The Moral Conflict of Law and Neuroscience

The Moral Conflict of Law and Neuroscience

PETER A. ALCES

THE UNIVERSITY OF CHICAGO PRESS   CHICAGO AND LONDON

The University of Chicago Press, Chicago 60637
The University of Chicago Press, Ltd., London
© 2018 by The University of Chicago

Published 2018
Printed in the United States of America

27 26 25 24 23 22 21 20 19 18   1 2 3 4 5

ISBN-13: 978-0-226-51336-2 (cloth)
ISBN-13: 978-0-226-51353-9 (paper)
ISBN-13: 978-0-226-51367-6 (e-book)
DOI: 10.7208/chicago/9780226513676.001.0001

Library of Congress Cataloging-in-Publication Data

Names: Alces, Peter A., author.
Title: The moral conflict of law and neuroscience / Peter A. Alces.
Description: Chicago ; London : The University of Chicago Press, 2018. | Includes
    bibliographical references and index.
Identifiers: LCCN 2017024311 | ISBN 9780226513362 (cloth : alk. paper) |
    ISBN 9780226513539 (pbk. : alk. paper) | ISBN 9780226513676 (e-book)
Subjects: LCSH: Science and law. | Law—Philosophy. | Neurosciences—Philosophy. |
    Responsibility. | Agent (Philosophy).
Classification: LCC K487.S3 A85 2018 | DDC 340/.112—dc23
    LC record available at https://lccn.loc.gov/2017024311

♾ This paper meets the requirements of ANSI/NISO z39.48-1992 (Permanence of Paper).

FOR BRADY CHRISTOPHER MACNEIL, ZOË PETE WILLIAMS, AND COLIN RAMON MACNEIL, IN HOPES THAT SOMEDAY THEY'LL REALIZE THAT THEIR GRANDFATHER, ONCE, HAD A FIRM GRASP OF THE OBVIOUS.

Contents

Acknowledgments, Debts, and Admiration

Surely the greatest joy of the scholarly life is experiencing the generosity of colleagues, students, and friends in the course of a project such as this. It is true that your work gets better only as those around you help you make it better. I have been surrounded by a wonderful extended community of people who were patient with my most unreasonable requests and indulgent of my most fantastic flights of fancy. It is not easy trying to convince smart people that much of what they have taken for granted is illusory, that common sense is importantly nonsensical. It is not surprising that I have not convinced them all, perhaps not even very many. But I have, maybe just a bit, gotten them to think along with me during my last three years on this book. No one has been more generous and patient and thoughtful than my friend and colleague Professor Paul S. Davies of the College of William and Mary Department of Philosophy. Paul is a wonderful, and extraordinarily patient, teacher (and, not incidentally, a brilliant scholar). I hope he will not be embarrassed to see this product of our many conversations over lunch and coffee. It is not, after all, his fault if I have proved uneducable.

I am indebted too to another wonderful friend and teacher-scholar, Professor George DeRise, who read parts of my manuscript with care and challenged me in many ways to appreciate and defend portions of my argument I had perhaps too cavalierly expected to carry inordinate weight. He also knew how much this book has meant to me and cared about that. His intellectual curiosity was an inspiration and humbling in its own right.

I am grateful to the many folks at the University of Chicago Press who have believed in the book from the outset. Christie Henry, editorial director, encouraged my initial enthusiasm for the project, and my original acquisitions editor, Christopher Rhodes, talked with me at length, carefully reviewed manuscript I sent him, and made this a better book than it could ever have been without him. I felt his loss, at far too young an age and with so much important work left to do, profoundly. I hope whatever good there may be in this finished product is, in some small way, a worthy memorial to his talent and professionalism. I am grateful to my editor, Chuck Myers; his assistant, Holly Smith; Christine Schwab; and Melinda Kennedy for the care and courtesy they have shown me as we, together, brought this project to a close. In a professional life that has provided much more bounty than I have deserved, a high point has surely been the opportunity to work with the professionals at this press.

Dean Davison M. Douglas of the College of William and Mary School of Law, my professional home for the last quarter century or so, has been supportive both materially and emotionally during my years on this project. He is among the last of a disappearing breed: the scholar-administrator. I am particularly grateful to the Rollins family for providing the funding to endow the Rita Anne Rollins Professorship in Law at William and Mary's law school. The family's commitment to the life of the mind generally and our law school in particular has been nothing short of astounding. I hope that they are pleased with this result of their generosity, even if they might not agree with every word in the book. It has been a great honor to have my name associated with theirs.

Many of my colleagues on the William and Mary law school faculty have generously read portions of the manuscript and taken the time to discuss their reaction with me, often at length. Professors Nancy Combs, James G. Dwyer, Adam M. Gershowitz, Vivian Hamilton, Laura A. Heymann, Paul Marcus, Nathan B. Oman, James Y. Stern, and Cynthia V. Ward have been particularly helpful, though, really, none of my colleagues has averted their gaze when they saw me coming to ask a question. And just as significant, my students in three years of Law and Neuroscience classes have taught me much more than I could have taught them but were consistently gracious enough not to make too big a deal of that. Our community is a very special place.

I benefited greatly from the work of an army of conscientious research assistants in the last several years, many of whom at first had no idea what they were getting themselves into but all of whom took my anxieties and foibles in stride, with good humor and the sense of professionalism that will serve them well in their legal careers. I am particularly indebted to Blaine Adams,

Christopher B. Anderson, Collin R. Atkins, James Bologna, Rachel M. Cannon, Tyler F. Chriscoe, Taylor L. Connolly, Justin D. Davenport, Eric R. Hammerschmidt, Jason M. Howell, Eric M. Loose, Paul-Michael R. Lowey, Shaina T. Massie, Walter M. Morgan, Jacob E. Mustafa, Lucas I. Pangle, Joshua M. Reynolds, Tyler J. Rosa, Gregory R. Singer, Janet M. Smith, and Tianye Zhang. Several other students helped along the way and were an inspiration too: Matthew P. Chiarello, Lauren M. Friedrich, Abigail M. Norris, Jenna M. Poligo, Elizabeth C. Smith, and Krista K. Wallace. It is understatement to say this book would have suffered without their help; more accurately, this book would not *be* without their contributions to it.

Many members of the broader legal academy also have made invaluable contributions to this finished product. I am grateful for the courtesy shown me by a distinguished group of (I assume) compatibilists at the September 2015 "Mind/Brain/Responsibility Roundtable" cosponsored by the University of Illinois College of Law and the University of Iowa Philosophy Department: Selim Berker, Richard A. Fumerton, Robert Van Gulick, Douglas N. Husak, Frank C. Jackson, William G. Lycan, Michael S. Moore, Stephen J. Morse, David Papineau, Dennis M. Patterson, Adina L. Roskies, and Katrina Sifferd, all of whom, I suspect, were unconvinced but all of whom were gracious. Several distinguished scholars also attended a conference at William and Mary's law school to discuss the normative ramifications of neuroscience for criminal law and theory in October 2015. At that gathering Christopher Conway, Matthew L. Hoag, Victoria McGeer, Adrian Raine, Francis Shen, and Nicole Vincent (as well as others noted earlier) commented on early drafts of chapter 2 of this book and made challenging as well as helpful suggestions.

In spring 2016 I was invited by Dean Luis E. Chiesa to present draft versions of chapters 2 and 3 of the book at a meeting of his SUNY Buffalo Criminal Law Center colloquium series. Professors Guyora Binder, Deborah W. Denno, Anders Kaye, Ken Levy, and Thomas Nadelhoffer pushed me gently, but firmly, and offered comments and observations that made this book (and maybe even me) better. (My participation in that session also led to my being invited back [virtually] to Buffalo to comment on Stephen Morse's presentation at the Buffalo Criminal Law Center, and that experience provided the genesis of chapter 8 of this book.) The Buffalo Criminal Law Center is unique.

Many friends in and out of the legal academy also responded to my requests (entreaties!) to read and share with me their thoughts on several chapters of the book: thank you, Kristen D. Adams, Mark E. Budnitz,

Carol Coghlan Gavin, Joseph G. Gavin, Michael M. Greenfield, Betsey J. Grey, I. Trotter Hardy, Nancy S. Kim, Adam Kolber, Susan Lyons, David Snyder, Bruce Waller, and Jonathan L. Williams (also a terrific son-in-law).

The Faculty Support Center at the William and Mary law school under the direction of Felicia Burton kept me in line throughout (sometimes literally), and Cody Watson made sure that the manuscript delivered to the press looked just right. I am grateful to Paul Hellyer for helping me conceptualize the framework for the index, and I am most profoundly indebted to James Curtis for composing the final index and making this book better therefor. Jim is a master craftsman.

Finally, it was Professor Martha J. Farah, through her Center for Neuroscience and Society at the University of Pennsylvania, who lit the fire under me. I told her at the end of the 2011 session of Neuroscience Bootcamp that the work she does changes lives. I am even surer of that now.

Preface, Premises, and Progress of the Argument

This book is, essentially, a thought experiment: What should law be in order to govern the affairs of human agents who do not have moral responsibility? It proceeds from the premise that human agents do not, in fact, have moral responsibility and that the mechanical nature of human agency is confirmed by neuroscientific insights that have revealed—albeit so far incompletely, perhaps even only vaguely—the chemical, electrical, and structural incidents of neural processes of the brain. And we are no more than our brains; we could not be. That conclusion entails hard determinism, the realization that we are the product of forces. Indeed, we cannot even say "the product of forces acting upon us" because we are the sum of the forces, not the object of their action. And that conclusion engages the contours of normative theory: even our understanding of our understanding.

The first chapter surveys the terms of that engagement, signaling the disruptive nature of the materialism that the thought experiment indulges. The chapter provides the necessary guide to the scope of the inquiry and describes the points at which the book's thesis joins the normative conversation. The focus is on the difference it would make to the law if things are not as they seem, if *we* are not as *we* seem. The book's argument is disruptive: I suggest that all, or virtually all, of our law largely depends on a gross misunderstanding of its subject—the human agent. The law often fails because the legal doctrine misunderstands what it means to be human. Further, extant comprehensive interpretive theories of law, theories that combine the positive and the normative, provide the arguments in support of the doctrine's misapprehension. It is the noninstrumental theories that make the fundamental conceptual error. Instrumental theories fail too, but their failure is largely attributable to empirical rather than conceptual error. Instrumental theory could take account of an authentic understanding of human agency; noninstrumental theory denies the materialism and the determinism that define human agency and so could not understand what it means to be human.

Chapters 2 through 7 proceed in pairs: Chapter 2 describes illustrative aspects of criminal law that depict the neuroscientific naïveté of the doctrine; chapter 3 explains the failure of normative criminal law theory to understand the authentic human agency the perspective vindicated by neuroscientific insights would reveal. Chapter 4, then, similarly presents illustrative tort law doctrine and chapter 5 describes the failure of noninstrumental tort theory, focused on corrective justice and civil recourse, to take account of an authentic conception of human agency. Chapter 6 treats the consent criterion in the contract law, primarily the operation of boilerplate in consumer contracts, and chapter 7 demonstrates the failure of noninstrumental contract law theory that relies on a misunderstanding of what consent and promise could mean to human agents, actors without moral responsibility.

Neither each chapter nor any pair of chapters is a self-contained whole. That is, the argument of the book progresses through the several chapters, emphasizing the portions of the argument that are best presented within the context of each of the doctrinal and theoretical discussions. There are some aspects of the determinism vindicated by neuroscientific insights that are best framed in the context of considering the retribution interest in the criminal law. Those observations may then be refined in the tort chapters and only appreciated in their full breadth in terms of the consent criterion in the contract law. The argument builds through the book to, ultimately, sustain the weight of the conclusion that the premise that founds much if not all of law—moral responsibility—is chimerical. Only at the end of the journey will the consequences of the argument emerge in full relief. That is the scheme, at least.

Chapter 8 then takes account of the arguments that might be (even anticipatorily have been) offered in response to critique of legal doctrine and normative theory that would rely on neuroscientific insights to deny the moral responsibility of human agents. The object is to join the conversation and also to suggest new lines of thought. The approach is heterodox, "scorched earth" in fact: Extant law, the orthodoxy, and apologies for it fail because the doctrine and theory misconceive human agency. So there is much work to be done.

CHAPTER ONE

Contours of the Conflict

The Question in Context, the Thesis

From the criminal law: A middle-aged school teacher rather suddenly began to solicit prostitutes and also to make subtle sexual advances to his prepubescent stepdaughter. After he was convicted of child molestation and then expelled from an inpatient rehabilitation program, a magnetic resonance image (structural MRI) revealed that the teacher had a large tumor on his orbitofrontal cortex, a portion of the brain involved in the regulation of social behavior. The tumor was removed, and he returned to his normal self. But the deviant behavior began again. It was discovered that the tumor had not, in fact, been completely excised and had grown back. When the tumor was then completely removed, the teacher was cured.[1]

From the tort law: A train approached an intersection in a Michigan town, and the operator observed a school bus entering the grade crossing and attempting to cross the grade by driving around the lowered gate. The train was traveling at sixty-five miles per hour, too fast to stop within the available distance. The train collided with the school bus. The operator thought the bus had been filled with children. It was not; only the bus driver was injured, though severely. The train operator suffered posttraumatic stress disorder (PTSD) and sought to avoid the defendant school district's interposition of a governmental immunity defense by alleging that he had suffered a serious impairment of bodily function, the PTSD. The court relied on a positron emission tomography (PET) scan to find the cause of the PTSD: "decreases in frontal and subcortical activity consistent with depression and posttraumatic stress disorder." There was a "bodily injury" to the operator's brain, "significant change in brain chemistry, brain function, and brain structure"; the PTSD was not just "in his head."[2]

And from the contract law: The decedent entered into a contract to sell approximately five hundred acres of land and a wheat crop. When the plaintiff-buyer brought a specific performance action to have the contract enforced and the sale effected, the appellate court relied in part on the testimony of a neurologist who examined the results of a structural MRI of the decedent's brain and "found evidence of brain shrinkage and hardening of the arteries . . . consistent with dementia." The court decided that the decedent was not competent to enter into the contract and refused to enforce the sale.[3]

The object of law is practical: to direct, even mold, human behavior; law is, therefore, normative. That is true whether you think law should be measured by its consequences or by realization of some more ethereal object. For law to work, to accomplish whatever goals, instrumentalist or noninstrumentalist, we have in mind, law must affect the human agent. So law must take the qualities of the human agent, what we are, seriously: What does it mean to be human?

Law relies on a conception of human agency; it must. Law takes for granted certain human attributes, both in prescribing and proscribing behavior. Indeed, for the last century the story of law has been the story of increasing acuity about the human condition: the legal realist movement of the twentieth century and the numerous "law and . . ." initiatives that followed thereon were designed to improve law by making it more responsive to what the "ands" (economics, sociology, psychology, etc.) revealed, scientifically or otherwise empirically. Economics, sociology, psychology, statistics, as examples, all can improve law by making it more consonant with revealed truths about the human condition, including our essential nature. Perhaps a natural development, even a culmination, of law's incorporation of insights from other areas of inquiry is a narrower focus on what it is that makes humans unique: our brain. Although all characteristics of sentient beings are points on a continuum, we may say with some confidence that what most certainly distinguishes humans from other life forms is our brain. We communicate, manipulate, and think about our own and others' thinking because of the particular way in which the human brain is organized and constructed. You may not believe that there is a supernatural reason for that uniqueness, but you cannot deny the uniqueness.

Within the last several years, likely owing to developments in our ability to look into the brain, research into how the brain defines what and who we are (as a species and individually) has given reason to reconsider what it means to be human. Surely we are more than the product of trillions of chemical and electrical processes—or are we? And even if something emerges that is more than such processes, can a better understanding of the underlying mechanics lead to a better understanding of human behavior and the role of mechanisms such as law that would affect human behavior? Can brain science, that is, neuroscience, affect law?

That question is now more than rhetorical. Certainly what we know about the brain has an effect on our law: We do not execute those who are profoundly intellectually impaired or even punish those whose apparently aggressive action was in fact the result of an epileptic seizure. So the criminal law at those margins surely is considerate of brain science. Similarly, law cares about state of mind in the imposition and measure of tort and contract liability: We do not impose civil liability in tort on those below a certain age, and we are comfortable reciting that contract liability will not lie if there has been no meeting of the minds. But those venerable examples of law's deference to empirical reality, cognitive limitations, are the product of a time when we knew less about the brain than we do now.

The question is how the law will (or should) respond to what developing neuroscientific insights have to tell us about the human agent. It may be in the first instance difficult to gain purchase on that inquiry in the most general terms, but surely we would all agree that there may be certain discrete criminal, tort, or contract law rules that would be subject to adjustment as we learn more about the human brain and its development. Recent United States Supreme Court decisions evidence willingness to take into account what the science reveals.[4] And if there is a way to objectively and certainly demonstrate emotional pain,[5] it is likely that courts will be receptive to such evidence and that legal doctrine will respond as well.

The object of this book is to take account of the current conceptions of the moral foundation of law, as revealed in illustrative aspects of the criminal, tort, and contract law, and compare those conceptions with human agency as revealed by the emerging neuroscience. The thesis here is simple: If emerging brain science reshapes what we understand to be the meaning of being human, then that same brain science must reshape our law, from the moral foundations up.

This introductory chapter describes, in broad strokes, the tensions engaged when we consider the effect that developments in neuroscience may have on law. Subsequent chapters chart a course through the doctrinal and theoretical thicket. The focus here is on the normative, or moral, underpinnings of the criminal, tort, and contract law: What does neuroscience reveal about the human agent that may affect the moral presumptions (and objects) of law? What happens at the normative intersection of law and neuroscience? Law, perhaps uniquely in human affairs, depends on morality: We would not brook immoral law; from at least one perspective, immoral law might even be an oxymoron. So the moral conflict of law and neuroscience is a worthwhile and, indeed, particularly important juncture at which to measure the impact of neuroscience on what it means to be human. As we shall see, just about all of the big issues, many summarily surveyed in this first chapter, are implicated.

The Received Wisdom

Neuroscience challenges the received wisdom, the sense we all have that we, as a species and even individually, are unique among the stuff of creation. We assume that we are more than mechanisms, more than the sum of our parts, and so not reducible to chemical and electrical processes. There is something that distinguishes us from inanimate and other animate objects and entities; we just *feel* it to be so. And religious as well as ethical precepts and practices reinforce that specialness. The sense of uniqueness may entail certain predispositions or moral commitments. For example, we believe that we and others think first and then act (and so are therefore responsible), that there is some homunculus inside that reviews the choices we confront and makes the decision for which we are accountable (thus that persistent internal monologue), that the mere fact that something exists does not make it right (the is–ought tension or naturalistic fallacy), and that we can infer the state of mind of others and respond to them on the basis of those inferences (folk psychology and theory of mind). That list is illustrative, not exhaustive, but it suffices to demonstrate how this felt sense of uniqueness manifests itself.

Neuroscience challenges that orthodoxy and so challenges conceptions of ourselves that have provided the moral foundation of law. Further—and this is crucial for the instant study—if normative theories of law, either instrumental or noninstrumental, depend on that received wisdom in ways that the neuroscience would undermine, then neuroscientific insights may challenge the very foundation of our law. Now, we may conclude that law is not based on a moral theory that depends on the received wisdom (or aspects of it), but then we would have to determine what the moral basis of law is, perhaps ultimately what morality is.

According to the received wisdom, morality has something of the ethereal about it: Morality is aspirational; it declares what we can be if we realize some object, perhaps making due allowance for reasons why we fail to realize that object. Our morality surely does not depend on the same forces that explain opposable thumbs or the ability to walk upright. We just know that morality is a uniquely human thing; your dog cannot be moral or immoral (except that she acts in a way that we would describe as moral if a human did it). As we shall see, neuroscience and the more empirical sense of morality that neuroscience suggests cuts into the received wisdom at this crucial joint. It is worthwhile to consider here, albeit summarily, the dichotomies revealed at the intersection of law and neuroscience, where law and neuroscience conflict. The chapters that follow will treat many of these issues in more depth. For now, though, in order to preview the argument of this book, it suffices to sketch in broad outline some important distinctions.

Is Naturalism Fallacious?[6]

It has become something of a truism that "is" does not entail "ought": We cannot reach a correct moral conclusion from an accurate empirical observation; might, for example, does not make right. Those who are fit may have better survival chances, but that does not give them a superior claim to survival. The so-called naturalistic fallacy just points out the difference between "is" and "ought." There is, though, a sense in which "is" may be a measure of "ought," and it is in that sense that insights offered by more empirical approaches to morality challenge the conclusion that the equation of "is" and "ought" is necessarily fallacious.

Consider your reaction to a child, perhaps your child or grandchild. We can identify a good evolutionary reason for the natural tendency or even desire to comfort that child when he is in distress. In fact, resisting the urge to come to the child's aid may make you uneasy, even physically uncomfortable. You extend your arms to him and embrace him, perhaps cooing soothingly as much for your own sake as for the sake of the child. We all understand that reaction, even on a physical level. But if we see an older man, destitute, homeless, curled up in a box on the street of our city, the reaction may not be, and in all likelihood is not, the same. Now that is not to say that as a moral matter (according to some coherent moral code) the reaction *should* not be the same. Indeed, the homeless adult may be in more distress and less at fault for his circumstances. The child may be crying because his diaper is wet; the homeless adult may be very ill, mentally or physically (a distinction, we shall see in due course, that ultimately proves specious).

The fact that we are attracted to the infant in distress at some emotional-physical level and not similarly affected by the homeless adult is a fact, an "is." We can certainly rationalize the divergent reactions; we can even weave an "ought" out of the emotional-physical responses. But we cannot deny that the two scenes affect us differently. Indeed, it would not be cynical or difficult to tell a story that makes some kind of sense of the different reactions, that reconciles the emotional and the moral. It may be that first we experience the emotional reaction and then rationalize it by embedding that emotional reaction in a moral rationalization.[7] The emotional reaction becomes the moral conclusion (and then maybe we codify the moral conclusion and call it law). So though we hesitate to say that the "is" (the emotional reaction) determines the "ought" (the moral conclusion flowing therefrom), we cannot deny the coincidence.

In *Principia Ethica*,[8] G. E. Moore argued that it is error to equate what *is* with what is *good*. Moral properties cannot be reduced to physical properties. That idea has been developed[9] and criticized.[10] Contemporary normative empiricists proceeding from a naturalistic perspective can assert, at least after a fashion, that "is" does equal "ought," but a good deal depends on what we mean by "is" and "ought."

Sam Harris, a noted atheist,[11] understood the challenge presented by Moore's identification of the naturalistic fallacy and observed that "Introspection offers no clue that our experience of the world around us, and of ourselves within it, depends upon voltage changes and chemical interactions taking place inside our heads. And yet a century and a half of brain science declares it to be so."[12] It is more than a bit discomfiting to reduce human agency to nothing more than physical reactions, albeit of awesome complexity. Harris asserted that all human normativity is based on human thriving but also recognized that that equation does not clarify much.[13]

Patricia Churchland offered resolution of the is–ought tension in naturalistic terms: "[M]orality can be—and I argue, *is*—grounded in our biology, in our capacity for compassion and our ability to learn and figure things out. As a matter of actual fact, some social practices are better than others, some institutions are worse than others, and genuine assessments can be made against the standard of how well or poorly they serve human well-being."[14] What inures to the net benefit of humankind, construed in evolutionary terms, that is, reproductive success, is *a* viable measure of goodness. But that naturalistic equation works only so long as what brings pleasure leads to evolutionary success, and it is not clear that *all* conceivable measures of human goods or capabilities result in reproductive success. It remains important not to simply dismiss out of hand noninstrumental perspectives of human goods and capabilities as quaint but wholly insubstantial. A thoroughgoing naturalism need not be so dismissive. It may be the case that the noninstrumental argument supporting naturalism's fallaciousness resides in the idea that there is a source of the good that goes beyond (in a sense, at least) human thriving.

The work of deontology is not complete with the demonstration that there is more to life, to life well lived, than reproductive success. It may be that noninstrumental appeals to not-obviously-consequentialist goals are not different, at the cellular level, from reproductive success. Aesthetic experiences may be different in kind from sex but no less grounded in neural composition. Deontology may demonstrate that a range of sensations may matter to human thriving. But that demonstration would not establish that there is any greater good than human pleasure. Indeed, naturalists can establish a connection between our ostensibly pure aesthetic sense and reproductive success.[15]

Even once we come to terms with the parameters and dimensions of the naturalistic fallacy, we need to appreciate the contours of a different, quite practical, challenge: the way we navigate the space between and among ourselves. Granting that we are social animals, need social stimulation in order to remain sane, how do we make sense of the relations among one another? Do we need to actually read minds? Or just act in ways that seem as though we can read minds? We will see that how we conceptualize our perceptions of one another matters to law, and that neuroscientific insights may affect law's assumptions.

Folk Psychology and Cognitive Neuroscience

Labeling a psychology "folk" is not to disparage it; folk psychology is not a term of derision. Folk psychology refers to what we engage in every moment of every day when we draw inferences about the thoughts and intentions of others from what we *imagine* to be going on in their minds. Indeed, there may be an essential identity between folk psychology and theory of mind,[16] our ability to look into the minds of others by using the inferences we draw from their appearance, words, and actions (as well as from what we know about what we would be thinking and feeling if we appeared that way, used those words, and acted similarly). Crucially, we infer intention and, accordingly, responsibility from certain behaviors, and such intention justifies a particular reaction in a moral sense (such as imposition of criminal or civil liability). You could think of folk psychology as akin to inductive reasoning.[17] We reason inductively when we infer a principle of general application, X, from the presence of A, B, and C; we engage in deductive reasoning when we start with the general principle and then confirm its operation by observation, empirically. Science, of course, depends on both forms of reasoning. The development of scientific theory relies on induction: After witnessing an array of phenomena, we reason to a cause of the phenomena. But applying science is a deductive process, distinct from theory. When the doctor prescribes an antibiotic, the doctor is not theorizing; instead, on the basis of her own and others' experience, she deduces that the correct, most efficacious response to the patient's high fever and inflamed throat would be administration of the medication that will respond to the cause of those symptoms. Deduction tests hypotheses and theories: It is not the source of them.

Folk psychology is theory building on a small, interpersonal scale: It endeavors to distinguish correlation from causation. So when we reach the folk psychological conclusion that the child molester assaulted the child, we infer that the reason for the assault was the actor's desire to bring about the result, a desire that is sufficient to establish responsibility, the criminal law's sine qua non of liability. In that way the standard criminal law attribution of blame—the basis of punishment—depends on the type of inductive theorizing that is typical of folk psychology. But folk psychology, like inductive reasoning, relies on observation of the phenomena from which we draw inferences and so is subject to error if the fundamental observations do not support the conclusions we draw from them. Responsibility is a conclusion inferred from folk psychology's dependence on intuitions about human agency; if folk psychological intuitions about responsibility and human agency are wrong, then folk psychological reasoning will lead to erroneous theory. Moral responsibility, the basis of blame and, accordingly, much of our law, may be just such an erroneous theory.

A simple analogy illustrates the point: You would not, really, *blame* your car for not starting one morning (though you might act as though the car were sentient). If you were to blame the car the way we blame recalcitrant people, you would try to modify its behavior, perhaps by punishing it, "sentencing" it to the garage for a week (or year) or two. You would, though, achieve a better result by having the cause of the car's failure to start corrected at the local garage. Indeed, if you were to incarcerate your car, there would be reason to question your capacity. Blame and responsibility are conclusions *on* which folk psychology relies and *to* which folk psychology reasons. Folk psychology infers blame and responsibility, theorizing from the coincidence of phenomena. And it works well, much of the time.[18]

Cognitive neuroscience may be distinguished from folk psychology by its more empirical basis and by the type of reasoning it represents. Though the folk psychologist might conclude that *A* molested a child at least in part because *A* was abused as a child himself,[19] cognitive neuroscience looks for the organic brain abnormality (*perhaps* the product of childhood abuse) that triggered the behavior. And where folk psychology is inductive, then cognitive neuroscience is analogously deductive, concluding from physical evidence. All that is necessary to make sense of human agents' actions from the perspective of cognitive neuroscience is an understanding of how the chemical and electrical networks within the brain work: There is no more than can meet the eye (broadly construed). Cognitive neuroscience does not rely on moral blame or responsibility; physical cause will be sufficient because all cause is, ultimately, physical. The attribution of normative substance to physical actions is not just erroneous; it may be misguided and ultimately undermine law's object.

To sum up, the difference between the two approaches—folk psychology and cognitive neuroscience—might be seen in their respective reactions to the same facts: Cognitive neuroscience seeks to identify the physical cause of the actor's behavior, the underlying neural aberration; folk psychology, while not inconsiderate of physical causes and willing to recognize excusing conditions, will generally find individuals responsible for their actions when those making the assessment believe they too would have been morally responsible under similar circumstances. The very idea of responsibility is treated differently: Responsibility means something for folk psychology that it does not mean for cognitive neuroscience. For folk psychology, responsibility has normative, moral valence; for cognitive neuroscience, it has only causal meaning. That distinction is crucial and matters across legal doctrine: It is a thesis of this book that the criminal, tort, and contract law proceed from the folk psychological perspective so that they can make sense of responsibility in a way that would not be accessible from the perspective of cognitive neuroscience. We must appreciate, though, that until we have a better grasp of the brain science, folk psychology makes sense as a second-best solution.

Certainly folk psychology, which is both the reason for and the product of the limitations on our understanding of human agency, accommodates a particular sense of the fit between the material and the immaterial. Because we have not yet figured out all we need to know about the physical, the material constituents of our being, we continue to rely on nonphysical conceptions to fill in the gaps. And that is fine, unless and until we deny physical explanations because we become too comfortable with the nonphysical ones. At some point, even the most comfortable fiction becomes pernicious.

There is, though, a real possibility that much of our normative sense, and so our morality and in turn our legal doctrine, is built on such a comfortable fiction. That persistent sense that we are distinct from the entity that we are in control of is familiar and powerfully comfortable. That sense is crucial to the law insofar as it informs conceptions such as responsibility, blame, and fault. Neuroscience confronts and unpacks the source and substance of that persistent sense and engages the meta-ethical tensions of dualism and monism.

Dualism and Monism[20]

The folk psychological view often, perhaps even necessarily, entails a dualistic conception of human agency: When A acts in some way that law proscribes, A is liable or subject to sanction because he did not "control himself." The core of dualism is that something exists that we cannot explain in purely physical terms—there is a "self" that some other entity or part of A controls. A made a choice to act or not act and is therefore responsible. There is, in that stark schematic, the stuff of dualism: the idea that humans are not unitary beings, the sum total of myriad physical properties, but, instead, are monitors of drives, forces, desires, and the like to which they sometimes succumb and which they at other times control. Modern dualism has evolved from the traditional Cartesian definition, which distinguished between the physical body and the nonphysical soul. Modern dualists, who almost certainly would deny that they are, in fact, dualists,[21] nevertheless distinguish the physical brain from the person or mind, the latter category defying physical explanation. The assertion is not that the mind or person is merely a useful concept, a placeholder for more nuanced scientific explanation (a fundamentally different concept).[22]

Insofar as law instantiates folk psychological precepts, it is dualistic.

That is the necessary conclusion once we distinguish dualist from monist perspectives. There is no such thing as a little bit of dualism: You either think we are nothing more than our brains or you think that we are also something more, something that is not captured by the physical brain. If you say "brains don't kill people; people kill people,"[23] you are a dualist by virtue of defining the source of agency as a nonphysical property. A monist concludes that all we are, all that we can be, is the product of chemical, electrical, and structural forces acting on and within the trillions of neural connections that compose our brains.

M. R. Bennett (a neuroscientist) and P. M. S. Hacker (a philosopher) offered a modern dualistic perspective in their 2003 volume, *Philosophical Foundations of Neuroscience*.[24] But they deny that.[25] So even those whose work reflects an intellectual commitment to the mechanistic can and do question whether all there is, *is* that which we can see, even if only under an electron microscope. Bennett and Hacker posited what they called the "mereological fallacy":[26] confusion of a part with the whole, that is, the brain for the "human being." They emphasized throughout the book that the brain does not (decide, act, promise, etc.); the "human being" does.[27]

More recently Pardo and Patterson, two law professors, reiterated the Bennett and Hacker argument.[28] While Bennett and Hacker concluded that the brain and the "human being" are distinct entities, Pardo and Patterson distinguished instead the brain and the "person."[29] That does not seem to be a substantial difference between the two pairs of scholars. (Pardo and Patterson did try to add something, though, by bringing the naturalism question into focus in the legal context.) Their conclusions relied on a supposed middle path between materialism and dualism by proposing that mind is merely the aggregate of a person's intellectual abilities, making it a function of the physical brain. Since mind is now a collection of properties rather than a separate *ousia*, they can claim to have avoided the dualistic premise of nonphysical substance. Nevertheless, they crossed the ontological divide by positing that the person who thinks and feels is not the brain that performs those concomitant functions (firing neural synapses, etc.). Personhood is the nonphysical entity linked to the now materially grounded mind. But on what grounds can this person be said to exist, if it is a nonphysical entity?

Bennett and Hacker, as well as Pardo and Patterson, described their arguments as conceptual,[30] relying on Wittgenstein's language theory to support their conclusion that the language we use conveys truths that naturalism would obscure.[31] But perhaps the language we use simply is evidence of general historical ignorance—a societal perspective that shifts much more slowly than scientific understanding. Here, critics of Pardo and Patterson could borrow from Nadelhoffer: "just because the criteria we traditionally relied on when *talking* about mental activities such as knowing, deciding, intending, and lying were behavioral, it doesn't follow that neural criteria could not possibly be adopted in the future in light of developments in neuroscience."[32]

A great deal depends on the *promise* of neuroscience to shift our understanding of behavior and, accordingly, the way we speak about it. Neuroscience already has begun to deliver on that promise, although the shift in our language, and the general conception of responsibility and choice, will ensue at a deliberate pace. That certainly promises to be a challenge for the law. For example, the legal resistance to the biomedical model for addiction, as recently articulated by attorney David L. Wallace, holds that "Brains do not smoke cigarettes; acting people do. . . . Law is about personhood, not biophysical function."[33] He argued that even an addicted person is assumed to be an otherwise reasonable legal person; a better understanding of underlying brain mechanisms is not the same as a legal cause. As the law stands, that is correct. But it would be foolish to cling too tightly to a framework that relies on this nebulous concept of personhood, a concept becoming more obscure as neuroscience increasingly points to the physical, mechanistic aspects of our thoughts and behavior.

Many popular and accessible neuroscientific sources also support the monist perspective (or, at least, provide substantial grounds to question dualism). For example, the books of the neurologist Oliver Sacks[34] clinically though compassionately describe the life experiences and consciousness of his patients trapped in abnormal states that are the product of neural anomalies. One reaches for his wife's head believing he is reaching for his hat,[35] another believes that her own limb is not her limb at all and wishes to remove it,[36] another is unable to recognize others (even the closest loved ones).[37] As we discover with increasingly fine precision, the source of all of Sacks's patients' misapprehensions, as well as those of others who suffer from similar pathologies, may be traced to chemical, electrical, or structural malfunction in their brains. Nonetheless, their consciousness depends on those misapprehensions, just as our consciousness depends on the observations that we are certain are accurate. Their behavior, their very experience of the world around them is not just colored by, but instead is actually *determined* by, such misapprehensions.

The false dichotomy between the mental and physical is further evinced in research by Vilanyan Ramachandran, who has worked with patients suffering from pain or discomfort in limbs that had been amputated (known as phantom limbs).[38] Ramachandran explored the brain's role in the experience of phantom limb pain and discomfort (as the brain is the source of all pain),[39] and theorized that reorganization in the primary somatosensory cortex after trauma causes the phantom limb phenomenon.[40] In other words, phantom pain is not merely *experienced* as real—it *is* real and has a physical basis in the brain. It is not difficult to appreciate how the reality of that experience could affect the consciousness of the victim; consciousness is revealed as the sum of physical experiences.

A final example in popular neuroscientific literature is the work of Antonio Damasio with his patient "Elliot." Elliot had significant ventromedial frontal lobe damage from a tumor and from the surgery to remove it.[41] Once a successful businessman and caring husband and father, after the surgery Elliot lost his job, depleted his savings, divorced, remarried, and divorced again. He could no longer make simple decisions. From a noninstrumental perspective, Elliot's behavior would be explained as irresponsible, perhaps attributed to some deficit of morality, self-control, or maturity. A dualist perspective might find fault with his behavior on account of a flaw in his mind or person. The monist perspective, however, reveals a glaring oversight, underappreciated by the noninstrumental, dualist explanation: Elliot's biology, including his neural damage, is not only one explanation, or a partial explanation, of his drastic behavioral shift—it is *the* explanation.

There are other illustrations of the wholly physical basis of what we consider to be the constituents of consciousness, and some will be treated in the chapters that follow. That appreciation of the physical basis of consciousness, and so of personhood and human agency, goes all the way down. We could not conclude that although a brain region, say the prefrontal cerebral cortex (the "executive center"), is physical, its processes are somehow the product of different stuff. The function of the prefrontal cortex is the product of chemical, electrical, and structural—inherently physical—systems. So if we identify a deficiency traceable to the work the prefrontal cortex does, we have located the source of a physical aberration. There is, then, no meaningful division between the physical and the mental, between the physical and the emotional, or ultimately even between the emotional and the rational that goes any deeper than the location[42] in the brain where the function (or malfunction) occurs.

The relationship between what we have taken to be distinguishable and even inherently distinct systems is important for appreciating what is at stake in resolving the dualist–monist tension. If there is only one system, comprehensible (though not yet fully comprehended) entirely by analyzing physical systems, then normative theory that depends on something more is at best misguided, and at worst insidious. Once we have identified inherent flaws in our normative framework that compromise its accuracy and efficacy, the only rational approach is to examine how we might improve it. Of course, even if we were able to certainly conclude, or at least agree, that all human experience pertinent to the normativity of legal doctrine may be reckoned in physical terms—in so many chemical, electrical, and structural actions, reactions, and interactions—we would have to acknowledge the limits of our current understanding and also appreciate and fashion law's function and operation in a physical world that remains enigmatic until we can lift the veil. We would still need a second-best solution, and would still need to discover what second best might be. Indeed, it *could* be the folk psychology framework. At the outset, though, it is important to appreciate the difference between what we do not yet know and what we cannot know and develop a framework that can adjust as our knowledge expands.

Empirical and Conceptual Limitations

It is an empirical conclusion to say that we do not yet understand the efficient neural basis of psychopathy; it is a conceptual assertion to conclude that psychopathy is inconceivable as the product of chemical, electrical, and structural processes in the brain. We can overcome empirical limitations, for example, by developing more precise imaging techniques and understanding how the brain works better than we now do, but we cannot overcome conceptual limitations: If human agents are something other than their brains, are not (merely?) the sum or cooperation of processes accessible to science, then no neuroscientific advance will ever formulate human agency.

A good deal of the current excitement about neuroscience is the product of optimism: the expectation that discoveries will in time lead to the revelation of important truths barely dreamt of in our state-of-the-art philosophy. There is a great deal that neuroscience cannot do, indeed, may never be able to do. And those who are most critical of irrational neurolaw exuberance[43] are able to find abundant examples of science's failures, false starts, and missteps. Further, even the best current science is misused, and that undermines the greater project. Attraction to the shiny, such as the colorful fMRI scan, has distracted those who would look for the important work that science can or might be able to do for law. Although it may be that neuroscience will not accomplish what its most enthusiastic proponents claim it can, it *seems* quite clear that neuroscience will be able to do more than it can today, and that that work will be very important.

There is little to be gained from dwelling on the empirical limitations of the current science. Candor and intellectual honesty compel recognition of the limitations of our understanding and appreciation of the consequences of those limitations for sweeping conclusions about neuroscience's potential impact on legal doctrine, but it is crucial as well to remain cognizant of the difference between what may be temporary empirical road blocks and what might be intractable conceptual obstacles. Emerging neuroscientific insights provide the means to distinguish them. Once we can see—literally, on a brain scan—what a brain lesion or abnormality can accomplish—physically—on the brain and then see (or infer) what behavior results (or may result) from a particular aberration, we will challenge our innate dualistic sensibilities and better distinguish the empirical impediment from the conceptual obstacle.

When we overcome empirical limitations, what appeared to be conceptual challenges may disintegrate, or at least revise in more tractable form. We can engage the work of Stephen Morse to demonstrate the nature of the empirical–conceptual divide.[44] The scenario he posited is drawn from the criminal law but pertains to responsibility and the normativity of legal doctrine generally. Professor Morse's description of the facts and appraisal of their significance bears repetition at length:

> Oft was a forty-year-old school teacher who was married and had a step daughter. He had an interest in pornography dating to his adolescence, but at the time in question he experienced a growing sexual interest in children and he collected child pornography and visited child pornographic Internet sites. He also solicited prostitution at "massage parlours," which he had not previously done. Oft tried to conceal his activities because he knew that they were unacceptable. Nevertheless, he continued to act on his sexual impulses because, he said, the "pleasure principle overrode" his restraint. Oft began to make subtle sexual advances to his prepubescent stepdaughter, who informed her mother.
>
> Oft was convicted of child molestation and ordered to undergo an inpatient rehabilitation programme instead of prison. Despite his desire to avoid prison, he solicited sexual favours from staff and other patients in his programme and he was expelled.[45]

Morse depicted an individual with insufficient self-control to restrain antisocial, even criminal, behavior.

Oft had always had such tendencies, but more recently something, apparently, had gone wrong: He has begun to act on those impulses. It is not clear whether the temptations have grown stronger over time or, colloquially, whether his "control" over them has diminished. Oft's change of behavior may represent the confluence of those two developments. So far, there is nothing unique about Oft: We can imagine any number of people who "change," who "give in" to forces that overcome their ability to avoid antisocial behavior. Indeed, that change could be no more than a reappraisal of the likelihood of detection, the evolution (or devolution) of social mores that makes the subject behavior more acceptable, or a general sense that there is less to lose by engaging in behavior that had been avoided, for whatever reason. The important point here is that Oft's behavior was not *obviously* solely the consequence or confluence of forces similar to those that many law-abiding people might confront and overcome in diverse contexts during the course of their lives. It is in fact the object of moral education to enable us to avoid behaviors that might afford immediate benefit but entail greater long-term costs. And no one had put a gun to Oft's head.

Morse continued:

> The evening before his prison sentencing, Oft was admitted to a hospital emergency room complaining of headache. Although no physiologic cause was suspected, he was admitted on psychiatric grounds with a diagnosis of paedophilia. He expressed suicidal ideation and a fear that he would rape his landlady. During neurologic examination he solicited female staff for sexual favours and was unconcerned that he had urinated on himself. He had various neurological signs, including problems with his gait. Oft was alert and completely oriented. His memory was intact, his speaking and reading skills were unimpaired, and he was able to inhibit motor responses on a standard test of this ability. Word generation was somewhat impaired. He did suffer from constructional apraxia, the inability to assemble a coherent whole from its constituent elements, as demonstrated by his inability to draw a clock or to copy figures. He also could not write a legible sentence. A magnetic resonance imaging (MRI) test was performed.[46]

Now Oft exhibited behavioral anomalies not as easily correlated with what we might consider to be a simple normative deficiency: His cognitive ability was impaired, and he demonstrated other, some relatively benign but objectively verifiable, inabilities to control himself. Although the inability to draw a clock is normatively neutral, it would seem to confirm that at least some aspects of Oft's deficiency were beyond his "control," in the familiar sense of the word. Apparently, Oft's behavior gave his physicians reason to suspect an organic rather than purely psychological cause.

Oft's case raises the issue of what constitutes control, or what constitutes legally significant ability to control. Indeed, that very formulation of the issue seems to invoke conceptions of dualism, the dominance of one agent over another, and conjures visions of a homunculus along with the attendant dualism. It may be, though, that the criminal law ultimately is not so concerned with control, though control would seem to be constitutive of both the *actus reus* and *mens rea* requirements. In the case of homicide, at least, the dominant test of legal capacity focuses on the ability to distinguish right from wrong, not self-control. The issue, as far as Oft's case was concerned, was not so much his ability to appreciate right from wrong as it was his ability to control his behavior.

Morse next described the discovery pertinent to control, if not moral awareness:

> Oft had a large orbitofrontal tumour. The orbitofrontal cortex is involved in the regulation of social behaviour. Lesions acquired in this region later in life are associated with impulse control problems and antisocial conduct, but previously established moral judgment is preserved. The tumour was surgically removed and Oft quickly recovered bladder control and normal walking activity. Two days post surgery, his neurologic examination was essentially normal. Oft then successfully completed an outpatient treatment programme for his sexual disorder. He was no longer considered a threat and returned home. About a year later, he experienced a persistent headache and again began secretly collecting pornography. MRI showed tumour regrowth and the new tumour was successfully removed.[47]

This part of the story connects Oft's antisocial behavior to an organic cause. Surely if the tumor was the *efficient* cause of the behavior, then punishing Oft for what the tumor "caused" would seem misguided, not to mention cruel and ultimately inefficacious. Oft did not want the tumor growing in his orbitofrontal cortex, and intentionally acted in no way to encourage its growth. Neither was there any suggestion that Oft made a lifestyle choice that accommodated the growth of the tumor. At one level, the tumor was wholly independent of Oft: It could be removed without impairing Oft's normal intellectual functioning; indeed, its removal improved Oft.

Morse was aware of the centrality of the control question insofar as moral, and therefore legal, responsibility is concerned:

> In their discussion of Oft's case,[48] the authors said that Oft "could not refrain from acting on his paedophilia despite the awareness that the behavior was inappropriate." They hypothesized that the problem was caused by a disruption of his somatic marker system, which led to a preference for short-term reward and thus impaired the "subject's ability to appropriately navigate social situations."
>
> Although paedophilia is not a sufficient mental disorder to support an insanity defense, it is not absurd to think that perhaps Oft deserved mitigation or excuse for his sexual deviance on the ground that he could not control himself. With respect, however, we do not know whether Oft could not—that is, lacked the capacity to—control his sexual behaviour, or whether he simply did not. Given the timing of the appearance of the sexual deviance and the tumour growth, we can be quite confident that the tumour played a causal role in producing and heightening his sexually deviant urges and in determining his inhibitory processes.[49]

Morse here demonstrated the potential confound between empirical limitations and conceptual judgments: It is one thing to say that we do not *yet* have the technology to determine the extent to which an organic anomaly rendered the subject unable to "control" his actions; it is quite another to conclude that once we do (and, for present purposes, assume that we will), that determination will not preempt criminal responsibility. It is clear that an act, committed when a gun is pointed *at* your head in order to get you to perform it, cannot be criminal; why would that change if the "gun" is actually *in* your head, and firing? "Control," that is, "self-control," would seem crucial to the imposition of criminal liability. But "control," as used by Morse, is a legal, not a neuroscientific, conclusion.

Morse did, though, reveal a significant problem for the law's reliance on mechanistic conceptions such as control. Taken to its extreme, an empirical measure of responsibility will undermine the free will assumption that is the necessary predicate for all our criminal law—indeed for all of law.

> The general legal question is how Oft is relevantly different from any other paedophile with similar urges and similar inhibitory controls? One assumption is that the sexual behavior is a mechanical product of the tumor and is thus just like the mechanistic sign of any other disease. The assumption begs the question of responsibility, however. Oft's desires may have been mechanically caused, but acting on them was an intentional action. An abnormal cause for

his behavior does not mean that he could not control his actions. This must be shown independently. We can reasonably infer that Oft had difficulty controlling behavior that harmed himself because he acted in ways he knew would have negative consequences. But this is true of all paedophiles and we do not excuse them. He may have had impaired executive function, but this may be true of many paedophiles and would again need to be established independently. Although there is reason to question whether Oft differs substantially from paedophiles generally, the temptation to respond to Oft differently is strongly influenced by the lure of mechanism.[50]

If we parse Morse's conclusion so far, we can see the manifestations of dualism: the idea that there is something other than the brain that acts, that a mind or the "human being" or "the person" is positioned to exert control over what the brain dictates. That is dualism and provides the foundation of Morse's analysis: "Oft's desires may have been mechanically caused, but acting on them was an intentional action." Whose intention? And what was its source? Morse recognized a dichotomy between the subject's desires and the subject's action on them. That conception is consistent with folk psychology: I desire a new car but decide not to act on that desire. There is an "I" that is crucially separate from the "desire," an "I" that acts on the "desire" that is somehow normatively and crucially different from me. Now, in fact, we can understand the desire and the decision to act on it as distinct properties of the same entity, which we partition into discrete attributive categories such as the brain, or mind, or human being, or person. So, in a not implausible colloquial sense, it may make sense to understand one part of consciousness as affecting another. Indeed, that metaphor is consistent with self-consciousness generally. I want something, but I decide not to buy it: "self-control," we call it, to pursue the metaphor.

What neuroscience reveals is that the metaphor is just that: a metaphor, a trope that enables us to make sense of the dualism that consciousness *seems* to entail. That trope serves us well. It enables us to distinguish uncontrollable pedophilia from the transient deviant tendency that is summarily dismissed. When we put the two on a continuum, the poles of the continuum are "controllable" and "uncontrollable." We would all agree that there is some point along that continuum when the imposition of criminal liability is appropriate, to serve one particular normative object or another (a point that depends on our normative commitments, instrumental or noninstrumental). Culpability is coherent only by reference to the normative object of legal liability.

Surely Morse would not argue that Oft would be or should be liable if somehow, without Oft's willing or knowing it, someone tampered with his orbitofrontal cortex. Robert Sapolsky makes this point: "If someone with epilepsy, in the course of a seizure, flails and strikes another person, that epileptic would never be considered to have criminally assaulted the person who [sic] they struck. But in earlier times, that is exactly what would have been concluded, and epilepsy was often assumed to be a case of retributive demonic possession. Instead, we are now a century or two into readily dealing with the alternative view of, 'it is not him, it is his disease.' "[51] Morse, though, did not appreciate the challenge to retributivist theory that neuroscience presents. His conclusion about Oft's criminal responsibility precluded his ability to overcome dualism. Morse's conclusion relied on conceptions of control and the dualism that reliance necessarily entails:

> We do know that Oft did not control his paedophilic and other sexual urges, including in circumstances in which it was unlawful or completely inappropriate to express them. Moreover, Oft understood that his behaviour was unacceptable and he reported that the pleasure principle overrode his inhibitions. It is reasonable to conclude based on common-sense inferences that Oft experienced substantial difficulty controlling himself, but how do we know that he lacked sufficient control capacity to deserve mitigation or excuse? . . . We do not know how firmly Oft resolved not to yield to his impulses or whether he took steps to restrain them. There is a hint in his comment about the pleasure principle that he took no such steps.[52]

Morse ultimately captured well the state of the law, and not just the criminal law. So when we appreciate the deficiencies in his analysis, we can see as well the deficiencies in legal doctrine across disciplines. Morse conflated the empirical and the conceptual: Because we cannot be sure (neuroscience has not *yet* provided the means) that Oft was not in control (a dualistic notion: what would be "in control"? A homunculus?), it is not appropriate to excuse his behavior. But "excuse" on the basis of what normative scheme: instrumentalism (deterrence) or deontology (retribution)?

The important point is that we do not have to answer all the empirical questions in order to begin to develop a sense of how they may be answered. It is enough if we can begin to infer that they may not be answered in the way that extant normative theories of law have assumed they would be answered. To do the work right, to make progress on the broad issues presented when we consider the neuroscientific challenges to our assumptions about the relationship between human agency and the morality of legal doctrine, it is necessary to carve out the sources of what is fundamental in the law, or at least what is fundamental to the primary legal categories.

The Primary Doctrinal Categories

This book considers whether extant noninstrumental theories of law cohere with emerging insights from the neuroscience about human agency. That inquiry necessarily entails a survey of representative theories as well as some consideration of the relationship(s) among the three primary areas of law: the contract, tort, and criminal law. Those three areas are not primary in the sense that they are more important than other doctrinal categories. Instead, the three areas are primary in the sense that they describe the legal relationships that are fundamental to all areas of law (in the way the primary colors are primary). Admiralty, trusts and estates, commercial law, products liability law, public and private international law, tax law—any area you might posit—is some amalgam of the constituents fundamental to contract law (the law governing consensual relationships), tort law (the law governing nonconsensual relationships), and criminal law (describing the power of the state to deny individuals and other entities freedom or property on account of an imposition on the public welfare). Indeed, for present purposes, it is not even strictly necessary that that assertion stand in such absolute terms; it is enough that the resolution of normative tensions in the contract, tort, and criminal law is typical of the resolution of normative tensions in other areas of law.

The contract law matters only when the parties consent to form a relationship that will be subject to legal analysis. That is, you must consent to enter into a contract, though what consent entails has morphed over time. It will be necessary to compare the assumptions about consent, on which the contract doctrine depends, with the reality of consent, revealed by recent discoveries (or realizations) offered by behavioral economics[53] and cognitive neuroscience.[54] Contracts depend on conceptions of agreement, bargain, consent, promise, the meeting of the minds, and basic assumptions of the contracting parties. We certainly have an accessible folk psychological sense of those determinants, but do they withstand the scrutiny of a more empirical cognitive neuroscience? And if they do not, will normative justifications of the extant doctrine fail as well?

The law of unintentional torts, the only type of tortious event considered in this book, depends on a "reasonable person" standard to measure
the responsibilities we owe to one another. The premise of much of tort
is that negligence represents a failure to act as one ought. It would seem
that such a conclusion should proceed from an appreciation of how the
actions of human agents are motivated and how they might be limited
or determined by circumstances, including neurological function. Duty, a
crucial constituent of negligence, may, like consent in contract, describe a
continuum calibrated to vindicate a normative object (or objects).

The tort law appraises injury by reference to accessible indicia that
might obscure rather than reveal real damage. For example, an important
component of tort damages is pain and suffering. We are familiar with the
objective indicia of pain and suffering, but we also are aware that those
indicia may be feigned. The same would be true of emotional injury. Tort
doctrine takes the uncertainty of pain and suffering and emotional harm
into account by limiting the recovery for each. It would be serendipitous,
though, were the limitations built into the doctrine in fact accurate guides
to appraise the injury. We do not yet have a certain basis for what we presently deem nonphysical injuries, and so we continue to consider pain and
suffering and emotional harm as nonphysical.

Neuroscience, though, provides at least the framework for discovering an objective basis of nonphysical damage. If we can identify the neural signature of pain (and even of emotional harm), then we can begin to
measure it, to compare it from one individual to the next. Then too, perhaps, once the physical–mental divide has been breached or at least compromised, we might discover a way to appraise and compensate for the
injury more accurately. Indeed, just identifying the physical basis of psychic harm would go some way toward overcoming bias (born of suspicion
and skepticism) against claims premised on injuries that are not readily
observable. It is neither likely nor necessary that neuroscience set out to
answer the legal question. There are sufficient incentives to ameliorate psychic harm to ensure that resources would be devoted to its amelioration
and could, in due course, reveal underlying organic causes. That revelation
would do as much to advance the tort law as it would medicine.

Similarly, tort distinguishes physical from mental attributes when it
draws distinctions between the competencies of putative defendants. That
is the case both with children of a certain age who are not deemed competent to commit some torts and with adults of diminished competence. We
do take defendants' physical limitations into account when considering cul-

pability and liability, but we more generally assume mental competence of defendants and in most cases do not insulate the mentally infirm from liability. If neuroscience confirms that *all* conditions are ultimately and necessarily physical conditions, then we would need to reconsider doctrine that assumes a strict divide between the mental and the physical.

Finally, the criminal law has so far provided the most fertile ground for appreciating (and anticipating) the changes to legal doctrine that neuroscientific insights may ultimately compel. To some extent that could be a function of research funding interests that are themselves, presumably, in turn a function of society's conclusion that there is much at stake in identifying and controlling those who would undermine the social order. Also, murder is just more compelling than breach of contract, as entertainment industry offerings regularly confirm.

So a good deal of experimental and scholarly energy attracted by emerging neuroscientific initiatives and discoveries has been devoted to the criminal law. The United States Supreme Court itself has referred to insights offered by cognitive neuroscience on brain development in considering the constitutionality of certain punishment regimes as applied to those who have not yet reached the age of majority and who have not realized the cognitive development that criminal law assumes of more mature offenders.[55] In the case of minors, the Supreme Court has even distinguished some normative choices from others, distinguishing the competence of minors to make a decision about abortion from their competence to make a decision to murder.[56]

The criminal law also is a primary focus of neuroscientific research because a consequence of criminal liability may be confinement. General deterrence may be accomplished by notoriety of jail or prison time, but specific deterrence—confinement of the criminal—certainly deters, at least during the term of incarceration. The sentencing decision, then, presents normative issues to which neuroscience might respond. Insofar as it is only necessary, and efficient, to limit the freedom of a criminal for the period during which he would commit crimes, at least as a matter of specific deterrence, neuroscience could inform decisions about the duration (and perhaps conditions) of confinement. Neuroscience also may provide a framework to support conclusions about the effect of sentencing on general deterrence objectives, as well as the fit between specific and general deterrence.

One of the challenges to which developments in neuroscience have been most pertinent is the identification of psychopathic personalities.

Psychopaths seem to lack the affective mechanisms that the other more than 97 percent of humans rely on to function. There often are correlations between psychopathic behavior and scores on psychological tests that reveal consistencies shared among those who demonstrate proclivities to engage in antisocial, even violent, behavior. But not all people who may be labeled psychopath from a high score on the Hare Psychopathy Checklist—Revised act criminally or commit acts for which they would be criminally liable.[57] It is not criminal to lack affect, or to fail to respond typically and compassionately to the fear or pain of others. Indeed, there may even be reason to believe that the neural mechanisms and functions that mark psychopathy are present in those who are very successful by conventional standards.[58] Not all psychopaths are criminals, but many criminals are psychopaths.[59]

So the ability to identify psychopaths before they commit a crime presents a conundrum, and here some would find a moral dilemma. In fact, the dilemma revealed resonates with the familiar empirical–conceptual tension. Surely if we knew, to a certainty, which psychopaths would commit a crime, particularly a violent crime, we would at least limit the freedom of such individuals and perhaps even incarcerate them. Indeed, we do this now with convicted child sex offenders, who may be required to spend Halloween at the local police precinct. And we would not and do not hesitate to quarantine those with dangerous communicable diseases, such as Ebola. As a *conceptual* matter, then, the answer is not difficult: preemptively remove from the general population those who certainly present dangers to others. Indeed, that is one of the justifications for specific deterrence. In the case of convicted criminals, however, to the extent that we incarcerate to specifically deter (and drawing lines between specific deterrence, general deterrence, and retribution is concededly problematic insofar as unitary sentencing does not delineate the three), we can justify the incarceration by pointing to the greater empirical certainty that the convict has demonstrated the propensity to commit a crime, perhaps the same one or one even more heinous in the future. That seems like a perfectly reasonable folk psychological conclusion, and one consistent with common sense, which is, after all, the stuff of folk psychology.

But in the case of someone who just has a troubling score on a psychological test, the reliability of which may be less than certain, there is reason to proceed cautiously. We can "see" Ebola in clinical tests; we have no similarly reliable determinant, yet, of dangerousness on account of psychopathic personality disorder. The problem is, again, an empirical one, not a conceptual one. Surely if we had the same confidence in our ability

to identify violent psychopaths as we do to diagnose Ebola, we might limit
certainly violent psychopaths' freedom. The fact that we quarantine those
with dangerous diseases confirms that there is no conceptual impediment
to what amounts to incarceration (though presumably in a less challenging environment than that presented by our prison system). Although it is
true that Ebola may be treated and the danger it poses to others thereby
minimized, the same is not *yet* true of psychopathic personality disorders.
We may achieve greater accuracy in our ability to identify psychopathy,
perhaps even by visible physical indicia, well before we have any treatment for it. Presented that starkly, reliance on neuroscience to inform the
operation and application of the criminal law seems quite problematic.

Consider as well the other side of the equation, which some may find
no less troubling. If we were to identify certainly a physical basis of psychopathy and then also develop a reliable treatment of it, we could treat
psychopaths and release them. Doing so would make sense on any coherent
normative basis: From the instrumentalist perspective, there is no work for
deterrence, general or specific, to do. Insofar as even retribution, a noninstrumentalist concept, is concerned, it would *seem* unfair to punish the
cured psychopath for crimes committed by him when he was sick, just as
it would be unfair to continue to quarantine the person cured of Ebola for
having been contagious before being cured. Although the example of psychopathy is particularly salient, at some organic level all actions, including
all criminal actions, are the product of neurological forces, chemical or electrical or structural features that may be adjusted by chemical or electrical
or structural interventions. And if the criminal law in its doctrine, including sentencing components, is inconsiderate of that neuroscientific conclusion, then its normative foundation is compromised, maybe ultimately
incoherent.[60]

Though the contract, tort, and criminal law would all be subject to
reconsideration in light of truths that neuroscience *might* reveal, it is important to appreciate why and how those three fundamental components
of the law would be differently affected. In this regard, a good deal may be
gleaned from comparison of the remedial objects of the primary doctrinal
categories.

Normative Object of Remedies

A brief description of the contract, tort, and criminal law normative objects supports comparison of them. The object of contract is not to punish

breaching parties or even to deter breach, really; the object is to provide the certainty of expectation that a system needs in order to accommodate exchanges that create welfare. That remains true even for those who would base contract on noninstrumental principles, and, for example, equate contract with promise in a Kantian (that is to say, deontological, duty-based) sense. The fact that those who breach contracts can be liable for damages would certainly serve to deter them from breaching contracts, but the object is not deterrence for deterrence's sake; it is to make the performance of promises more likely by removing (or at least limiting) the benefits to be realized from breach.

Contract liability is strict liability: If you breach, we do not care why you breached. Unless some excusing event intervened, you are liable for whatever measure of damages will enable the nonbreaching party to realize the benefit of her bargain. It is possible, though, to conceive of contract in noninstrumentalist terms, and to award damages on noninstrumentalist bases. Such normative theories would rely on conceptions of the human agent and depend on conclusions about the nature of the affront that breach accomplishes. Those conceptions would have to be consistent with whatever neuroscience reveals about human agency. It may be the case that there is a consequence of breach that is not compensated by the award of expectation damages. Unless it is a conceptual error to do so, the law would have to take account of the residual harm; neuroscience could confirm that such harm in fact occurs, and perhaps even ground an empirical basis to measure it. But it could be error to assume some compensable dignitary affront on the basis of armchair speculation.

The bottom line, though, is that extant contract doctrine, insofar as damages are concerned, focuses on vindication of the expectation interest (only considering alternative measures of damages when confidence in the expectation measure is lacking). Although we may have reason to look to empirical evidence to determine what the expectation is and how it might be vindicated, for purposes of comparing contract with tort and criminal law, at least insofar as punishment and remedies are concerned, contract's focus is on compensation of the nonbreaching party, not on punishment of the breaching party or retribution for any harm caused to society generally. That formulation works well enough to distinguish the remedial object of contract from the criminal law.

Contract is distinguishable from tort with regard to the basis of liability. Contract liability is, for the most part, strict liability, and so not dependent on deterrence or retribution premises to provide a normative foundation, whereas tort law is more generally based on fault. That is true even in many cases of so called "strict products liability."[61] Tort liability is imposed on those who (at least) negligently harm the interests of others. That negligence is framed in terms of breach of a duty, a failure to act reasonably. Though the tort law discourages, and so deters, unreasonable conduct and the imposition of tort liability redresses the imbalance[62] resulting from tortious conduct, we again do not require deterrence or retribution premises to found a normative explanation of tort law. The law works when the victim of a perpetrator's negligence is made whole, but in no better position than she would have been in had there been no tort.

Indeed, the curious bilateral nature of tort damages[63] makes clear that victim recovery is not normatively related to perpetrator culpability. If A, a very wealthy individual, negligently (but not recklessly or willfully) drives his expensive new Ferrari into B's ten-year-old Honda Civic and causes $5,000 of damage to the Civic and $100,000 in medical expenses for injuries to B, A may not be substantially deterred from engaging in such behavior by B's recovery of the $105,000 from A if A chooses to pay it from his checking account. Indeed, the same result largely obtains if A is not wealthy but has sufficient insurance. There is no reason to be confident that any increase in A's insurance rate would be sufficient to cause A to take any more care than he otherwise would just taking account of his own interest in not damaging the Ferrari. Now we can certainly imagine circumstances in which the imposition (or threat of the imposition) of tort damages could encourage greater care and so deter less careful conduct. But the law does not depend on that, and any deterrent effect will often be diluted if not obscured by the availability of insurance against tort liability.

Keep in mind also that unreasonable behavior alone, not followed proximately by injury, does not expose the unreasonable actor to liability. If the object of the tort law were strictly to deter unreasonable conduct, the imposition of liability would not depend on the existence of a victim who suffers compensable harm. Indeed, the requirement that we find a sufficiently direct victim also undermines any normative explanation of tort based on retribution. Arguably, per corrective justice, even those not directly harmed by tortious behavior have suffered harm (lack of respect? contempt?) by the unreasonable actor's demonstration of indifference toward their well-being.[64]

Any arguments that corrective justice and civil recourse theory[65] provide a normative (and noninstrumental) explanation for the tort law ultimately may be controversial and depend on folk psychology assumptions that lack empirical support, or at least rely on incomplete folk psychology accounts. Although, arguably, entertaining ratiocination may posit a noninstrumental account of tort, it is clear that the doctrine in no way depends on it. Surely the law could be changed to ensure results consistent with corrective justice or civil recourse theory, but that would not provide a positive account of the extant doctrine. Tortfeasors who injure others in person or property may discharge their liability by paying an amount of money sufficient to compensate their victim but insufficient to either deter future conduct or to impress on them the moral affront they have caused. (Indeed, it is not clear that there is a moral affront when injury results from negligence: Even a dog, it is said, knows the difference between being kicked and being tripped over.)

We need know less about the human agent in order to impose contract or tort liability in a manner consistent with an instrumentalist normative object: Contract affords the nonbreaching party the benefit of her bargain so that transactors may rely on the future value of promises in fixing risk (which is what contracts do); tort puts the injured party back in the position she would have been in had there been no tortious conduct, all that is required to avoid the consequences of unreasonable actions that result in harm. So once we know that contract and tort need only be concerned with compensation (by reference to the expectation interest in contract, and restoration of the status quo ante in tort), we can reach some conclusions about the nature of human agency pertinent to the normative inquiry in both contexts. From that perspective it becomes clear that pure instrumentalism can do any necessary heavy lifting: fix liability and damages in such a way as to vindicate expectations (contract) or restore status quo ante (tort). We can arrive at the sufficient liability and damages conclusion without much reference to the moral responsibility of the defendant actor. Indeed, insofar as contract liability is strict liability, there is no requirement that we find that the breaching party has done anything wrong. The breaching party may well have gone to extraordinary lengths to attempt to avoid breach or to minimize the consequences of breach. We may still find liability notwithstanding such heroics. Similarly, in tort, there is nothing necessarily morally repugnant about being negligent. You could even conclude that virtually everyone is negligent in some way virtually every day. The reason that there is not more tort liability is that most negligence does not result in compensable injury, or injury for which it makes sense to seek compensation. And certainly in some tort settings, inherently dangerous instrumentalities and manufacturing defects, the defendant may be liable notwithstanding the exercise of the utmost, even extraordinary care. The object of tort is to compensate the wronged, not to identify and punish a wrongdoer.[66]

None of the foregoing is meant to suggest that the contract and tort law could be indifferent to the nature of the human agent. In both bodies of law we will need to answer questions comparing the actions of the parties to those of typically situated actors: Both contract and tort rely on conceptions of reasonableness.[67] Insofar as the contract and tort law take into account and make assumptions about the cognitive characteristics of human agents (and the ranges and vicissitudes of those characteristics among human agents), neuroscience may contribute to the rationalization of the apposite doctrine. But the criminal law is importantly different from contract and tort in ways that are particularly salient with regard to the normative commitments of the three foundations of legal doctrine. Useful comparison may be made between contract and tort, on the one hand, and the criminal law, on the other.

The object of the criminal law is either deterrence or retribution or some combination of the two. The criminal law is not concerned with vindicating expectations or restoring the status quo ante: It may rely on contract and tort to do that.[68] In fact, then, we may see the three bodies of doctrine as normatively complementary: Each fills gaps left by the others and may assume that the others fill gaps it leaves.

The criminal law provides abundant room for normative considerations based on deterrence and retribution interests to operate. The focus of the criminal law is on the defendant. Now that might not be the sole concern of the criminal law, but it is certainly the distinguishing constituent of the criminal law's normative object. The consequence of a criminal conviction is correction, both for the sake of the defendant (at least ostensibly) and for society. The criminal defendant is to be corrected; his relationship with society and his victim is to be corrected. Once we recognize the criminal law's interest in correction, it becomes clear that deterrence and retribution may be constituents of that correction.

Consider first deterrence: We punish criminal behavior in order to accomplish both specific and general deterrence goals. A convict incarcerated is not free to commit further crimes (specific deterrence) and others considering that same or similar behavior have a concrete example of the consequences (or at least potential consequences) of such behavior (general deterrence). From those simple premises, it is clear that we could, at least theoretically, calibrate the specific or general deterrence (the fine or

sentence or other limitation on the defendant's freedom) to best effectu-
ate our object. There are certainly obstacles to an accurate calibration:
There is no reason to believe, much less assume, that the single sentence
(e.g., ten years in a medium-security facility) will realize both the specific
and general deterrence objects. The two may be in tension, perhaps irreconcilable tension.

It is difficult to know what particular type or quantum of specific deterrence is necessary to accomplish the goal:[69] Certainly incarceration will keep
the defendant "off the street," but there may as well be potential victims
"on the inside." So even that limited object of specific deterrence may be
frustrated; the sentence is, at best, a guess. Further, with regard to specific
deterrence, to the extent that the threat of a sentence (i.e., fine or incarceration) is designed to dissuade the potential perpetrator from engaging
in the proscribed conduct, we would need to know something about the
cognitive and emotional characteristics of the subject before we could calibrate the threatened punishment accurately. That target may be a difficult
one to hit even approximately: We would need to appraise the potential
perpetrator's ability to forecast affect,[70] to appreciate genuinely the effect
that incarceration for a particular term or the imposition of a fine in a particular amount would have. It does not take much reflection to appreciate
that it would be very difficult, likely impossible, to get the mathematics
just right.[71] And you could point to high rates of recidivism to confirm that
we often get the mathematics wrong. Surely one-size-fits-all minimum sentences could accomplish accurate deterrence only serendipitously. Even
if we were, by luck or superhuman prescience, able to calibrate the deterrence accurately, we would need to be able to factor in a detection discount, the unlikelihood that the criminal activity as well as the identity of
the perpetrator would be discovered.[72]

The object of general deterrence is to send a message, to impress on
those other than the particular defendant the consequences of actions that
violate the law in order to discourage such violations, to make an example
of the defendant. The efficacy of general deterrence is a function of many
variables, including its salience generally and for the third party who is or
becomes aware of the consequences of legally proscribed actions. Those
consequences, of course, may be both legal and extralegal.

There are certainly many reasons to question whether we can do the
consequentialist mathematics accurately, at least with regard to deterrence
by punishment. The case is no easier, and likely a good deal more difficult, if we focus on retribution. What, exactly, are we trying to accomplish

through retribution? There is a sense (perhaps an Aristotelian sense)[73] that justice is done when balance is restored. A criminal act knocks things askew, and the criminal law redresses the imbalance by restoring the non-instrumental status quo between the perpetrator and the victim(s) of his crimes. That may well include consideration of (even instrumental) communitywide values compromised by the criminal actions.[74]

The more moving parts, the more victims we identify, the more we need to know in order to do the retribution mathematics. However we approach the calculus, it should also be clear that we need to know a good deal about the cognitive constituents of the perpetrator (and perhaps of the victims as well). We cannot accomplish the accurate rebalance without knowing at least (1) what to put on the scale, (2) the "weight" of the constituents ex ante, and finally (3) how to redistribute whatever needs to be redistributed in order to return the parties (the community, the cosmos?) to the positions they were in before the commission of the crime.[75]

Now whether or not retribution contemplates consideration of more than the mental states of the perpetrator and victims of a crime, it would be difficult to make sense of a normative system, particularly one cast in non-instrumental terms, that was inconsiderate of the mental and emotional well-being of the parties. In order to take account of well-being so construed, we would have to come to terms with human agency. Certainly any construction of human agency must involve the nature of human agency generally as well as the unique combination of attributes that determines the human agency of the parties concerned, arguably all parties concerned. Neuroscience may inform both phases of the inquiry; it at least endeavors to reveal its contours.

So both deterrence and retribution in the criminal law, and then necessarily the coincidence and potential cooperation of the two, require that we understand human agency as accurately as we can. It is difficult to make normative sense of criminal law that is inconsiderate of human agency and does not require thoughtful consideration of what it means to be human. It is no less true that though we may be able to do some of the mathematics approximately, we could not have any confidence that we are getting it right from the perspective of either deterrence or retribution even most of the time.

Similar challenges undermine the normative coherence of the contract and tort law as well. These areas of the law do not depend on conceptions of deterrence and retribution to an extent approaching that of the criminal law, but they do rely on conceptions of the normatively pertinent characteristics of human agency no less sophisticated than those assumed by the criminal law. The point, then, is that we cannot escape the fact that an understanding of human cognitive processes is crucial, even indispensable, to the formation of a conception of human agency that would inform an appraisal of the normative commitments of legal doctrine and the operation of normativity in the course of human events. Our reliance on folk psychology and theory of mind both confirms that and, once examined, reveals the deficiencies of extant conceptions.

Nevertheless, there remains . . .

The Specter of Phrenology

There is a very real possibility that our fears will outpace our science, that we will be tempted to draw conclusions on slender, even erroneous, evidence and limit the freedom of those who would in the fullness of time present no threat to anyone, even themselves. Certainly the history of science, medicine, and law is replete with examples of partial truths that turned into great lies. Contemporary critiques of neuroscience liken it to the worst episodes of scientific misdirection, even eugenics. A troubling comparison is with phrenology: the study of skull configuration to discover personality and cognitive characteristics.

Although phrenologists were much more wrong than they were right,[76] their work may well have been a necessary step on the path toward the modern neuroscience that has improved lives. We certainly now know that neural function is not as localized as it might at first seem (from a perspective redolent of phrenology), but within the networks of neurons that produce cognitive function, *where* chemical or electrical activity occurs matters. The brain is not an undifferentiated mass but a system of neural properties with localized capacities and competencies. We also know that if your hippocampus is compromised, your memory will be impaired; if there is damage to your amygdala, your affective function will suffer; and if a thirteen-pound iron rod enters just below your eye socket and pierces your frontal lobe before exiting through your skull, your personality will change, dramatically, but the insult would not necessarily kill you.[77]

There is a persistent tension between what we know about the operation of the parts of the brain and what we are coming to understand about the cooperation among the parts that results in the experience of mind. New discoveries and our evolving understanding are encouraging greater confidence in the ability eventually to figure it out. But there remains considerable disagreement about what that "it" is, and there remains room for skepticism whether *the* answer or meaningful answers will be forthcoming in the near future. Indeed, it often seems to be the case that the more we learn, the more we have to reevaluate what we have learned before and surrender what seemed to be progress.

Imaging techniques have quite literally revealed much about the brain's operation and suggested as well applications of brain science that could matter to law. There is no shortage of those who make bold claims based on slender evidence (sometimes for commercial gain) and in doing so threaten the integrity of the science.[78] It is at least as important to appreciate the limitations of neuroscience as it is to applaud its ambitions. We can use the science to look into the brain, but we cannot (yet) read minds, at least not in any way that would have a real effect either on the law or on an appraisal of the doctrine's normativity. But we need not actually read minds to draw a line from a neural state to its translation into the type of actions that matter to the contract, tort, and criminal law. Though the contract law, for example, is ostensibly concerned with finding agreement at a meeting of the minds, neuroscience may provide reliable bases to infer competence to contract and authentic consent even if we cannot actually yet see beyond the parties' representation of their understanding and intentions. Similarly, if we can identify a reliable neural signature of pain or emotional distress, we can overcome doctrinal limitations to tort recovery for them. And though certainly the criminal law would benefit as much as any other area of the law from the ability to appraise with confidence the veracity of defendants and witnesses against them, more directly connecting neural state with resulting actions could go a long way toward informing responsibility determinations. Indeed, I shall argue that even with what (little?) we know now for sure, there is ample reason to reappraise the role and substance of the responsibility calculus across all three fundamental doctrinal areas. I conclude that we have misunderstood responsibility all along, and across all areas of the law.

The discussions of the neuroscience in the chapters that follow will take account of what the science does not yet reveal as well as what it does. Though the science is, by some measures, young, it is not naive. We do not need to know what each and every neuron is doing and precisely how networks of neurons cooperate in order to appreciate challenges to conceptions we have taken for granted, but which may warrant reconsideration in light of enhanced empirical understanding. The science is headed in a direction, and it need not reach its destination before we can begin to see why it may be necessary to reevaluate where we are now.

Ultimately, though, there is a challenge to our invocation of neuroscience that may prove resistant to our best efforts, a challenge that at least for now is daunting.

Consciousness: A Final Frontier?

Our awareness of our awareness, our consciousness, is inscrutable. Though we distill neural functions into their component subatomic parts, there still seems to be something there that we cannot fathom. We *feel* as though we are both spectators and actors responding to what we perceive. That feeling persists even after we become aware of the contingency of our perceptions and awareness.

Neuroscience has not solved consciousness. Although we are able to identify degrees of consciousness among human actors and across species, we have not come very close to appreciating the substance of consciousness. Until we do, a mystery will remain for which science of the mind does not have a solution. So, for those who are skeptical of neuroscience *ab initio* (who remember phrenology), the persistence of the consciousness mystery provides grounds to question the sufficiency of neuroscience as a comprehensive conception of human agency. If neuroscience cannot support such a comprehensive conception of human agency, then what it means to be human may not be a matter of neuronal architecture and operations. Indeed, if there is a something to consciousness that neuroscience cannot reveal (or may even obscure), then there *may* be something neuroscience misses that is crucial to our ultimate conception of human agency.

Yet even were we to establish that consciousness has some substantial independent existence, that consciousness is more than brain function, the portion of consciousness that matters for the type of interpersonal relations that are the subject of law may be contingent (and even manipulable) in such a way that the more ephemeral sense of consciousness just does not matter to the normative objects of legal doctrine. It is one thing to say that there is a sense of consciousness that empirical inquiry will not (even cannot) reveal; it is wholly another to say that that sense of consciousness is normatively pertinent to the operation of legal doctrine. For example, though we know that the normative objects of retribution and deterrence depend, in large part, on what may be idiosyncratic aspects of the human subject's rational capacity, we do not need to know much about consciousness to reach some helpful neuroscientific conclusions about the efficacy of retribution and deterrence in the particular case. So even though consciousness may at bottom (for now), be mysterious, that mystery does not certainly frustrate our efforts to make use of neuroscientific insights. The limits imposed by the challenges to our understanding consciousness are not an impermeable barrier to reconsidering the normative object of law from a perspective informed by neuroscience.

Determined, Not Free

Finally, books have been written that focus on the free will–determinism opposition and the compatibilist accommodation of the two. This is not another one of them. If you believe that humans have free will, an assumption on which much (all?) of law is based, then you believe that at some level we must be morally responsible for the choices we make. If you are a determinist, at least in the strictest sense, there are no uncaused causes in human agency; our thoughts, our actions, our feelings are the product of forces beyond our control. And if you are compatibilist, you believe that humans are determined creatures but that there is enough that is free-will-like to support impositions of moral responsibility based on free will conceptions.

Implicit in the materialistic, physicalist approach to human agency that neuroscientific insights vindicate is the conclusion that we are determined creatures who have nothing like the libertarian (in the philosophical, not political, sense) free will that is, in fact, indispensable to deontology and noninstrumentalism generally. That means that if we in fact do *not* have free will, are wholly determined in our actions (including our choices), then a normative philosophy built on free will is incoherent, at least as something usefully applicable to the morality of human agents.[79] Insofar as neuroscience demonstrates the essentially mechanical (chemical, electrical, and structural) nature of our human agency, and mechanical entities are determined entities, lacking the *sui causa* free will of deities, free will is certainly a failed hypothesis (though it may make for an entertaining, or at least occasionally comforting, theology).

The perspective of this book and the premise of the arguments developed here are resolutely deterministic. That conclusion is based on empirical findings such as those developed and reported by Adrian Raine, whose important work is considered at length in chapter 2. It does not depend on the experiments of Benjamin Libet or even the most sophisticated observations and conclusions of Daniel Wegner. Libet devised experiments intended to demonstrate that we make the decision to act before we are conscious of that decision: So we must not have free will because only actions consciously chosen could be free.[80] Wegner reasoned that our consciousness is an illusion.[81] Criticisms of Libet's method seem plausible.[82] But Libet's findings are not crucial to the deterministic conclusion: That is, even if Libet is wrong and we *do* consciously choose when we choose to act (including when we decide), that would say nothing about the forces that determined that conscious choice. Raine, as we shall see, provided the data to suggest (if not establish) that we are the product of forces that act on us—and *only* the product, not the producer of such causes. So it would not matter if something like choice intervened when the human agent acts: If that choice is determined by nature and nurture, there is no basis for moral responsibility independent of those forces. And since we are not morally responsible for the formative forces that determine us, there is no "we" responsible for the consequences of those choices. The law assumes that moral responsibility, and so it is no surprise that legal doctrine often fails. It is based on a false premise.

The majority view, barely,[83] among philosophers is "compatibilism," the view that human agents have *sufficient* free will to justify the imposition of moral (rather than merely causal) blame even though we are determined creatures. I find that view incoherent; none of the extant justifications of it are convincing. But the space limitations of this project preclude thorough consideration of the failure of compatibilism. It may be worthwhile, though, to offer two potentially pertinent observations: First, of that bare majority of philosophers who subscribe to compatibilism, it is likely that only a small number agree to any particular iteration of that perspective. In that way, compatibilism is like theology: A majority of people may believe in a supernatural deity, but their views are ultimately irreconcilable (and in some cases have led to the bloodiest wars in world history). Second, it is probably true, though I am aware of no source that would confirm this, that most, if not virtually *all* professional philosophers are deontologists, at least in their conception of the morality of human agents. And so their arguments about what human agents do and should do necessarily entail deontological (often, Kantian) precepts. Well, if human agents are determined agents, agents without libertarian free will, then such moral perspectives (and the professional careers based on them) are vacuous, just another form of supernatural theology. Further, it would not be surprising that the (bare) majority of moral philosophers are compatibilist. After all, most contemporary moral philosophers were trained before neuroscience began to provide the materialistic reconceptualization of human agency. But once we come to understand that human agents are determined, not possessing free will, normative philosophies positing stuff like moral realism are revealed as vacuous, and human institutions such as legal doctrine, founded on and justified by those normative philosophies, are undermined. This book is about that undermining and tries to answer the "what if" question: What if neuroscientific insights, broadly construed, confirm that human agents are determined creatures, without free will in any meaningful moral sense? What happens to our legal doctrine when the normative premise of our legal doctrine, noninstrumentalism, fails?

The Plan

From the premises sketched in this chapter, the book offers a review of the normative commitments of the legal doctrine as representative normative theories present them. The several chapters that follow develop the course of an argument. The next six chapters of the book present, in turn, the criminal law doctrine and criminal law theory, the tort law doctrine and tort law theory, and the contract law doctrine and contract law theory. Of course it is not possible, in any work of determinate length, to survey all the doctrine in those three primary areas of the law and all the normative theory that would interpret the doctrine. In each of the three doctrinal chapters, I focus on fundamental aspects of the doctrine that make the materialistic critique offered by neuroscientific insights most salient. The treatment of the doctrine, then, supports critique of noninstrumental (generally deontic) interpretive theories that would make sense of the doctrine. We shall see that the description and prescription offered by those interpretive theories founder.

The order of the presentation is significant: The criminal law has provided the most accessible and typical setting in which commentators and courts have considered the effect that neuroscience may have on the law. So the points made in the course of examining criminal law doctrine and theory set the stage well for the subtler, but no less telling, effect that neuroscience would have on the tort law and theory. Arguments developed in the course of discussing the criminal and tort law then build on and are built on in the contract law doctrine and theory chapters to question the coherence of our pervasive law of even consensual relations.

The final chapter is something of an anticipatory reply to critics, cast as an assault on the strawmen set up by those who would be most discomfited by the materialistic critique of our law that an appreciation of the neuroscientific vindication of determinism entails.

CHAPTER TWO

Neuroscience and Criminal Law Doctrine

Though only seventeen years old, Christopher had long fantasized about killing someone. He often talked about it with friends. One night during his junior year of high school, Christopher Simmons murdered Shirley Cook.

It was around 2:00 a.m. when Christopher reached his hand through an open window to unlock Shirley's back door. With the help of a friend, Christopher bound Shirley's wrists and ankles with electrical wire and placed duct tape over her mouth and eyes. Loading Shirley into her own minivan, Christopher transported her to a state park where he applied more duct tape, covering her entire face with it, then tossed Shirley off a bridge and into the water below, where she drowned.

Christopher was proud of his first murder. He bragged about it at school. When he was taken into police custody, Christopher waived his right to legal counsel and offered to reenact the murder for the officers. Within a year, Christopher was convicted and sentenced to death. His defense attorney asserted adolescence as a mitigating circumstance that called into question his culpability. Using Christopher's age against him, the prosecutor also pointed to Christopher's minority status, but suggested that all the years in the world would not alter the fact that, at his core, Christopher was a monster.

In 2005 the Supreme Court heard Christopher's argument that it would be cruel and unusual punishment to execute a defendant for a crime he committed as a minor.[1] At the invitation of Christopher's attorneys, the Court contemplated the analogy between the diminished culpability of the mentally handicapped and the diminished culpability of unthinking teenagers. Indeed, the adolescent brain does produce somewhat more reckless and impulsive behavior than the adult brain. And, according to neuroscience, the (not-so-bright) line between the adolescent and adult brain should be drawn closer to twenty-five than to eighteen.

The Court ruled against application of the death penalty to crimes committed by minors. Writing for the majority, Justice Kennedy cited behavioral research studies indicating that adolescents are more impetuous and reckless than adults, more susceptible to peer pressure, and unable to realize the benefit of fully formed personality traits.[2] The justice explained that "From a moral standpoint it would be misguided to equate the failings of a minor with those of an adult, for a greater possibility exists that a minor's character deficiencies will be reformed."[3] He also cited neuroscientific findings.

Introduction

As the body of law that governs the power of the state to exact punishments for transgressions, the criminal law sees issues of moral responsibility and the essence of human agency loom larger than in any other primary doctrinal area. The criminal law assumes that human agents are responsible in a normative sense, an assumption inherent in both noninstrumental and instrumental conceptions of desert. First, we punish those we punish because they deserve it, and in so doing, we assume human agents freely choose the criminal act. Second, we punish those we punish to minimize the incidence of such behavior in the future and thereby assume individuals choose, in some sense, not to act criminally. It may well be that such noninstrumental and instrumental objectives could both be served by the same, unitary punishment decision, but there is nothing inevitable about that result. In fact, serving both instrumental and noninstrumental goals simultaneously may be impossible in most cases. Realization of one goal may necessarily entail frustration of the other in all, or virtually all, cases.[4]

The basis of noninstrumental punishment is retribution, which its champions might distinguish from revenge.[5] The idea is that there is some, perhaps ineffable, quality of the criminal act that warrants, even demands, societal response in the form of punishment. The basis for such desert is elusive. If there is something that just feels right about exacting punishment, is that something laudable or contemptible, efficacious or futile? The tension is venerable, or, at least, durable.[6] And if we do decide on some normative basis that punishment is deserved, how do we determine

the quantum of punishment? To what extent would idiosyncratic characteristics of the perpetrator or the particular circumstances of the act matter
in the normative calculus? Is such normative mathematics even possible?

With a focus on deterrence, instrumental bases of punishment may
avoid some of the problems presented by noninstrumental premises, but
the mathematics is no easier. We must still achieve (or at least assume) a
level of normative acuity that is quite artificial with the current limitations
of even the best neuroscience. It may be impossible to know what specific deterrence (deterring the antisocial behavior of a particular defendant) is required or how general deterrence (deterring others) would be
calibrated to serve rather than frustrate the underlying normative object.
Although this book focuses on the doctrinal incoherence of noninstrumental, primarily deontological, normative principles, it is worth noting
that in the criminal law, the limits of human knowledge may render instrumental objects just as elusive.

Criminal law is premised on conceptions of normative rather than (or
in addition to) causal responsibility. If your car does not start one morning, your car is, in the causal sense, responsible for your being late to work.
If the reason your car does not start is that you failed to put gas in the
tank, then you, in both a causal and at least colloquial normative sense, are
responsible for the car's not starting (and for your being late to work). We
are used to that important normative distinction in our common attribu-
tions of fault: when it makes sense to blame yourself, when it does not make
the same sense to blame your car.

Normative responsibility and its limits are the province of the criminal
law. We aim to punish those who are responsible in the normative sense;
we do not punish inanimate objects or even people who act as inanimate
objects because such punishment would not serve any noninstrumental or
instrumental purpose. Accordingly, the criminal law must engage issues of
responsibility in ways that other doctrinal areas need not. For the most part,
the criminal law is premised on the coincidence of a *mens rea* (a culpable
state of mind) with an *actus reus* (the criminal act), and the parameters of
each of those constituents are subject to layers of nuance and scholarly
inquiry.[7] For our purposes, it is sufficient to interrogate, in a general manner, how conceptions of human agency in the current criminal law rely on
assumptions that emerging neuroscientific insights could undermine.

By focusing on the criminal law's punishment and sentencing doctrine,
this chapter will demonstrate the uneasy fit between noninstrumental and
instrumental objects to reveal how neuroscientific insights might expose

the ultimate incoherence of the extant doctrine. Although no corner of the criminal law is immune to reconsideration in light of developments in neuroscience, certain contexts present the dominant tensions in more revealing relief. The criminal law as it relates to juvenile justice offers such revelation. We seem to feel, even profoundly, the difference in the normative calculi when we juxtapose punishment of a child with punishment of an adult. Regardless of which normative concept of responsibility we embrace, the criminal law is put to the test when it comes to the doctrine's treatment of adolescents and children. We might intuit that preadults are less culpable for crimes committed, but, apart from culpability, we might also feel compelled to protect preadults from the consequences of their own actions. That is human nature, after all.[8] In turn, when we investigate our intuitions about the punishment of juveniles, we need to better understand both juveniles and ourselves and in terms that neuroscience may explain, if not vindicate.

The issues surrounding juvenile justice generally provide fertile ground to review the doctrine in light of neuroscientific insights, but punishment issues more generally provoke such introspection. We need the neuroscience to help us better understand what punishment does to the criminal, whether and how punishment works at the neuronal level. If we are, after all, nothing more (or less) than the sum of certain chemical, electrical, and structural incidents of our brains, then we need to know how punishment acts on those incidents in order to know whether we are serving or frustrating our normative object(s).

This chapter will proceed through three principal parts to demonstrate the deficiencies of the criminal law doctrine. First, and primarily, the sense of responsibility that emerges from the doctrine depends on conceptions of desert that may be undermined when we better understand the contingent relationship between mind and brain. If mind is just a label we assign to some constellations of brain function, then criminal responsibility premised on state of mind ultimately makes sense only in terms of cerebral mechanics. Second, the chapter considers the doctrine's inconsistent sensitivity to limitations of cognitive capacity, often in defiance of neural reality, by examining the variances in sentencing for juveniles, adults with congenitally diminished intellectual capacity, and adults with acquired cognitive impairments resulting from trauma. Third, this chapter will show that, once we better understand the effect of nature and nurture on the neural state that determines responsibility, the doctrine's inability to account for such complicating factors prevents the criminal law from successfully asserting any normative foundation, noninstrumental or instrumental.

"Responsibility," Retribution, and Deterrence

Whether based on noninstrumental or instrumental premises, any theory of criminal punishment encourages society to punish only those who are, in some normatively coherent sense, responsible for their actions. That, in fact, is the basis of our requirement that there must be coincidence of *mens rea* and *actus reus* in order to impose criminal liability. We do not punish merely violent thoughts without subsequent action, and we do not (criminally) punish the epileptic whose grand mal seizure causes his arm to strike a bystander.[9] Criminal liability requires the coincidence of mental state and voluntary action.

There are, though, degrees of volition. Even those who believe in free will must admit there are degrees to which our actions are freely undertaken or determined by outside influences. Criminal liability, or the punishment thereof, might be mitigated by factors that undermine free will. Not even the most idealistic believer in free will (and so too, the compatibilist) denies that at least some of our actions are the product of forces beyond our control. And the determinist (or incompatibilist) just goes one step further, citing the omnipresence of external influences to conclude that *none* of our actions are the product of free will in any sense of normative responsibility.[10] Yet the criminal law (indeed, like all of our law) is committed to the reality of free will. But if there is no such thing as free will, there is no meaningful normative sense of responsibility. That is, if all we are is the product of forces acting on us, material alterations of neural chemical, electrical, and structural properties, where is the "I" that deserves punishment?

Once we appreciate the precariousness of a distinct I, we must question the premise of retribution. Retribution is based on desert (as it must be)[11] and requires that desert be measurable in the normative sense. And, while desert itself is ephemeral, courts persist in treating retribution as a concrete concept. *State v. Kirkbride*[12] presents the typical conflation. The defendant committed a particularly heinous crime,[13] and the sentencing court specifically contemplated punishment based on retribution, notwithstanding the fact that the statutorily articulated sentencing policies did not include any reference to retribution. On the basis that retribution had traditionally been an object of the criminal law, the Montana Supreme Court confirmed the validity of the sentence in spite of Montana's statutory sentencing guidelines, finding it sufficient that the concept of retribution was an ostensibly integral part of the criminal law to insulate the pronounced sentence from review.[14] The court offered no suggestion how the appropriate retributory effect of the sentence might be measured, though the court did point out that the interests of deterrence also were served by the retributory sentence imposed.[15]

The criminal law may have two goals—retribution and deterrence—but there is only a single sentence for any given crime, and there is no immediately obvious way to determine the extent to which a sentence serves either purpose. Further, it would be serendipitous in the extreme if the same number of days, months, or years of incarceration (or community service, for that matter) would simultaneously accomplish both purposes. Indeed, it is more likely, perhaps even inevitable, that the same sentence would necessarily work at cross-normative purposes. The very idea behind deterrence, especially general deterrence (an instrumentalist idea), is antithetical to that of retribution (a noninstrumentalist commitment): The noninstrumentalism of Kant's deontology prohibits us from using anyone as a means to an end rather than as an end in himself.[16]

Melding retribution and deterrence results in fundamental incoherence. The United States Sentencing Guidelines[17] demonstrate this truism:

> The court, in determining the particular sentence to be imposed, shall consider—
>
> 1. the nature and circumstances of the offense and the history and characteristics of the defendant;
> 2. the need for the sentence imposed—
>   A. to reflect the seriousness of the offense, to promote respect for the law, and to provide just punishment for the offense;
>   B. to afford adequate deterrence to criminal conduct;
>   C. to protect the public from further crimes of the defendant;
>   D. to provide the defendant with needed educational or vocational training, medical care, or other correctional treatment in the most effective manner.[18]

The phrase "just punishment" seems unnervingly ambiguous, but it is clear that the inquiry contemplated by (A) must entail normative factors different from those contemplated by (B) and (C). (A) demands that the court consider noninstrumental conceptions of retribution (maybe even revenge), and (B) and (C) ask the court to focus on deterrence (general and specific). The use of the phrase "just punishment" in (A) at best invokes something like corrective justice[19] and at worst sanctions revenge (to the extent that revenge is somehow different from retribution).

Though many courts have reviewed the "parsimony provision,"[20] very few have adequately analyzed the definition of "just punishment" set forth in Section 3553(a)(2)(A). In *U.S. v. Wilson*,[21] Judge Cassell[22] asserted that "Just punishment means, in essence, that the punishment must fit the crime."[23] To Cassell, however, this vague standard did not depend on the opinion of each individual judge: "'just punishment' requires the court to consider society's views as to appropriate penalties."[24] To justify the use of the Sentencing Guidelines, he argued that there is a strong correlation between the guidelines' recommended sentences for particular crimes and the public's opinions of the sentence an offender deserves.[25] The use of public opinion fails to consider that laypeople do not engage in a philosophical inquiry to determine what sentences should be administered. "Just deserts" seems to lend support to a retributive justification of punishment, but obscures the fact that the US public has long supported rehabilitation as the primary goal of the criminal justice system.[26]

The latest iteration of the Model Penal Code[27] (MPC) attempted to impose a logic on the melding of retribution and deterrence in sentencing guidelines. With regard to juveniles in particular, the new code recommends that sentencing be guided by two utilitarian priorities. First, sentencing should serve "the purposes of offender rehabilitation and reintegration into the law-abiding community." Second, the new code suggests that "priority may be given to the goal of incapacitation" for high-risk violent offenders. Yet the MPC also demands that these juvenile sentencing priorities be considered in the context of the code's general sentencing provisions, which provide, in pertinent part, that punishment be calibrated according to "the gravity of offenses, . . . harms done to crime victims, and the *blameworthiness* of offenders."[28] The comments that accompany the MPC make clear that placing primacy on such retribution-based, noninstrumental proportionality serves to protect society from the hazards of sentencing focused on instrumental goals, which apparently include both radical rehabilitation efforts and seemingly lenient (albeit effective) punishment: "Deontological concerns of justice or 'desert' place a ceiling on government's legitimate power to attempt to change an offender or otherwise influence future events. So too, an appeal to utilitarian goals should not support a penalty that is too lenient as a matter of justice to reflect the gravity of an offense, the harm to a victim, and the blameworthiness of the offender."[29]

That raises the question: if a lenient punishment is instrumentally effective, why would justice require a harsher punishment? And how much harsher? Insofar as the doctrine would have us compare noninstrumental and instrumental goals and then combine them, the doctrine simply makes no sense unless it can describe the contours of that calculus and amalgamation.[30]

For present purposes, it is particularly noteworthy that the Model Penal Code distinguishes the role of retributive concerns in sentencing juveniles from the role such noninstrumental principles play in the sentencing of adults. According to the MPC, utilitarian purposes should guide the court more in sentencing juveniles than in sentencing adults, intimating that the hazards of a lenient sentence are somehow reduced in the case of juveniles. Neuroscience generally supports such line drawing, because the minds of minors are quite different from the minds of adults in a variety of ways, but it is not immediately clear that the doctrine draws the lines correctly. To the extent that the Model Penal Code reflects doctrine,[31] we must understand why the lines have been drawn where they have been drawn heretofore, and whether emerging neuroscientific insights compel us to return to the drawing board. Further, an inquiry into more scientifically valid line drawing may cause us to conclude that the entire doctrine must be reworked and may support the realization that retribution is ultimately an imprecise and unhelpful concept that serves only to prevent the criminal law from achieving coherence—in the case of both juveniles and adults.

Measures of Capacity in the Doctrine

The criminal law recognizes the normative infirmity, or at least the inefficacy, of imposing punishment without accounting for the mental capacity of a convict. That is, just as mental illness may call into question the *mens rea* requirement for the imposition of criminal liability,[32] the criminal actor's mental capacity may affect the imposition of a sentence. The following policy statement supporting the United States Sentencing Guidelines[33] explains how the *fact* of diminished capacity may militate in favor of a reduced sentence as long as the public safety is not compromised:

> A downward departure may be warranted if (1) the defendant committed the offense while suffering from a *significantly reduced* mental capacity; and (2) the significantly reduced mental capacity *contributed substantially* to the commission of the offense.... "Significantly reduced mental capacity" means the defendant, although convicted, has a significantly impaired ability to (A) understand

the wrongfulness of the behavior comprising the offense or to exercise the power of reason; or (B) control behavior that the defendant knows is wrongful.

However, the court may not depart below the applicable guideline range if (1) the significantly reduced mental capacity was caused by the voluntary use of drugs or other intoxicants; (2) the facts and circumstances of the defendant's offense indicate a need to protect the public because the offense involved actual violence or a serious threat of violence; (3) the defendant's criminal history indicates a need to incarcerate the defendant to protect the public; or (4) the defendant has been convicted of [obscenity, sexual abuse, sexual exploitation and other abuses of children, or crimes of transportation for illegal sexual activity and related crimes].[34]

Notably, that statement reflects a relatively coherent merging of instrumentalist considerations related to specific deterrence with the noninstrumentalist concern of blameworthiness. Permitting blameworthiness to calibrate sentencing does not render the punishment scheme incoherent because blameworthiness is equated with mental capacity. Viewing retribution as determined by the gravity of the offense alone, without reference to the capacities of a given "I," is what remains hazardous. Here, the gravity of the offense is considered in conjunction with specific deterrence and public safety; the particularities of the "I" involved are not as easily overlooked. Yet, even if it seems intuitive to make assumptions about the risk to public safety posed by certain perpetrators as shown by the particularities of the crime committed, those intuitions and assumptions may well be flawed.[35] Clearly, unexamined bias should not be permitted to orient the doctrine or its application, and neuroscience should be employed to reveal such prejudice.

Blameworthiness considered via mental capacity may be the key to creating coherence within the criminal law. Allowances made for those with diminished capacity constitute a coherent sentencing scheme, integrating both instrumental and noninstrumental purposes, because determining desert depends on establishing mental capacity. Neuroscience, rather than prejudice or assumption, can facilitate conceptualization of pertinent blameworthiness more accurately.

Measures of Capacity in the Neuroscience

Neuroscience enables us to make generalizations about the development of cognitive capacity and function from infancy through adulthood, to reach conclusions about the comparative intellectual and social abilities (and

disabilities) of typical individuals, and to compare any particular child, adolescent, or adult with the "norm."[36] Accordingly, neuroscience enables us
to derive normative judgments *in light of* neural development. It would not
be fair, in some sense, to expect as much of a child as we do of an adult when
the two do not have the same cognitive capacity. Likewise, adolescents are
not inconsiderate of others on account of some moral deficiency; they are
less considerate than mature adults[37] because they are wired differently.[38]
Neuroscience enables us to *see* chemical, electrical, and structural brain
states and differences and then challenges us to determine the normative
significance of those material differences.

The criminal doctrine does distinguish, at least in the capital punishment setting, between individuals who are intellectually disabled and those
who are not, by way of intelligence quotient test scores,[39] an objective measure. But there is no easily administrable test to determine how a particular adolescent's cognitive capacities compare to the adolescent norm. Many
rightly question the probative value of comparing group data (e.g., the age
at which *most* adolescents development other-regarding sensibilities) with
evidence that a particular adolescent matured sufficiently (e.g., became
himself sensitive to the perspective of others). Of course, there will also be
those for whom such realization never occurs. Indeed, given the potentially
low level of confidence in any results, we might decide for instrumentalist
reasons[40] that the normative benefits of drawing the line are not worth the
costs entailed. Such inquiry is already baked into the law, in the elements
of particular crimes,[41] the terms of affirmative defenses,[42] and the aggravating[43] and mitigating[44] factors pertinent to sentencing decisions. We may
not adequately understand the neural premises behind those distinctions,
but the fact that the doctrine embraces them demonstrates the extent to
which the law encourages us to employ neuroscientific insights to evaluate
how responsible a particular criminal is for his crime. The challenge will
be to determine what is left of normative responsibility once we have
exhausted its simple causal sense.[45]

Hooks v. Thomas[46] well illustrates the tensions that arise when responsibility is interpreted in terms of mental capacity. The *Hooks* court
applied the Supreme Court's decision in *Roper v. Simmons*,[47] the case
narratively noted at the beginning of this chapter, in which the Court
prohibited imposition of the death penalty for crimes committed before
age eighteen. Hooks characterized his criminal actions[48] as the product
of frontal lobe dysfunction,[49] arguing that the death penalty should be off
the table because of ongoing research into the correlation between such

dysfunction and antisocial behavior.[50] If the Court decided *Roper* on the basis that, for developmental reasons, juveniles and adults have different mental capacities, then the Court has, by implication, encouraged *any* criminal defendant to introduce evidence of a neural abnormality that might render the defendant the normative equivalent of an adolescent. But the *Hooks* court denied the defendant's request for the type of brain scan that would have established his diminished normative capacity on account of frontal lobe dysfunction. The court clarified that *Roper* was based on "categorical rules to define Eighth Amendment standards."[51]

If Hooks indeed did suffer from a frontal lobe dysfunction that would render him the intellectual and normative equivalent of a juvenile, what *non*instrumentalist reason could there be for *not* reaching the same Eighth Amendment "cruel and unusual" conclusion? The basis of that constitutional limitation, insofar as the death penalty is concerned, may be considered desert,[52] a retributory concept. If the adolescent is less blameworthy because of diminished capacity, then adults with similar diminished capacity also should be less blameworthy. The retributivist, then, cannot explain the court's distinction. But the instrumentalist easily rationalizes differential sentencing for the dysfunctional adult and the developing adolescent on the grounds that an adolescent brain holds the promise that a perpetrator might grow out of his violent criminal behavior.[53] Underlying the focus on the perpetrator's potential to outgrow his violent behavior is a realization that he will literally be a different person, neurologically speaking, in the future. For some, however, the interplay between their genetic makeup and their surrounding environment undermines the possibility of that instrumentalist rationale.

Conceptions of Capacity and the Curious Case of Psychopathy

We are neither *only* what our genes determine us to be nor *only* what our environment molds us into; we are about 50 percent of each.[54] Our genes may *predispose* us to particular behaviors, and our environment may precipitate certain choices, but we are not wholly determined by either. That indeterminacy leaves sufficient room for parents to avoid all the blame (and precludes their taking all the credit) for how their children turn out, and it affects normative conceptions of responsibility. Surely there are decisions that we make, but if those decisions are the product of chemical, electrical, and structural neural accidents that are themselves products of nature and nurture, then at some point we lose an entity that can take the credit or the blame.

Neuroscience confirms that events occurring after conception but prior to birth may affect the capacities of a newborn and the effects may persist for years. Intuitively, hindrances to fetal brain development in utero may handicap brain development long-term. What is breathtaking is the ultimate significance of even the most ostensibly insignificant chemical, electrical, and structural effects. Insofar as a child's impeded development will lead to behavioral differences that, in turn, elicit different reactions from others (including parents) to the child, we can begin to appreciate the rich nature-nurture interaction that determines the intellectual and emotional characteristics of the fully formed person.

It is reasonable, then, to inquire into whether events that affect the development of certain brain structures and processes may ultimately affect behavior, broadly construed. For example, we know that the amygdala is the primary neural area "subserving fear conditioning."[55] A study published in the *American Journal of Psychiatry* followed 1,795 children from age three to age twenty-three and demonstrated that early deficits in fear conditioning were associated with criminality later in life. Because the amygdala is rarely vulnerable to injury or illness, the study may support the conclusion that dysfunction of the amygdala is more a product of nature than nurture.[56] But that rendition of the born criminal would be both disturbing and perhaps overbroad. We do not know, for a fact, how nurture might affect the relationship of one area of the brain with another—such as the cooperation of the amygdala (emotional)[57] and the prefrontal cortex (executive). So the jury must still be out, but neuroscience is helping us ask better questions.[58]

Perhaps the most compelling context in which responsibility questions might be asked today concerns the criminal liability of psychopaths. Are psychopaths born (or conceived) or made? Should that distinction affect how the criminal law treats offending psychopaths? Even Stephen Morse, a leading skeptic of neuroscience's potential effect on the law,[59] has suggested that psychopaths are not responsible for their criminal actions, at least not in the way that nonpsychopaths might be.[60]

Roughly 1 to 3 percent of people are psychopaths, and the condition is more prevalent among men than women.[61] Significantly, not all psychopaths are criminals; some are (even very) highly functioning. Psychopathy describes a continuum: Some people are more psychopathic than others. To justify the label, you must reach a certain level (or constellation) of psychopathic tendencies, meaning there are a variety of ways to be classi-fied as such.[62] The general deficiency displayed by the psychopathic person-ality is the lack of empathy that would alert him to the fact and degree of fear or pain felt by another.[63] Frequently associated with psychopathy is a certain charm, an ability to manipulate others, often (though not always)[64] in order to make them victims of a crime.

An exemplary case is *In re Martenies*.[65] The appellant had been con-victed, after a guilty plea, of "intrafamilial sexual abuse in the first degree" and was incarcerated therefor. His criminal actions, directed toward his stepdaughter, were particularly brutal. Mental health professionals testify-ing at the appellant's trial concluded that while he knew the consequences of his actions, he was unable to control himself. All experts agreed that the perpetrator had psychopathic personality disorder, but all experts also con-cluded that psychopathy was not the same as mental illness. As it did not qualify as a mental illness, psychopathy could not the basis for any excuse that might have resulted in commitment rather than criminal incarceration.

The law has trouble with psychopaths: The condition seems to have cer-tain and reliably ascertainable indicia, but is not a mental illness, as such, so does not provide for mitigation of criminal punishment.[66] Psychopathy is not a protected status in any way: Psychopaths may be criminally liable notwithstanding their aberrational psychological profiles, which can now be confirmed by brain imaging.[67] The normal brain registers emotional cues in a way unavailable to the psychopath. Harming another person would actually, in a very real way, *hurt* the 97+ percent of the population who are not psychopathic. Psychopaths do not feel pain caused to another. When asked to describe the emotion revealed by someone in great fear, psycho-paths might well respond that they do not know what the expression dem-onstrates, but they recognize it as the expression on their victims' faces.[68] That does not mean that psychopaths do not have methods by which to conform their behavior to prevailing social standards. Indeed, many psy-chopaths rely on other cues to exhibit the charm that attracts their victims.[69]

We can determine whether someone is a psychopath by his performance on the *Hare PCL-R* and by brain imaging. Psychopathy is a verifiable and very material brain condition.[70] As imaging techniques and technologies ma-ture, it may become increasingly easy to detect psychopathy. Accordingly, a worthwhile thought experiment emerges: If we knew, actually and certainly *knew* beyond the shadow of a doubt that someone was (or would become) a psychopath, what should the law's reaction be? This is the *Minority Report* problem:[71] When we can certainly predict crime, should we prevent it?[72]

The easy answer is that our neuroscience is currently too imprecise to resolve that conundrum. But, assuming we do have the technology to discover accurately the psychopath and to distinguish the violent psychopath from the merely unpleasant psychopath, *should* the criminal law intervene? The responses may be varied and quite rich in their nuance, involving moral calculi similar to those employed in determining criminal liability for attempted criminal acts that never come to full fruition.[73] *Mens rea*, we know, is not enough: There must be coincident *actus reus*. But what is the normative significance of that act requirement?[74] And would prosecution of someone for *being* a violent psychopath before he commits any violent act amount to the imposition of criminal liability on account of status? The constitutional implications of that would be profound and uncertain.[75]

Even now, we do limit the freedom of those whose status threatens the common welfare. Statutory provisions permit detention of the physically ill whose presence in open society threatens to cause an epidemic.[76] From a victim's perspective, what is the difference between being infected with a deadly virus by patient zero and being murdered by a psychopath if both are the product of that victim merely being in the wrong place at the wrong time? This is not to suggest that there might not be important moral bases that distinguish the two issues. Even still, as neuroscience grows more sophisticated, the line between ostensibly mental (psychopathy) and physical (antibiotic-resistant tuberculosis) threats blurs, and the criminal doctrine may be ill-equipped to maintain a distinction between the two.

Even though being able to measure a person's mental capacity for empathy and other such faculties may lead to normative dilemmas, it also would allow for greater sensitivity in punishment. A clear view into a person's mental hardwiring would allow for a greater understanding of how various neural activity affects one's capacity for understanding, particularly in cases involving impaired or damaged brains. Such detailed measurement would ultimately shift the relevant analytical question from what caused the impairment to its potential neurological effect.

Impairment

What makes impairments pertinent to punishments is the fact that certain forms of them involve the type of neural consequences that have normative resonance. This section considers a range of impairments to conclude ultimately that the criminal law will achieve normative coherence only by evaluating responsibility in terms of the *sources* of rather than the *effects* of impairment. Focusing on the constituents of the impairment that has given rise to criminal behavior may rob responsibility of all its normative meaning and force us to abandon retribution (and attendant concepts like blameworthiness and desert) as the basis of criminal law and punishment. That, then, would leave us with only instrumental, utilitarian goals as the basis of punishment (and might well put us on the road to a *Minority Report* reality).

Human agents are the product of nature and nurture, and no two agents have the same nature and nurture, not even twins.[77] Accordingly, two actors may be equally responsible in the causal sense (for pulling a trigger), but no two actors are identically responsible in the normative sense. That truism presents a formidable challenge to the law generally, but it may be most troubling in the criminal law. If two criminal defendants do not have the same neural capacity, it would seem harsh (if not fundamentally unfair and inefficacious) to subject them both to the same punishment or even to the same elements of criminal liability *ab initio*.[78]

In exploring the effect of neuroscience on the notion of responsibility, the discussion that follows owes a debt to Adrian Raine (2013), which describes and integrates the proliferation of studies, many conducted by Raine himself, that reconceptualize our understanding of human agency. We are, indeed, *just* the sum of forces—and we cannot even say "the sum of forces acting *upon us*," because there is no "us" independent of those forces.

Bred in the Bone: Nature

Each cell in the human body contains chromosomes that consist of long DNA molecules, which constitute a person's genetic makeup or genotype.[79] A gene is a segment of DNA that may synthesize a single type of polypeptide, consisting of amino acids, to create a functional protein, such as an enzyme or antibody.[80] For every gene, there are alternative forms called alleles. The alleles of your parents' genotypes combine to determine the alleles present in your own genotype. The location of a given cell (e.g., brain or liver) and your individual genotype (i.e., allele variation, mutation, etc.) will influence the kind of protein produced by a cell, as well as the rate of production and the level of functionality of the proteins produced.[81] Genes are integral to the functioning of the human body and the smallest variation or mutation may affect behavior and well-being.[82]

For example, consider the transforming growth factor beta 1 (TGFβ-1) gene. Located on chromosome 19, this gene results in the production of the TGFβ-1 protein, which regulates cell proliferation (growth and replication rates), differentiation (division of labor among cells), motility (range of movement), and apoptosis (scheduled cell death for mature cells soon to be dysfunctional and susceptible to pathology).[83] Within the TGFβ-1 gene, there are several polymorphisms (alternate alleles). For an individual recently diagnosed with cancer, the form of her TGFβ-1 gene may provide important information about the potential growth rate of her cancer cells or the metastatic potential of her cancer cells.[84] Just as researchers study polymorphisms within the TGFβ-1 gene to inform a cancer-treatment protocol, researchers also study the genetic polymorphisms that affect neurotransmitter regulation to understand how certain pathological antisocial behavior moves from generation to generation.[85] This is epigenetics.

Perhaps it is unsurprising that studies suggest that violent parents produce violent children. To make sense of this finding, it might be tempting to rely on the explanatory power of learned behavior. Children mimic what they see and if they witness adults, particularly those closest to them, navigating life violently, then children follow suit and become violent themselves. Our parents are our first teachers, for good or ill. And if such learning is the efficient source of violent behavior, we might be able to curtail violence by reteaching maladapted children or, even better, by teaching their parents to act nonviolently.[86] But if the source of a propensity for violence is genetic, then it is more resistant to teaching, and violence as a social problem is even more intractable than previously thought.

Studying adopted children never directly exposed to the antisocial behavior of their biological parents sheds stunning light on the extent to which genetics predict criminal tendencies in general and violent behavior in particular. In a sample of more than 14,000, a 1998 Danish study found that adopted sons were at an elevated risk of criminal conviction if at least one of their biological parents had one or more convictions. Of adopted children convicted of one violent offense before the age of eighteen, 45 percent had a biological parent who had been admitted to a mental hospital and, among adoptees convicted of two or more violent crimes, the figure was 68 percent.[87] Theories premised on learned behavior offer no solace for those results.

In similarly disturbing fashion, a 2007 study found that the presence of particular genes may predispose adolescents to delinquent (including violent) behavior.[88] In a study of more than 2,500 US adolescents, researchers found an association between violence and the TaqI polymorphism in the DRD2 gene and 40-bp VNTR in the DAT1 gene. This research suggested that such genetic differences may account for as much as twice the likelihood of serious delinquency. That means that even if A and B grow up in essentially similar circumstances, each confronting the same challenges and having the same opportunities,[89] A and B may be differently susceptible to encountered circumstances and influences, and thus the likelihood of avoiding antisocial behavior may be very different for A than for B.[90]

If we are to develop a coherent concept of criminal responsibility—from either the noninstrumentalist perspective (desert) or the instrumentalist perspective (deterrence)—then these findings demand that we integrate into our normative calculus the genetic inheritance that determines certain aspects of cognitive function at the moment of conception. Criminality as a function of genetic inheritance is frighteningly dissonant with our society's aspiration that all should be equal under the law. The only way for the law to treat us equally is by recognizing the variety of ways in which offenders are uniquely unequal—products of importantly different nature and nurture.

Developmental Impairment: Nurture

Just as each of us is the product of genetic predisposition, our destinies also may be determined, at least in part, by our mother's prenatal diet and environment. Something as ostensibly benign as a pregnant woman's avoiding seafood out of fear for potentially attendant neurotoxins (such as mercury)[91] may cause an expectant mother to experience Omega-3 fatty acid deficiency, which is predictive of lower IQ scores, less intrauterine growth, impaired fine motor skills, slower information processing, and irreversible deficits in serotonin and dopamine release.[92] Such malnutrition may lead to persistent alterations in the neurotransmitter systems responsible for the regulation of norepinephrine, serotonin, and dopamine, resulting in abnormal uptake in adulthood, which, in turn, may correlate with antisocial behaviors.[93]

Confirming, with a vengeance, that "the sins of the [mothers] are visited upon their sons,"[94] studies show lifelong deleterious effects for children with mothers who consumed tobacco or alcohol during pregnancy. Such children are at a cognitive disadvantage vis-à-vis their peers.[95] That effect may be seen even at lower levels of alcohol or tobacco consumption.[96] You might be tempted to surmise that consumption of alcohol and tobacco during pregnancy would also indicate a lack of more conscientious parenting skills and that the latter may explain observed cognitive differences, but studies that control for other aspects of nurture confirm the observed effects.[97]

Even those mothers who ingest the recommended amount of Omega-3 fatty acids and avoid all alcohol and tobacco consumption during pregnancy may find themselves powerless to prevent deficits owing to environmental exposure to toxins. For instance, there are many ways a young child might be exposed to lead.[98] A child exposed to lead will have lower brain volume as an adult and studies show a dose-response relationship between the amount of lead exposure and the degree of brain volume decrease in the ventrolateral prefrontal cortex and the anterior cingulate cortex. Such decreases correlate with deficient fine motor function and a greater tendency toward antisocial behavior (not surprising given the emotional regulatory function of those two brain regions).

Children are in significant part also a product of their social environments, even though the extent to which they are susceptible to certain social forces may be the product of genetics, in utero nutritional deficiency, malnutrition during infancy, and so on. And indeed, the best, most effective cure for adolescent antisocial behavior may be growing out of adolescence. Most youthful offenders are one-time offenders, never again offending either in youth or adulthood.[99] But those who spend their normatively formative years in threatening environments are more likely to develop long-term antisocial skills that make them more able to survive in antisocial environments than in prosocial environments.[100] The state may be complicit, in many ways, for exposing impressionable youth to such environments, when adolescents "enter the system."[101]

Certainly there are sound public safety reasons for limiting the freedom of adolescents who would hurt others; the costs of *not* incarcerating those who would harm others may be considerable for victims in particular and for society more generally.[102] But we are coming to realize that some types of punishment themselves entail costs that are not immediately obvious, yet nonetheless quite real. Neuroscience provides objective ways to discern the neural effect of punishment in particularly compelling contexts. Isolation, or solitary confinement, may serve important correctional functions (e.g., protection of the detainee and protection of those whom the detainee might harm), but there may be attendant developmental costs.[103]

Solitary confinement is an incredibly pernicious kind of imprisonment. It quite literally destroys minds. The practice is particularly destructive when imposed on adolescents whose brain development requires socialization crucial to the formation of neural networks that promote normal social functioning. Denying the adolescent brain social interaction is akin to blinding the immature eye by denying it the visual phenomena that ensure development of visual acuity.[104] Solitary confinement causes brain damage, no less surely than a blow to the head with a tire iron.[105] The damage caused to the adolescent brain is far greater than that caused to the adult brain placed in isolation.[106]

The instrumental purpose underlying the criminal law demands that we calibrate sentencing according to the efficacy of a given punishment. Our evaluation of the efficacy of punishment must be more than an inquiry into whether a sentence is an efficient deterrent or impotent deterrent. We also must ask whether a sentence can be instrumentally counterproductive, causing harm by reinforcing antisocial behavior (even by causing brain damage or stunted brain development) and thereby increasing costs to society.

Trauma

Since the time of Phineas Gage's (in)famous misadventure,[107] we have known that physical insults to the brain may manifest as profound alterations in personality. Modern courts have been invited to take into account the effect of physical trauma on the culpability of criminal defendants.[108] Surely if mental deficiency resulting from nontraumatic neural aberration, say profoundly low IQ, would be pertinent to determining criminal responsibility and related consequences,[109] then a more obviously physical injury, whether the result of accident or parental abuse, also should be part of the desert calculus.

That is the assumption explored in an important article by Nita Farahany.[110] Professor Farahany considered the significant implications of the United States Supreme Court's opinion in *Atkins v. Virginia*,[111] which held that executing people with severely diminished intellectual capacity violates the Eighth Amendment. Farahany argued that the *Atkins* holding should not be limited to those born with diminished intellectual capacity but should apply as well to defendants demonstrating a similar level of disability when deficits are the result of some other cause—such as traumatic brain injury, dementia, developmental disorder, or central nervous system dysfunction. "Through the Court's new jurisprudence, a new disproportionality has emerged—a capital defendant who suffers traumatic brain injury at age twenty-two, and exhibits all of the same behavioral manifestations as a medically diagnosed mentally retarded capital offender, can be subject to the death penalty while one with early onset mental retardation cannot. These legislative enactments[112] are now ripe to be challenged on equal protection grounds."[113] Farahany's provocative conclusion resonates with the need to evaluate normative responsibility in terms of demonstrable cognitive deficits without differentiating deficits according to their origin.

Farahany illustrated the injustice she theorized by discussing at length the decision of the Louisiana Supreme Court in *State v. Brown.*[114] The defendant was convicted of two counts of first degree murder and was sentenced to death. Years before the double homicide, Brown had been shot in the eye in the course of committing armed robbery, and the resultant traumatic brain injury precipitated his serious intellectual disability. Several experts testified that the areas of the brain damaged directly related to impulse control, the ability to interpret nonverbal stimuli, and the aptitude to think rationally in general. The court's opinion curiously recited facts that it believed indicated Brown did not suffer from the mental deficiency his injury (and expert testimony) suggested, even affording some weight to the fact that the defendant acted as his own counsel.[115] The apposite state statute required that, to constitute mental retardation, the condition's "onset must occur before the age of eighteen years."[116] Further, the statute provided that "traumatic brain damage occurring after age eighteen . . . does not necessarily constitute mental retardation."[117] Although the Louisiana statute may not be perfectly considerate of the neural inquiry that should proceed from evidence of impaired cognitive function, the court certainly *could* have found sufficient reason, per *Atkins*, to question the imposition of the death penalty.

A year after the *Brown* decision, the Louisiana court considered these issues again in *State v. Anderson,*[118] where another defendant suffered a traumatic brain injury that had caused serious, demonstrable cognitive impairment. Ruling against defendant's sentencing appeal, the court concluded that Anderson failed to prove with certainty that the brain injury directly caused his criminal behavior. But why should the mentally deficient minor have to make the same showing of causal connection between injury and impairment? In such cases, the *cause* of the mental deficiency (nature, nurture, or trauma) is actually probative of nothing. It is the *fact* of the impairment that matters. The moral reasons (noninstrumental or instrumental) for *not* punishing the person whose cognitive function is impaired for one reason do not differ when another person suffers the same cognitive impairment for a different reason.

In *U.S. v. Candelario-Santana*,[119] the federal court for the District of Puerto Rico put the burden on the defendant, a former boxer, to establish that any impairment of cognitive function he suffered as a result of his twenty-seven years in the ring had been the cause of his having committed capital murder. Candelario-Santana was convicted of twelve murders. He had struggled in school but was ultimately able to hold down a full-time job and even obtain a license to operate heavy equipment. There was, though, evidence of brain trauma: He stated that he suffered approximately ten to twenty concussions during his boxing career as well as a head injury in a motorcycle accident.[120] The court, in finding no cognitive impairment bar to the defendant's death sentence, focused on the fact that experts testified Candelario-Santana was of average intelligence, "had proper average intellectual resources to work and produce."[121] His IQ score was 75, above the mental retardation threshold at the time.[122]

Similarly, the court in *State v. Stanko*[123] determined that the defendant's impaired cognitive function was not the normative equivalent of mental retardation for sentencing purposes. The evidence of brain damage was compelling:

> An expert in physiological psychology testified that Appellant suffered damage to the frontal lobe of his brain from two separate incidents. The first incident occurred during Appellant's birth when his brain received a reduced oxygen supply. The second incident occurred during Appellant's teen-age years when he received a blow to the back of his head from a beer bottle, driving his brain forward. Appellant presented psychiatric testimony that he had diminished or lowered function of his brain in the frontal lobe areas, and that there "could" be a causal connection between diminished function in the frontal lobe and mental illness. . . .
>
> [One] expert testified that the damaged lobe of Appellant's brain played an important role in impulse control, judgment, and empathy. In fact, he stated, "This abnormally low function or abnormality or injury would significantly compromise and impair an individual's ability to exercise judgment, impulse control, control of aggression."[124]

The court, though, seemed more moved by the fact that the defendant had an IQ of 143. Stanko was able to communicate and care for himself, and his "behavior before and after the . . . murder, demonstrated an ability to formulate and execute deliberate plans."[125] The court conflated intelligence and sufficient cognitive capacity for sentencing, and Eighth Amendment, purposes.

More recently, the Supreme Court of Missouri, in *State ex rel. Clayton v. Griffith*,[126] considered the capital sentence imposed on a defendant who had suffered a severe brain injury when he was struck by a piece of wood that lodged in his head and had to be surgically removed. The surgery "resulted in the loss of nearly eight percent of Clayton's brain and 20 percent of a frontal lobe."[127] But that accident occurred twenty-four years before the murder. The change in the defendant following the accident, though, was of Gageian proportion: Clayton had been a part-time pastor and evangelical before the accident and thereafter his marriage dissolved, he drank alcohol excessively, and he demonstrated an antisocial personality.

The court recognized that *Atkins* applied, but also recognized that intellectual disability, according to *Atkins*, is a matter of state law. So the apposite Missouri statute was dispositive: " 'intellectual disability' or 'intellectually disabled' refer to a condition involving substantial limitations in general functioning characterized by significantly subaverage intellectual functioning with continual extensive related deficits and limitations in two or more adaptive behaviors such as communication, self-care, home living, social skills, community use, self-direction, health and safety, functional academics, leisure and work, which conditions are manifested and documented before eighteen years of age."[128] Because the injury did not occur until the defendant was thirty-two years of age and the altered behavior did not begin until after the accident, the court found that Clayton was not within the scope of the statute's protection. But the court's confusion, normative if not statutory, is not without precedent.

Recall *Roper v. Simmons*,[129] introduced at the beginning of this chapter. In that case, the United States Supreme Court held that it was unconstitutional, a violation of the Eighth Amendment, for a state to sentence to death someone convicted of a murder committed before the age of eighteen. Writing for the majority, Justice Kennedy took account of neuroscientific insights into adolescent brain development. The Court did not consider the state of Simmons's brain in particular, but instead relied on studies of neural development generally. Justice Scalia found reliance on that evidence troubling, at least insofar as the Court had relied on contrary evidence when earlier weighing the intellectual maturity of adolescent women making the decision to have an abortion.

In that earlier case, *Hodgson v. Minnesota*,[130] the Court considered whether two-parent notification of a minor's intent to have an abortion and a forty-eight-hour-notice requirement violated the Constitution. In arguing against the constitutionality of the state law, the American Psychological Association maintained that minor women were intellectually and emotionally competent to make the decision to have an abortion without parental intercession. But in *Roper* that professional association argued that a minor who commits murder does not have the same emotional and intellectual maturity as an adult, and so sentencing the minor to death for murder would violate the Eighth Amendment proscription of cruel and unusual punishment. Dissenting in *Roper*, Justice Scalia noted the apparent inconsistency:

> [T]he American Psychological Association, . . . which claims in this case that scientific evidence shows persons under 18 lack the ability to take moral responsibility for their decisions, has previously taken precisely the opposite position before this very Court. In its brief in [*Hodgson*], the APA found a "rich body of research" showing that juveniles are mature enough to decide whether to obtain an abortion without parental involvement. . . . Given the nuances of scientific methodology and conflicting views, courts—which can only consider the limited evidence on the record before them—are ill equipped to determine which view of science is the right one. Legislatures "are better qualified to weigh and 'evaluate the results of statistical studies in terms of their own local conditions and with a flexibility of approach that is not available to courts.' "[131]

Leaving aside the dubious conclusion that legislatures are any better than courts at considering neuroscientific evidence, and whether there is anything particularly local about the normative calculus, Justice Scalia's ignorance of the neuroscience is understandable if not excusable.

It may well be the case that different decisions, different actions, relate to different forms of cognitive competence. While we (or at least Justice Scalia) may conflate normative capacity to appreciate the consequences of actions among violent adolescents with normative capacity to do the delicate balancing between having a child and seeking a legal abortion, it is just not clear that the two decisions entail the same intellectual and emotional capacities. That is the point made by Laurence Steinberg: "The skills and abilities necessary to make an informed decision about a medical procedure are likely in place several years before the capacities necessary to regulate one's behavior under conditions of emotional arousal or coercive pressure from peers."[132] For our discussion, the extent to which those decisions are similar is not the point. Rather, our takeaway should be the power of neuroscience to reveal line drawing in the criminal law that is empirically invalid.

As neuroscience makes the ineffectiveness and arbitrary nature of those lines even clearer, it will highlight the need for the law's normative understanding of responsibility to account for an individual's specific neural capacity. The focus of that specificity, however, is critically important. Neuroscience confirms that the proper focus is on the existence of a cognitive deficit and not on whether the deficit resulted from genetic predisposition or was later acquired from a traumatic event. Trying to maintain a distinction between cognitive deficits rooted in causation is based on a dangerous misunderstanding that is as invalid as the arbitrary lines it precipitates.

Acquired Normative Impairment

If there is good reason to draw lines based on mental capacity, we would imagine that those same lines should, at least presumptively, be drawn when we consider the objects served by punishment decisions about those whose intellectual or emotional competence has been impaired by trauma. And the neuroscience bears out the connection between traumatic events and impairment of social functioning. The evidence, though, is mixed and so may intimate empirical limitations of the brain science, at least so far. Nonetheless, it is worthwhile to consider some of the studies Raine presented in his meta-analysis that may shed light on brain trauma and the normative inquiry.

Three studies described by Raine[133] are particularly provocative. Koenigs et al.[134] concluded that damage to the prefrontal cortex makes it more likely that victims will engage in utilitarian (rather than deontological) reasoning. The investigators presented fifty hypothetical scenarios to six patients with focal bilateral damage to their ventromedial prefrontal cortex (VMPC) and to a control group. The study found that the differences between the patients and the controls were most pronounced when participants considered whether to smother a crying baby in order to save a group from fatal detection.[135] That choice is wrenching for anyone with normal sensibilities, but the reaction of those with VMPC damage was less wrenching and more utilitarian.[136] That is not to say that those with damaged VMPCs were *less* competent to make the moral decision.[137] Indeed, a very good case could be made that, in the provocative thought experiment, the more utilitarian choice was the better choice. For present purposes, though, what matters is the direct connection between the traumatic brain injury and the way the decision was made. Although there was not a general deficiency in the capacity for moral judgment as a result of the injury, the injury certainly affected the decision-making processes of the participants.

A study conducted by Raine and an associate[138] considered cases of "acquired sociopathy [synonymous with psychopathy]," in which traumatic injury to the VMPC in adulthood results in pseudo-psychopathic, disinhibited, antisocial behavior, bad decision making, and lower anticipatory skin conductance responses to stimuli predicting negative outcomes. Injuries to the adult brain manifest differently from injuries to the developing brain. Unlike acquired sociopathy (adult-onset sociopathy), developmental sociopathy is associated with significant impairments in moral reasoning and judgment.[139] Another study found that infants who sustain damage to their prefrontal cortex are more impaired than adult-onset lesion patients. The early-onset patients were never able to acquire complex social knowledge and so were more impaired, presenting more like psychopaths than adults who suffered similar cognitive injuries later in life.

The differential effect of the same trauma on the developing brain versus the adult brain may be analogous to differential capacity of children and adults to learn a second language or master a new musical instrument. But we need not delve into the nuances of neuroplasticity here. For present purposes, it suffices to recognize that the timing of trauma may be determinative of the cognitive consequences of the trauma. And that arms us with another reason the criminal law should evaluate normative responsibility in terms of demonstrable cognitive deficits and should not put primacy on the particular cause of diminished capacity.

Conclusion

This chapter has reviewed exemplary criminal law doctrine in light of the realities exposed by the evolving neuroscience, broadly construed. What emerges is a sense of mismatch: The doctrine seems to assume a being very much unlike the human agents neuroscientific research, including the studies recounted by Adrian Raine, reveal each of us to be. We are just not responsible in the way we would have to be for the law to make much normative sense. And the great damage done by the doctrine's misconception is not just to individual victims of the law's misunderstanding;

the greater damage is to all of us who maintain and defend a system that is actually at odds with what is likely our own conceptions of what is moral. Think of the shame, perhaps real pain, you feel when you wrongly blame, or accuse, or punish someone: That is the shame our criminal law system has made us all party to. And that is compounded by the fact that many, perhaps most, of the actions the doctrine takes in pursuit of providing a safer environment for us all has actually made our society less safe.

The brief survey provided by this chapter supports the review and critique of noninstrumental normative criminal law theory that is the focus of the next chapter. And this is where the action has been so far in most of the literature. Indeed, it seems safe to say that if neuroscientific insights unravel noninstrumental theoretical approaches to criminal liability, then the noninstrumental, particularly deontological, moral theory generally is undermined. Because if we are wrong about what it means to be human in the criminal law, we are wrong about what it means to be human throughout the law (and much else, too). We shall see that desert and blame and retribution are incoherent: Human agents are not morally responsible, at least not in the way the law would have them be in order for the doctrine to make normative sense. So the next chapter engages the apologies and apologists for the noninstrumental status quo in the criminal law and demonstrates that it is based on an inauthentic, even wholly fictitious, account of the human agent. Noninstrumental theory has actually aided and abetted the moral failure that is our criminal law. Much of what the next chapter develops will work to cast profound doubt on the conception of human agency on which the tort and contract law rely as well. The dominoes fall.

CHAPTER THREE

Neuroscience and Criminal Law Theory

Introduction

The neuroscientific challenge to legal doctrine is most salient in the criminal law context. Neuroscientific insights require a fundamental reconsideration of human agency, testing the responsibility criterion upon which the criminal law is founded. The goals of the criminal law will remain out of reach if they misrepresent human agency. The first part of the chapter focuses on the folk psychology that is the foundation of criminal law and the normative work it must do to be coherent. That inquiry considers folk psychology tenets as manifested in the criminal law's conception of human agency. Folk psychology ignores the reality that we are, in important ways, more like mechanical devices, like cars, than we are like the entities idealized from the libertarian (not in the popular political sense) perspective. By refusing to acknowledge that human agents are the products of forces beyond our own control, the folk psychology depiction of human agency actually *undermines* normative objectives of the criminal law. The second part of the chapter then confronts retribution and the arguments in support of normative conceptions that would distinguish it from revenge. Retribution and revenge are ultimately indistinguishable and arguments in favor of retribution ultimately founder on the same shoal as noninstrumentalism, particularly deontology, generally. The failure of the perspective that would vindicate retributionary principles is demonstrated by returning again to the problem of psychopathy. Finally, part three of the chapter focuses on *moral* responsibility, generally and particularly in the criminal law, and concludes that the insights provided by the materialistic perspective neuroscientific insights vindicate make the case *against* moral responsibility, even the case against the morality of a system based on the supposed moral responsibility of human agents.

The Folk Psychology of Human Agency

Normative responsibility is the basis of criminal liability. Before an actor may be subject to criminal penalties, the prosecution must establish "beyond a reasonable doubt" that the actor is "guilty" of the charged crime. The jury must deem the actor culpable. Only then do criminal penalties make sense: Whether the object of criminal punishment is instrumentalist or noninstrumentalist, the criminal law serves its normative object only when punishment is a response to culpable behavior. Desert, the basis of retribution (and, in an important way, of deterrence too) provides the foundation of criminal punishment, whatever form that might take. A necessary predicate, then, of the normative responsibility calculus is an accurate conception of the human agent.

It would be absurd and ultimately inefficacious to understand as moral actors those entities that we consider purely mechanical objects. We do not punish cars that do not start or dishwashers that leave watermarks on glasses. In the case of the recalcitrant car, we fix it, if we want it to start. That might entail replacing the battery or starter, but it does not entail beating it with a stick or sentencing it to isolation in the garage.[1] We could respond to the problem with the dishwasher by changing detergent, installing a water softener, or adding a rinse agent. Whatever you do, you endeavor to correct the problem, in an instrumental way.

Our intuitive sense of human agency, though, resists such a pure instrumentalist response. Indeed, there is a sense that we somehow deny the humanity of criminals if we treat them as objects to be fixed rather than as sentient moral beings normatively responsible for their actions and answerable in terms of desert.[2] We must be free to make bad, even criminal, decisions and answer therefor in order to be truly free in this intuitive sense. Nothing less than the fundamental sense of humanity is at stake.

Neuroscientific insights, however, strain that noninstrumentalist perspective by conceiving of human agency in mechanistic terms. The challenge is profound, and unsettling. Essentially, at the most extreme level, the mechanistic, physicalist conception of human agency understands the difference between household appliances and the human agent as a matter of degree, and so that conception grates. The result of that tension between the two conceptions of human agency—at the extremes, between human agent as determined mechanism and human agent as just below angels on the scale of divinity—is criminal law doctrine that at times confounds. That was the conclusion of the preceding chapter, which considered the doctrine. It is worthwhile, even necessary, then, to appreciate the relationship between normative responsibility in criminal law and neuroscientific insights into the nature of human agency.

This part will focus on the difference between folk psychology—premised on the essential reality of beliefs, desires, intent, and motivations—and cognitive neuroscience to describe the significance of that distinction for a normative appraisal of criminal law doctrine. The distinction carries a great deal of metaphysical weight and considerable philosophical baggage as well. Much is at stake: Folk psychology depends on there being something ineffable, or at least not *completely* accessible from the materialistic, physicalist, monist perspective.[3] Cognitive neuroscience is vindicated by Francis Crick's "astonishing hypothesis":[4] All we are is the sum total of physical stuff. We may not (yet) understand how the physical stuff manifests itself in things that folk psychology labels "beliefs, desires, intent, and motivations," but ultimately, folk psychology describes the manifestations that may be reduced, also a term of art,[5] to physical phenomena not beyond apprehension (at least no more so than any other physical phenomenon is beyond our apprehension).

Conception of Human Agency: Cars Distinguished

Understanding human agency depends on our coming to terms with the challenge that cognitive neuroscience presents to the assumptions of folk psychology: If we are no more than interacting physical entities, as cognitive neuroscience maintains, then normative systems based on folk psychology will *often* fail, and fail in ways that even apologists for folk psychology would acknowledge. Prerequisite to appreciating the contours of the challenge is formulation of folk psychology in terms that resonate in the criminal law doctrine and draw starkly the difference between folk psychology and cognitive neuroscience. What is the relationship between them that supports normative critique of criminal law doctrine? That is, *how*, precisely, does folk psychology err in just the way that entails its normative infirmity? We shall see that the normative difference between folk psychology and cognitive neuroscience is just like the difference between your child and your car: The two are normatively distinct because what is efficacious with regard to performance modification of one would be incoherent with regard to performance modification of the other. Now that is not to say that there would not be useful analogies between the proper care of both—food for your child, gasoline for your car—but treating one just as you would the other (and hoping for normatively coherent results) would be absurd.[6]

Folk Psychology: Beliefs, Desires, and Intent in the Criminal Law

Folk psychology posits the essential irreducibility of beliefs, desires, intents, and motivations. No one disputes that folk psychology animates the criminal law. Stephen Morse, one of the most prolific defenders of extant criminal law doctrine against the neuroscientific critique, has quite correctly acknowledged that folk psychology provides the basis of criminal law doctrine: "The criminal law is a thoroughly folk psychological enterprise that is completely consistent with the truth of determinism or universal causation."[7] Morse intended that statement to be an accurate empirical observation, and it is. Perhaps more precisely it is an accurate empirical statement ("thoroughly folk psychological enterprise") and likely an equally accurate evaluative one: "completely consistent with the truth of determinism." But, as we shall see, that might be to damn the criminal law with faint praise.

Morse's conclusion that the criminal law doctrine is a product of folk psychology is confirmed by the language of the criminal law, focusing on *mens rea*, and even the very terms of folk psychology: beliefs, desires, intent and motivation. (It is not much of a leap to conclude that beliefs and desires are intentional too.) Although intent is a foundation of folk psychology, it would be wrong to conclude that intent does not exist for cognitive neuroscience. Intent is every bit as real for cognitive neuroscience as it is for folk psychology;[8] indeed, in the criminal law, it is important to note, intent may be more real for cognitive neuroscience than it is for folk psychology because cognitive neuroscience cares about the source and substance of intent in ways that are elided by folk psychology. That is the point, really: The criminal doctrine, consistently with folk psychology, generally focuses only on the fact of intent (but for limited affirmative defenses),[9] but cognitive neuroscience inquires into the substance and constituents of intent, enabling an evaluation of the extent to which intent is normatively operative. Though this will be developed further later in the discussion, understand that folk psychology significantly reduces to cognitive neuroscience and so cognitive neuroscience reveals the deficiencies, in the normative sense, of folk psychology.

If Morse were to have been satisfied confirming folk psychology as the basis of criminal doctrine, he would have made a worthwhile contribution to the conversation about the neuroscientific integrity of the criminal law. And, in fact, a generous reading of his work could conclude that all he is saying is that the extant doctrine in fact reflects folk psychology, not that it is normatively coherent in doing so. Morse might even go so far as to say (and in fact he has said) that we are predisposed, if not hardwired, to find the normative premises and conclusions of folk psychology attractive and comforting, because of the way human agents are constituted.[10] To his great credit, Morse left open the possibility that further work will vindicate what he would describe as neuroarrogance, the propensity of some commentators (including, perhaps, the author of this book) to overclaim by suggesting that neuroscientific advances will undermine folk psychology and thus undermine the normative bases of the primary doctrinal areas.[11] But he has argued that we do not *yet* have empirical support for the conclusion that folk psychology is conceptually infirm. There is room for disagreement.

Folk Psychology as Alchemy

What if folk psychology were wrong, or even just so incomplete as to mislead profoundly? Even before that, what would it mean for folk psychology to *be* wrong? Surely, there are such things as beliefs, desires, and intent. There are reactions, rational or affective (if you think those two need to be distinguished), that we describe as beliefs, desires, and intentions. Just as there is the color blue, there are the incidents of folk psychology, and so folk psychology is real. But continue the analogy between a color such as blue and the reality of blue.

The color blue is certainly at some level a state of mind. It does not take much imagination to indulge the speculation that no two people *see* blue in just the same way, because no two people's vision system structures, at the finest neuronal level, are just the same. Significantly, though, close enough is good enough in the case of color labels (perhaps, ultimately, in all things). Indeed, even were two people's visual systems identical at the most basic neuronal level, you also can appreciate that insofar as color is the reflection of light (wave or particle) and that reflection is subject to spatial differences and no two (four) eyes could occupy precisely the same space, your blue car could never, at the finest neuronal level, be my blue car. Again, though, for all practical intents and purposes, we can reach sufficient consensus to label our two sensory experiences the vision of blue.

So mere disjunction at even some particularly very fine level of acuity does not undermine the social and even general clinical utility of labeling the sensation created by reflected light at a particular wavelength (or range of wavelengths) blue. We can say, then, that color exists, even if its basis depends on social convention. But in a real way, color does not exist. The car is not, strictly speaking, blue: The light reflected by the collection of solid shapes that compose the car is within the blue wavelength range (around 475 nm).[12] But the convention works because it does not mislead. That is, although color in an important way reduces to wavelength, nothing significant is lost or misrepresented in most cases by the shorthand. As a matter of fact, a good deal is gained by agreeing that the car is blue rather than calibrating the wavelength that each of us perceives in terms of nanometers.

Folk psychology, then, with its dependence on beliefs, desires, and intents and the like does no harm and even helps insofar as social conventions converge sufficiently around what a belief, desire, or intent is. And they generally do. But there is a crucial difference between what folk psychology describes and what color labels describe: Wavelengths, the basis of color and distinctions, certainly exist, are certainly real wholly independently of the perceiver, but beliefs, desires, and intent do not, at least not in the same way. That is jarring, and it is jarring for reasons that evolutionary psychology can explain.[13] Folk psychology describes fictions that are useful at some levels of acuity for some purposes (just like colors) but that are pernicious at the levels of acuity that matter to the calculus of responsibility. Paul Churchland made that counterintuitive leap clear when he likened folk psychology to alchemy, on the way to endorsing an "eliminativist materialism."[14]

The problem with folk psychology, the measure of its deficiency, is found in the degree of its inaccuracy. You could depict the sun as a chariot moving across the sky, but you'll make real mistakes if you take that metaphor literally, or even too seriously. Churchland described "the major philosophical positions on the mind-body problem . . . as so many different anticipations of what future research will reveal about the intertheoretic status and integrity of folk psychology."[15] The philosophical differences (or confusion) are the consequence of conjecture in the face of insufficient empirical evidence.

The identity theorist optimistically expects that folk psychology will be smoothly reduced by completed neuroscience, and its ontology preserved by dint of transtheoretic identities. The dualist expects that it will prove irreducible to completed neuroscience, by dint of being a nonredundant description of an autonomous, nonphysical domain of natural phenomena. The functionalist also expects that it will prove irreducible, but on the quite different grounds that the internal economy characterized by folk psychology is not, in the last analysis, a law-governed economy of natural states, but an abstract organization of functional states, an organization instantiable in a variety of quite different material substrates. It is therefore irreducible to the principles peculiar to any of them.[16]

So cast, the measure of folk psychology and the disagreement among theorists over the ultimate efficacy of folk psychology is the extent to which folk psychology predicts and will depict the reduction of mental states such as beliefs, desires, and intentions to neural states. If there is something about mind that will not reduce to brain, then issues of responsibility will not reduce to brain state either because neuroscience will necessarily miss something that is normatively essential. The eliminative materialist takes the most extreme (or confident) view of what neuroscientific insights will ultimately reveal: "The eliminative materialist is . . . pessimistic about the prospects for reduction, but his reason is that folk psychology is a radically inadequate account of our internal activities, too confused and too defective to win survival through intertheoretic reduction."[17] But what could inform that pessimism about folk psychology (optimism about what advances in neuroscience *can* reveal)?

The ultimate difference between the two extremes—the dualist and the eliminative materialist—is a difference of opinion about what future discoveries will vindicate. Now, were the difference a matter of degree, which it might seem to be, we could expect to find the truth somewhere along the continuum between the two. But that is not possible: Either monism or dualism is right, at least in terms of our current understanding of how human agents might be constituted, and there could not be degrees of either or both. Either folk psychology is a metaphor (sometimes good, sometimes not so good) or it is undermined by neuroscientific insights. If the eliminative materialist is right, at the end of the day folk psychology will prove to be a deficient, even problematic metaphor because it will support incoherent normative judgments that would actually frustrate human thriving. So folk psychology would make no more sense than alchemy, and functionalism, which attributes some sense to folk psychology, seems as dangerous as alchemy too.[18] Functionalism, which Churchland attributed principally to Putnam,[19] rejects folk psychology as incorrigible but does not abandon the beliefs, desires, and intents of folk psychology as viable functional idioms, ultimately irreducible. Eliminative materialism is more radical: "the correct account of cognition, whether functionalistic or naturalistic, will bear about as much resemblance to [folk psychology] as modern chemistry bears to four-spirit alchemy."[20] The eliminative materialist points to the developments in neuroscience specifically and science generally to support the intuition that when we figure it all out, we will have figured it all out in terms that demonstrate the incoherence of folk psychology.[21] We are trapped by the terms of folk psychology because they so comfortably confirm our ignorance, just as our forebears made causal sense of the path of the sun by reference to gods in chariots: "Eliminative materialism thus does not imply the end of our normative concerns. It implies only that they will have to be reconstituted at a more revealing level of understanding, the level that a mature neuroscience will provide. . . . [I]t is important to try to break the grip on our imagination held by the propositional kinematics of FP [folk psychology]."[22]

You can get the sense of what eliminative materialism would accomplish, as a philosophical theory of mind, if you appreciate why it would be ludicrous to blame your car for not starting one morning, as if it didn't want you to get to work on time. Now we use such terms all the time: "my car failed me again"; "I hate the car"; "that's the last time I trust that thing." We use those same terms when talking about people, but that does not mean we are using them the same way, attributing human characteristics to so much plastic and metal. And no rational person would think that we were using the same terms the same way. Similarly, when we use the term *love* in the following three sentences, we are using it in importantly even if perhaps in subtly different ways: "I love my daughter"; "I love my dog"; "I love my car." The term works in each declaration, but its operation, its meaning, is not the same. Nonetheless, the term works well enough within the contexts to communicate effectively. It is because we know the evaluative context that we have no trouble making sense of the term *love*. If the nature of the inquiry changed, if we needed to reach a normative conclusion that would be confounded by the use of the same term in those three contexts, then the term would be problematic *precisely* because it would obscure (perhaps even undermine) that normative judgment.

So materialism does not depend on our purging our vocabulary of all terms that do not have an immediately accessible and certainly discernible neural correlate. We could still use terms such as *belief, desire,* and *intent* as long as we understand what they do not denote as well as what they can connote. But what materialism does certainly entail is a skepticism that such terms do the normative work the criminal law doctrine would have them do. And that skepticism will prove more than fair in contexts in which the criminal law relies on retributive principles in guiding as well as punishing behaviors. Make no mistake: A thoroughgoing materialism challenges criminal law doctrine, and appreciation of what the perspective entails would profoundly reorder the morality that criminal law doctrine instantiates.

Evolution of the Doctrine

In an important contribution to the conversation, Greene and Cohen concluded that although neuroscientific insights are unlikely to affect the operation of the criminal law in the near term, those insights will, over time, effect adjustments of our normative perspectives that will ultimately be manifest in changes to criminal law doctrine.[23] Neuroscientific insights tell us more and more about what it means to be human: We learn more about consciousness (but frustratingly not enough);[24] we can *see* the difference between the adolescent and adult brain (which makes responsibility determinations more salient);[25] we better appreciate the effect that neural networks in emotion centers of the brain have on decision making and impulse control.[26] Although we all know and have known, probably from the beginning of time (or, at least life, intelligent or otherwise) that adolescents are less mature than adults, make worse decisions that involve deferred gratification, it was the contribution of neuroscience that supported Justice Kennedy's conclusion in *Roper.*[27] So Greene and Cohen, writing before *Roper,* seem to have been prescient: "The legitimacy of the law itself depends on its adequately reflecting the moral intuitions and commitments of society. If neuroscience can change those intuitions, then neuroscience can change the law. . . . The fact that people are tempted to attach great moral or legal significance to neuroscientific information that, according to the letter of the law, should not matter, suggests that what the law cares about and what people care about do not necessarily coincide."[28] According to Greene and Cohen, then, the normative failure of criminal law doctrine[29] will increasingly become manifest as the science outpaces the law and application of the doctrine leads to results that are morally indefensible. Solitary confinement, particularly in the case of minors, might be a particularly stark example of that consequence,[30] as might developments in restorative justice initiatives that neuroscientific insights could confirm in accessible empirical terms.[31] The findings of studies such as those cataloged by Adrian Raine[32] and introduced in the preceding chapter make it easier to feel compassion rather than animosity toward those who commit even violent crimes.

The nature of the intellectual gestalt shift that that type of normative reappraisal and reorientation triggers may be captured most provocatively in an extension of the metaphor suggested by Robert Sapolsky:[33] Does it make sense to equate cars and humans for purposes of determining the normative nature of human agency? If it offends us, in some very fundamental sense, to liken mechanical systems to human systems—and not as a matter of analogy—we need to appreciate why we are offended. Is the comparison inapt—does it miss a fundamental and normatively significant difference between people and cars?—or is it just upsetting because it obscures something unique and uniquely special about and to us as human agents, something supplied, perhaps by what is ineffable about consciousness?

It is not necessary to discount the significance of complexity. Your modern car is *not* the same thing, exactly, as your toaster. We would miss something important about cars if we said that they were just toasters with wheels. But is that *something* the kind of thing that has normative significance? You would not punish your car but simply repair your car even though your car is a more complex mechanism than your toaster. Certainly the nature of the repairs you would effect on each would differ, would respond to the mechanical differences between the two products. But in both cases you would repair; you would not punish retributively.

If human agents are distinct from cars as a function *only* of their relative mechanical complexity, then it would make no more sense to punish humans on some retributionary basis than it would to punish your car and repair your toaster *solely* because the car is more complex than the toaster. Ultimately the object should be to remedy the deficiency in the toaster, car, and criminal. If we are to make sense of punishment from the materialistic perspective, it must effect such a remedy, or there must be some other normative reason for the punishment, a reason we would not find when considering the malfunctioning toaster or car. Where would we find that reason, that something *other than greater complexity* that distinguishes the car from the human agent? The search for that something is the inquiry pursued by scientific reductionism.

Reductionism: Something in the Gap?

We could posit that psychology *reduces* to biology, which *reduces* to chemistry, which *reduces* to physics, which *reduces* to mathematics (maybe more specifically, statistics). Appreciating that serial reduction requires us to make sense of the relationship between each level of reduction, and particularly to focus on what reduction reveals or obscures about the reducing science and the reduced science as well as the essential relationship between them. The reductionist dialogue and debate are vibrant, with much sound and fury, but the substance of it for the dualist–monist inquiry pertinent to the mind–brain relation may be succinctly formulated: "If qualia-concepts or concepts of conscious states pick out mental states in terms of features that cannot be fully explained in terms of the vocabulary of the reducing science, the explanation and hence the reduction is not successful. Even if our mental terms and neural terms refer to the same states, if the former pick them out using mental concepts that cannot be neurally explained, we will not have reduced to [*sic*] the mental to the neural. There will be a residual explanatory gap."[34] Now what fills that gap *may be* quite important; it could be everything. If will or consciousness resides in that gap, and will or consciousness is pertinent, even fundamental or essential, to the responsibility of human agents, then that something ineffable is conceptually distinct from the physicalism that generally informs science, cognitive neuroscience more specifically. And if that intuition, inferred from the ineffability of qualia and will or consciousness is correct, then we will not be able to make sense of human agency in terms that the law needs until we make sense of will or consciousness: a daunting prospect. That seems not too different from concluding that we cannot conceptualize human nature until we understand God or some other supernatural (in the literal sense) entity.

A metaphor first offered by Greene and Cohen and then treated by Michael Moore captures well the limits of reductionism, or the sense that reductionism *has* limits:

> At some time in the future, we may have extremely high-resolution scanners that can simultaneously track the neural activity and connectivity of every neuron in a human brain, along with computers and software that can analyse and organize these data. Imagine, for example, watching a film of your brain choosing between soup and salad. The analysis software highlights the neurons pushing for soup in red and the neurons pushing for salad in blue. You zoom in

and slow down the film, allowing yourself to trace the cause-and-effect relationships between individual neurons—the mind's clockwork revealed in arbitrary detail. You find the tipping-point moment at which the blue neurons in your prefrontal cortex out-fire the red neurons, seizing control of your pre-motor cortex and causing you to say, "I will have the salad, please."[35]

Moore's response was dismissive:

In Greene and Cohen's imagined string of neural firings determining a decision of salad over soup for lunch, for example, there is no room for persons, selves, or moral agents, on nonreductionist premises. We—our agency—would have to be something extra, a ghostly commander leading the blue neurons (for salad with blue cheese dressing?) to victory over their red (for tomato soup?) competitors. And this is silly—it would make the soup versus salad decision at the neural level like the battle scenes in Kurosawa's film, Kagemusha, with all the roles in the scenes filled before selves enter the stage.[36]

One might respond to Moore that the *metaphorical* (a subtlety missed) neural battle over salad or soup need be nothing more mystical, or willed or conscious, than the electronic device that turns on your furnace or tells your automatic garage door that there is something in the way. There is no self missing when your thermostat starts your furnace, any more than there is a metaphysical self necessary for learning. Neurons do not have to be animated by consciousness to do the work they do.[37] At least we do not have to posit will or consciousness for that work to be done. So although something may, indeed, be missing that is crucial to human agency, something conscious that is more than the propagation of chemical and electrical signals, Moore's response to Greene and Cohen offered no sense of what that might be and how it could relate to a normative calculus.

It would be something of a leap of faith (maybe literally) to conclude that that something ineffable about will or consciousness precludes the advance of the neuroscientific inquiry in any way pertinent to legal liability and doctrine. That is, we need not know how everything works before we can develop helpful insights into how some things work. We do not need to understand all of quantum theory even to do something as ambitious as sending a person to Mars. Those who rely on the limitations of reductionism, who may even conclude that reduction fails in some perhaps significant way sometimes, place a great deal of tension on that gap when they infer therefrom that folk psychology captures a truth that cognitive neuroscience will never reveal.

The point of those who leverage the limitations of reductionism into a refutation of a neuroscientific reappraisal of human agency relies on the conclusion that what is in the gap, what is lost in translation, represents a conceptual barrier between what materialism can ever demonstrate and what provides the necessary normative foundation of human agency. For them, then, the limitations of neuroscience are just not empirical, not the kind of thing that may be overcome by bigger magnets and more refined software in fMRI machines, for example. Will or consciousness, for them, is not just the final frontier; it is the impregnable obstacle.

Those more sanguine about the promise neuroscience holds for our better understanding of human agency appreciate the limitations of the current science, recognize that we are just at the beginning of what is certain to be a long journey, the contours of which we can only dimly see now, but they are able to focus on the direction of science—since roughly the beginning of scientific inquiry—and find reason to foresee that we will find answers to some of the vexing questions pertinent to the relationship between human agency and law. The limitations of the current science are, for this group, so far epistemic, *not* conceptual, limitations. They imagine that as we get stronger magnets and the software to support them (as well as the theory to direct the inquiry), fMRI and similar devices will reveal more of what we need to refine our understanding of human agency in terms that matter to legal doctrine and practice.

Note that it is not necessary to understand *everything* before law can take advantage of neuroscientific insights to fine-tune law's normativity. We did not have to split the atom to realize that burning witches and punishing epileptics as demonically possessed was not normatively coherent: The gap need not be bridged completely, and all that we will learn we do not need to learn all at once. So those who are currently dubious should probably not put too much weight on the conceptual side of the balance. Indeed, even the most conspicuous skeptics already could acknowledge that neuroscience, even in its current relatively nascent stage of development, can inform responsibility calculi in ways that are fundamental to broad swaths of the criminal law.[38]

Philosophers and some neuroscientists, though, do strenuously insist that the limitations of neuroscience are conceptual, that no amount of scientific inquiry into the natural or material world would be able to bridge the gap. They insist that there is a quality of human agency that entails more than just the brain, and say things like "Brains don't kill people[,] People kill people,"[39] and "Brains do not convince each other; people do."[40] So for at least some (and perhaps even all) of the skeptics, there is something essentially human that neuroscience misses, must miss. There is something to the human agent other than brains, the neural system.[41]

Although a comprehensive recapitulation of the philosophical schools of thought about reductionism would be beyond the scope and space limitations of the instant inquiry, it is possible to posit the fundamental tension in accessible terms that do not rely too much on jargon:

> An epistemological reductionist concerning a science or theory S holds that in fact, we are (or at some point will be) able to reduce S to a more fundamental science. It is a thesis about what we can epistemically achieve. The ontological reductionist is, in this respect, more modest: She just holds that in fact, there is just one sort of objects and properties out there in the world; however, due to our cognitive limitations, we might never be able to actually carry out all the reductions that would be appropriate given the actual ontological structure of our world.[42] . . .
>
> On one interpretation, the non-reductive physicalist opposes the idea that we can and should in fact reduce high-level sciences; we need them for epistemic or pragmatic purposes.[43] We need a plurality of autonomous theories and frameworks. However, the non-reductive physicalist accepts that what is actually out there might very well be all of one kind, at least in some ultimate sense such that ontological reductionism is true.[44]

Now for present purposes, it does not matter which form of nonreductionism one endorses: The work of Adrian Raine and other physicalist perspectives has shown that nonreductionist positions generally would fail as fundamental normative justifications of extant criminal law doctrine (though the epistemic nonreductionist would likely not offer that conclusion as a justification of the doctrine in the first place).

There may be some level of empirical and perhaps even ontological acuity that we, as human agents, cannot appreciate or perceive. Even were that the case, though, even were it true that there are "more things in heaven and earth . . . than are dreamt of in [our] philosophy [or physical sciences],"[45] concluding that the stuff essential to human normativity resides within those ineffable, or unexplained, or inexplicable somethings is quite a leap. It is just the type of leap that noninstrumentalist normative theory needs in order to make sense of human agency in libertarian or compatibilist terms.

There is, then, a good deal at stake in the noninstrumentalists' efforts to identify the gap, the something lost in translation, as we work down from the familiar and comfortable appearances provided by folk psychology to the more concrete and elemental neural stuff that neuroscientific insights reveal. If noninstrumentalism fails, if Paul Churchland was right in likening folk psychology to alchemy, then deontology *as well as* normative systems, including legal doctrine based on folk psychology, are, in fact, no better than alchemy.

Moore, perhaps better than any other noninstrumental moral theorist writing today, has understood the profound significance of the neuroscientific challenge to extant (largely deontological) theory: If the materialism neuroscience would vindicate is right, much of noninstrumentalism is wrong. Moore has recognized that the relationship between willing and action is importantly pertinent to the debate and so has engaged the challenges to conscious will presented by the work of Benjamin Libet and Daniel Wegner. The contributions of each are central to the free will–determinism contest and familiar to most students of the colloquy. Succinctly, Libet established experimentally that resolution to action occurs *prior* to consciousness of the intention to act,[46] and Wegner comprehensively defended the thesis that "we are intrinsically informed of how our minds cause our actions by the fact that we have *an experience of causation* that occurs in our minds."[47] Moore described Wegner's conclusions as relating to the epistemic: We are not infallible "knowers of when we acted."[48] So while we might have unique access to our sense of consciousness, that sense is often wrong: Our privileged access is neither transparent nor incorrigible. That epistemic challenge is profound: It might leave us with an "epiphenomenal will."

Moore, then, in the course of considering the challenge of a merely epiphenomenal will, explained what is at stake: "If our choices, intentions, and willings truly do not cause our voluntary actions, that challenges directly the folk psychology assumption of autonomy,"[49] as well as, I would add, the noninstrumental normative basis of legal doctrine based on that assumption. That would be a reasonable construction of the challenge presented by Libet. Wegner's thesis too, according to Moore, would be disruptive of the normative status quo: "This denial of privileged access is more challenging than it might seem at first blush. Such privileged access is arguably a main marker of the boundaries of self and personhood. . . . If this line between the actions we do as persons, and the subroutines our bodies do ('without us,' so to speak), is eroded, that would be a challenge to our sense of selves."[50]

It may well be correct that if the Libet and Wegner challenges either separately or in combination stand up to a critique, dualist or otherwise, that would preserve in a meaningful normative sense human agency in the terms that animate folk psychology and noninstrumentalist theory, then much of our legal doctrine would be subject to reappraisal. But is it necessary that consciousness be compromised in the terms offered by Libet or Wegner in order for the materialist critique to undermine popular normative conceptions of the human agent? No, it is not.

It is certainly true that, if our consciousness does not "control" our thoughts and actions in a normatively significant way (in a way that would provide a meaningful sense of human *agency*), then there is no such thing as responsibility in anything more than a mechanically causal sense: like "the cue ball collided with the eight ball after the cue ball was struck by the cue stick." Such merely causal responsibility could not establish fault, much less culpability or desert, the basis of retribution. So if Libet and Wegner are right in the way free will libertarians and compatibilists fear, then there is no moral responsibility: full stop. And that is true whether we are talking about actions temporally remote or those proximate to the conscious motivation or will to take them. If your actions precede your willing them, then you are importantly not responsible for them, and you do not become responsible for them anywhere along the cascade. Indeed, it may not be wrong to understand human agency as you would the operation of a car: mechanical, automatic. So it is not surprising that Moore—who resists, strenuously, the determinism and monism of Greene and Cohen and Sapolsky—describes the Libet-Wegner challenges to noninstrumental conceptions of human agency as fundamental. If things are as Libet and Wegner suggest they are, we are not what we thought we were, and not what deontology and extant legal doctrine assume we are: And therefore, there is no basis for normative desert and retribution.

But it also is true that even if Libet and Wegner are wrong, if there is *some* scintilla or even more of conscious control over our choices and actions,[51] that does not save legal doctrine that takes insufficient account of the limitations on that control. Indeed, legal doctrine that is not considerate of different levels of control in determining desert or any noninstrumental determinant of punishment is at least insufficiently coherent.[52]

The findings of Adrian Raine and the others whose work he recounted has provided evidence at the most fundamental level that, contrary to the law's operating presumption, all human agents are *not* normatively equal. That is not to say that all people of typical intellectual and emotional capacity are not similar enough to be subject to the same proscriptions and prescriptions to keep the trains running on time and achieve a desirable level of interdependence and cooperation. But it is to suggest that taking the operating presumption too literally, assuming that we all start from the same place when we do not, and ignoring the fact that we are subject to normatively significant forces beyond our control (and, perhaps, even understanding) will result in legal doctrine that undermines rather than serves the normative object of the doctrine. Another automotive analogy: If you are having trouble getting the car started, continuing to depress the gas pedal will just flood the carburetor (assuming you drive a classic); it will not start the car. Similarly, subjecting the youthful offender to isolation because "he deserves it" may only increase the dangerousness of the child, rather than reform him.[53]

So neuroscience need not demonstrate the total failure of responsibility, in the normative sense; it is enough if it reveals the constituents of responsibility. And here the salience of neuroscientific evidence may have rhetorical power. Certainly it is true, as Morse and others have reminded us, that we know and have known for a very long time that teenagers are immature and do immature things, sometimes incredibly brutal immature things.[54] But the moral calculus changes, and is perhaps enhanced, in ways Greene and Cohen predicted when we can see, vividly on an fMRI scan, the corporeal evidence of that neural development or underdevelopment. And here is where the power of the critique of folk psychology emerges: To the extent that folk psychology recognizes moral blameworthiness, normative responsibility, on the basis of something less mechanical, less physical than the operation and cooperation of neurons, folk psychology undermines rather than serves the normative object of legal doctrine. Folk psychology suggests that there is something normatively important, crucially important, in beliefs, desires, and intents that is inaccessible to mere physical investigation of human agency. But what if there isn't? What if what is *currently* beyond our understanding has no normative significance beyond the normative significance of *any* physical system? Although the computer running your car is more complex than the carburetor that ran your grandfather's car, it is no less mechanical in a way that would have any normative force. It would be as absurd to talk about the beliefs, desires, and intents of your 2015 Mercedes as it would to talk about the beliefs, desires, and intents of his 1966 Mustang. Mere complexity does not change the nature of the two cars' agency, at least not in any normatively significant way: They are both just cars, even if one or the other or both sometimes seem to have a mind of their own. But some whose perspective would clash with (if not specifically reject) the mechanics of folk psychology seem to hold out hope that by undermining wholly physical explanations of normative decisions they leave important room for noninstrumentalist moral theory that challenges wholly instrumentalist accounts.

The Nonfalsifiability of the Inscrutable

An accessible and well-received example of that perspective is Selim Berker's "The Normative Insignificance of Neuroscience."[55] Berker wrote the piece in response to articles by Joshua Greene[56] and Peter Singer[57] specifically that relied on brain imaging[58] to support their conclusion that characteristically deontological[59] normative perspectives are less reliable (indeed, wrong) vis-à-vis characteristically consequentialist perspectives. Greene[60] performed a series of experiments based on "the trolley problem,"[61] which investigates the bases of normative distinctions in settings not obviously normatively distinguishable. That is, we seem to (maybe intuitively) react differently to scenarios in which we act[62] in distinguishable but arguably not normatively distinct ways to bring about the death of one innocent to avoid the death of five innocents. From a strictly consequentialist perspective, the question is not difficult: take the one life to save five.[63] Indeed, as Berker noted, deontologists, or some of them at least, may be able to reach the same conclusion.[64]

Significant for Greene, though, was the fact that something in us, perhaps in all of us, at least hesitates when the scenario is adjusted to require more direct intervention of the bystander-human agent to bring about the death of the one to save the five. For Greene, it seemed clearly a normative mistake to *not* cause the death of one to save the five, and so if he can identify what is going on in the brain, very mechanically, when that mistake happens or that is not happening when the mistake is not made, in other words, when the correct instrumental conclusion is reached, then he will have demonstrated that neuroscience can be used to appraise the correctness of moral decisions in a way that would be apparent to (and maybe even convince) deontologists.

Berker revealed deficiencies in Greene's method[65] and so compromised Greene's conclusion. But beyond those valid methodological criticisms, Berker reached important conclusions about the nature of moral reasoning and so the cogency of anything about it that we may glean from investigation of cognitive mechanics. Moral reasoning, for Berker and others,[66] is just not the same thing as the type of arithmetic reasoning we do when we are confounded by such circumstances as the salience bias that Kahneman and Tversky's work revealed:[67] how badly we do mathematics under some conditions, how heuristics fail us in predictable settings. But Berker and others have maintained that the constituents of moral inquiry are distinguishable from the constituents of the computational biases identified in behavioral economics.[68] So we should not be surprised if the equation Greene suggested fails when the nature of the question changes so dramatically. Greene, by those lights, in fact rigged the test when he equated the correct moral conclusion with the utilitarian moral conclusion, obscuring distinctions that can and, deontologists would argue, *should* matter to the normative calculus. That is the important takeaway from Berker's critique.

But does that critique support the conclusion that neuroscience is normatively insignificant, or does the claim exceed the proof? Recognize that there are two parts to the claim. The first part relies on mechanical neuroscience itself: What does what in the brain, and how does it do it? The second part relies on the nature of inference: What may we infer about moral reasoning from what we *now* know about the neural mechanics? Neuroscience may provide the means to track the path and progress of reasoning by reference to the brain regions engaged in the course of decisions, moral or otherwise. And we know that we do not know all (maybe not even very much) of what we would need to know to reach reliable conclusions about how the brain functions when it approaches certain cognitive tasks: What does it mean when a brain region lights up on an fMRI? How functionally discrete are apparently discrete cognitive centers when the brain weighs the constituents of any decision, normative or otherwise? Does the neuronal network supporting a cognitive function operate in ways not predictable from the sum of the neurons firing in different brain regions? And there would certainly be many other questions that the current science cannot yet resolve.

At the same time, we do know a good deal more about the brain than we knew even just a generation ago, and there is no reason to imagine that our knowledge of the brain will not increase apace, perhaps even growing geometrically. For the time being, though, neither those who are most sanguine about the promise of neuroscience to resolve normative questions nor those most skeptical, like Berker, can do much more than claim they know how the story will end. Are we really prepared to say that neuroscience will ultimately prove insignificant? There are reasons to believe that neuroscience has already demonstrated sufficient significance to (at least help) convince even the most skeptical.[69] Though it is clear that we do not yet know all that we will know, the state of the science continues to tell us something about the constituents of the decisions human agents make. Further, we can identify the kind of things that go wrong, at the neural level, when human agents err.

It may well, then, turn out that, when we trace neural activity through brain regions and brain states, we will confirm how and when the brain makes mistakes such as the biases Kahneman and Tversky discovered. That was just the type of thing Greene was trying to do. The fact that Greene's method was deficient is not enough to disestablish his hypothesis. At the end of the day we are just where Greene found us: Because we do not yet know enough to distinguish how the brain certainly functions when it errs from the way it functions when it succeeds, we can infer nothing certain about the normative valence of brain activity from neural process. Berker's conclusion, keep in mind, likely would have been just the same if Greene's methods were unassailable and if Greene in fact discovered what he thought he had discovered about brain function: Berker and those who proceed from his noninstrumentalist perspective would still say that normative reasoning is not the same as arithmetic calculation, and so error in one setting cannot be conflated with error in the other. That, of course, is the sum and substance of Berker's contribution. Though Berker cited[70] the brief popular article by Adina Roskies and Walter Sinnott-Armstrong, he seemed to miss the sense of the subtle and important point they made: "Future studies should explore the distinctions that the current literature roughly characterizes as emotion versus cognition, and deontological judgment versus utilitarian judgments. Further clarification will come with a more precise specification of which functional processes constitute the controlled cognition that is supposed to cause utilitarian moral judgments. Clearly, more work needs to be done."[71] There would seem to be two important points to be made here that Berker's analysis and conclusion have obscured (as does Greene's analysis and conclusion): First, the emotion versus cognition category fails if the object is to delineate discrete neural processes. It is no more true to say that emotion and cognition (or rationality) are distinctly constituted in the brain than it would be to distinguish decisions made in the head and heart, though that might make the stuff of good poetry. Second, there is no reason to believe that different brain systems reliably and always track different normative systems: We do not have a deontological lobe and a utilitarian lobe that are in tension.

The battling armies image offered by Greene and Cohen and rejected by Moore is an analogy or metaphor, at most: The point is not that armies of neurons battle in our brains between soup or salad or consequentialism and deontology; the point is that conclusions we label as soup or salad or consequentialism or deontology have neural referents. They must, if they are to be "of the brain," and so ultimately "of the mind" (should that distinction retain currency). Berker's rejection, then, of Greene's misleadingly simplistic demonstration of the consequentialist and deontological tension in the brain was just a rejection of what could already be an analogy or metaphor as illustrative of physical fact. It is as though Greene posited real micro-armies waging war, and Berker revealed that there really are no such armies, discoverable by nanoscience or otherwise. Greene was wrong to reduce consequentialism or deontology to particular neural signatures, and Berker was wrong to dismiss out of hand neural bases of evaluating, or at least describing, normative commitments. So neither Greene's nor Berker's analysis could do much to advance the ball, as is evidenced by Berker's conclusion that the case against deontology would not have been proved *even if* Greene had been able to accurately demonstrate that we err when we think deontologically in just the same way we err on account of the cognitive biases described by Kahneman and Tversky.[72]

Understood that way, the conflict between Greene and Berker sounds a lot like the familiar challenge to normative systems based on a supernatural deity or deities: If God ordered a believer to do something that the believer considered to be immoral, would it be immoral? That is, would the morality be found in the agency of the believer or in the pronouncement of the deity? Well, the believer could respond that what the deity prescribed could only be moral (is *automatically* moral), and so if the deity ordered it, it would not be immoral. In the same way, noninstrumentalists could say that if the neurons were to fire when confronting a moral question in a way that would be erroneous if the agent were considering a mathematical question, then that would establish a truism: Moral questions are not mathematical questions, and so we should not expect them to yield (merely) mathematically correct answers.

Keep in mind, though, that neuroscience need not resolve the ultimate instrumental versus noninstrumental controversy in order to be significant, and we have no reason to believe that Berker would be uncomfortable with that conclusion. Within the scope of the doctrinal challenges described in chapter 2, is there room for neuroscientific insights to improve the law, in the sense of making it more consonant with its own normative commitments? That is, would it not be the case that neuroscientific insights could guide development of the law in ways that would appeal to noninstrumentalists and instrumentalists alike? Could neuroscience take the next step and illuminate better the questions about which the two perspectives would disagree? Yes, and there may be no better place to begin than with banishing hobgoblins.

Retribution

Punishment by reference to retributionary principles is necessarily noninstrumental. Indeed, if retributionary punishment has any instrumental effect, that effect would be inconsistent with retributive principles: at least if we take Kant as the source and measure of the deontology upon which retribution depends, as we shall for present purposes.[73] A Kantian would not punish A to teach B a lesson, for to do that would be to treat A as a means to an end rather than as an end in himself.[74] It would be wrong, immoral, from the noninstrumentalist perspective, to do that to A. Concomitantly, the instrumentalist would at least likely have no interest in retribution, might deem it no more moral (or ultimately efficacious, for that matter) than actions taken in revenge, and so would resist any punishment administered to effect retributive rebalancing.[75]

It is difficult to formulate certainly, uncontroversially, the normative substance of retribution. At least doing so, in terms even reasonably unassailable, seems to have been elusive for noninstrumentalists so far. Indeed, it may be problematic to distinguish retribution from revenge, though the connotations *seem* different. In any event, it is clear that a comprehensive review of retribution, even just as an apology for the criminal sentencing doctrine, would fill a thick volume or two.[76] It would, of course, be necessary to distinguish retribution from instrumental theories of punishment. If your argument for retribution (whether you acknowledge it is an argument for revenge or not) depends on some psychic or emotional benefit realized by the direct victim of criminal activity or even society at large, that is ultimately a consequentialist argument: Any such benefit to the victim or society will have to be weighed against harm visited upon the criminal defendant, those who care about and may depend upon him, and those in society generally who feel psychic pain when they see what incarceration does to those convicted of crimes. And we may feel that same distress no matter the heinousness of the crime that led to imprisonment. Even those who can justify torture do not justify it as a punishment.[77] For a theory of retribution or revenge to work noninstrumentally, it must rely on no instrumental object.

There are problems, of course, with all punishment calculations, some problems that neuroscience may help us address, but other problems as well that do not admit of resolution in the reasonably near future from any projection of neuroscience's potential. That is true whether our object is retribution or deterrence: The same punishment that would be just right for defendant *A*, in terms of either the instrumentalist or noninstrumentalist perspective, would necessarily present two insolvable problems: (1) retribution and deterrence are necessarily inconsistent, and (2) the experience of punishment would not be the same for any two criminal defendants, *A* and *B*, because the two are not just alike. Imposing the same punishment on both would be to miscalibrate the sentence imposed on one or the other (if not, as is more likely, to miscalibrate the punishment imposed on both, in light of the practical impossibility of calibrating punishment correctly given the crudeness of the tools with which we would measure retributive or deterrent effect, *even were we to agree on what the appropriate retributive or deterrent effect should be*).[78]

Blame, Desert, and Culpability

Michael Moore, after dismissing other theories of retribution,[79] offered a justification that depends on a basis that neuroscience can recognize, or at least with which neuroscience can gain some purchase. Moore would found retribution on guilty feeling, the emotional reaction. The fact that we feel guilt when we act in a certain way both justifies retribution therefor and also obligates the state to exact retribution; according to Moore, criminals are owed retribution by the state. After recognizing that retribution is based on desert,[80] he concluded that in cases where retribution is justified, "one emotion . . . predominates, and that is the emotion of guilt. A virtuous person would feel great guilt at violating another's rights by killing, raping, assaulting, etc. And when that emotion of guilt produces the judgment that one deserves to suffer because one has culpably done wrong, that judgment is not suspect because of its emotional origins in the way that the corresponding third person judgment might be."[81] Now, laying aside the conclusion that an emotion, a neurological state, can reliably justify a particular normative response without inquiry into the neural foundation of that state in the particular case, it is necessary to discover what work the emotion is doing in Moore's normative calculus. Is retribution appropriate, justified, because the perpetrator feels guilt or because the perpetrator should feel guilt? Clearly it must be the latter, because

Moore posited the virtuous person. We are left, then, to take account of the guilt that would be felt by the virtuous person, a particular reasonable person standard. And the guilt that would be felt is to be determined by (in the sense of "limited by") desert. Moore's focus on the emotion of guilt seemed to rely on the emotion as the reliable test for the rectitude of retribution: We can be confident that retribution is the proper (the only proper) response to criminal actions because we each feel guilt when we act criminally. The emotion, the feeling, confirms the rectitude of the punishment, and, perhaps the nature and extent of the punishment.

Moore cited extensive sources that confirm the reliability of the emotions as heuristics, indicators of appropriate behavior and response to others' behavior. Our emotions are our main heuristic guide to finding out what is morally right: "We do both them and morality a strong disservice when we accept the old shibboleth that emotions are opposed to rationality. There is . . . a rationality of the emotions that can make them trustworthy guides to moral insight. Emotions are rational when they are intelligibly proportionate in their intensity to their objects, when they are not inherently conflicted, when they are coherently orderable, and instantiate over time an intelligible character. We also judge when emotions are appropriate to their objects; that is, when they are correct."[82] So an emotion, guilt, is ultimately the heuristic that guides and justifies retribution. We can know that retribution is the appropriate response because it feels right, the way actions consonant with any emotional reaction feel right. But Moore recognized that, as heuristics, emotions are subject to the same shortcomings as are all heuristics, just as the biases identified by Kahneman and Tversky may actually undermine rather than serve decisions. Now that is not to say that emotions or biases are necessarily unreliable, but it is to say that emotions cannot be the determinate of their own rectitude.

We know that the cognitive biases unveiled by Kahneman and Tversky mislead, are wrong because they in fact encourage behavior that is inconsistent with the human agent's avowed object: act efficiently,[83] take only the efficient amount of risk,[84] no matter how the terms of the risk calculus are cast.[85] When we are focused on efficiency, we can be confident of when the emotions are reliable guides and when they mislead because their object is certainly discernible. Not so, though, when we would use emotions as a guide to something that is not amenable to certain calculation, like the appropriateness of retribution or revenge. So Moore's analysis, when he identified the heuristic value of guilt, obfuscated in the same way that Berker's response to Greene did: There's just something about noninstrumental responses to moral dilemmas that is ineffable and cannot be reduced to an instrumentalist calculus (just as cognitive biases could be confirmed by comparing what the agent wanted to accomplish with what operation of the heuristic did accomplish). Ultimately, Moore's rationalization (almost literally) of retribution in terms of the emotion of guilt is no more productive (perhaps because no more falsifiable) than Berker's invocation of noninstrumentalism in response to Greene's instrumentalism.

But there may be an even more substantial failure attributable to Moore's conception. To be clear, Moore would not rely *solely* on the fact of the guilt reaction to confirm the measure, at least the rectitude of retribution; Moore would rely on guilt only to the extent that it is the appropriate reaction under the circumstances. So, ultimately, the work is not done by the emotion; the work is done by whatever means help us determine whether the emotional action is appropriate. It is just not clear how Moore's equation of the guilt emotion with rectitude and measure of retribution really helps, at least on the ground. And if Moore's point is only to defend retribution as a punishment principle, then it would seem that more work has to be done than his introduction suggests in order to determine whether guilt is well founded, or deserved in the deontological sense.

What would it take for someone to deserve the guilt they feel? Certainly we may feel guilt, the emotion, whether we deserve to feel it or not in any real normative sense. You could feel guilt because you have disappointed someone whom you admire, or someone who admires you. Surely that is not the guilt that Moore's equation of guilt and desert of punishment contemplates, but it is not clear how that guilt would differ, unless, of course, we consider other determinants of whether retribution is appropriate. Further, two people could perform the very same act but one might (and by some independent and objective measure, should) feel guilt and the other might not (and by the same independent objective measure, should not) feel guilt. The real work is being done by the application of whatever standard we employ to determine whether (and perhaps the extent to which?) the feeling of guilt is appropriate to the actor and the circumstances.

Adrian Raine's metastudy,[86] though, made clear that the devil is (almost literally) in the details, or the particular circumstances. Whether you feel guilt, or even whether you should by reference to some independent objective measure feel guilt, is a function of genetic and environmental forces beyond your control (and even if you believe that human agents have some modicum of control sufficient for particular normative purposes, you likely recognize that some determinants are beyond the agent's control, at least those genetically or epigenetically programmed in). The point is illustrated quite vividly in the case of psychopathy, and the problem is posited and resolved correctly by someone whose work has generally been skeptical of the effect that neuroscientific insights might have on the criminal law doctrine. It is worthwhile at this juncture to appreciate the ramifications of Morse's understanding of psychopathy, even if Morse does not seem to appreciate all those ramifications himself.

The Special Case of Psychopathy

Recall that psychopaths, to various degrees, lack a capacity that the other 97+ percent of the population possess: the moral emotions of empathy and guilt. There might be many reasons, from the selfless to the selfish, that explain why nonpsychopathic people do not harm others. Certainly the law deters such conduct—criminal prosecution would not be pleasant—and just the social stigma accompanying a reputation for violent lawlessness would discourage antisocial behavior. But perhaps (indeed, we would hope) for more than external, obviously instrumental reasons to avoid criminal activity virtually all of us avoid violent criminal behavior because it would cause us pain: We actually hurt when we see others in pain, and that empathy would likely be enhanced were we the cause of the pain. Another manifestation of empathy is that we feel guilt after the fact for having caused the pain, and guilt makes us uncomfortable (the source of its normative power, at least).

Psychopaths, though, to some degree lack such empathy and guilt; theirs could not be the reaction that would demonstrate the moral rectitude of retribution. Moore's theory accounted for that, when he posited that the guilt that vindicates retribution is the guilt felt by a normally constituted human agent, and a psychopath would not be such a normally constituted human agent. So in the case of psychopaths, Moore's theory might support the criminal doctrine as it is now formulated.

Recall that Morse too recognized that criminal law doctrine can accommodate many, perhaps all, of the insights that neuroscience has provided so far or might in the near future.[87] Free will is not necessary for the imposition of criminal liability; some form of compatibilism, and maybe even determinism, would do just as well.[88] While the moral claims that the doctrine would make would differ depending on whether libertarian free will or compatibilism or determinism best described the animating normative assumption of the doctrine, the nature of the moral claim would not impact the internal integrity of the doctrine: The criminal law would be coherent, if not unassailable by reference to some normative standard.

Morse, though, has recognized that neuroscience conceivably *could* change our understanding of what it means to be human, *could* revise profoundly our conception of human agency.[89] He just does not think that neuroscience has done that yet, at least not comprehensively. But it seems that in one crucial context, the case of psychopathy, neuroscience certainly *has* revealed enough about human affective function to cause Morse to question the moral coherence of the doctrine: "I believe the [extant] law's assessment is morally incorrect and should be reformed. Psychopaths are not morally responsible and do not deserve blame and punishment."[90] He then offered two reasons for that conclusion, one he described as specific, the other as general. Specifically, Morse maintained that the best reasons people have for refraining from criminal activity is their recognition that it is wrong to do so and their feeling of empathy for those who suffer. Normal people will rely on that moral sense, built from an understanding of what is wrong and from the feeling of empathy and refrain from criminal acts. Morse cited no authority for his conclusion that the moral sense, so constituted, provides the best reason for not violating the rights of others, but that omission does not preclude our following his argument, even if it might cause us to hesitate to follow it with nodding approval. He did, though, double down on that thesis (which would seem to be empirical and, so, testable): "Internalized conscience and fellow feeling are the best guarantors of right action. The psychopath is not responsive to moral reasons, even if they are responsive to other reasons. Consequently, they do not have the capacity for moral rationality, at least when their behavior implicates moral concerns, and thus they are not responsible. They have no access to the most rational reasons to behave well."[91] It would seem that what are "the best guarantors of right action" might well vary from one setting to the next, and that the deterrence provided by the positive criminal law might, at times, provide the "best" guarantor, particularly when the level of personal animosity is high enough to obscure such fellow feelings but still within the range of normalcy and the risk of detection is high as well. Further, there are many what we might call high-functioning psychopaths who well understand the prudential reasons for acting civilly even if they do not feel the imperative in quite the same way.[92]

Morse then endorsed the thesis of Paul Litton, to the effect "that psychopaths are [not] rational at all because they lack any evaluative standards to assess and guide their conduct."[93] So as a general matter, "severe psychopaths are out of touch with ordinary social reality"; they "have a general diminished capacity for rational self-governance that is not limited to the sphere of morality."[94] Litton's argument is especially provocative because it breaks down, or at least assaults, the distinction between moral and rational thought. For Litton, the psychopath's moral blindness is just symptomatic of the psychopath's general cognitive impairment. And it is at least, as an empirical matter, certainly the case that psychopathy and intellectual deficiency are frequently coincident.[95] But most provocatively, Litton's thesis presents a fundamental challenge, indeed a challenge that would confront the neat compartmentalization of the emotional and the rational. We can identify brain areas that activate brain states we describe as emotional (i.e., the amygdala) and areas that we identify with rational thought, such as deferred gratification (i.e., the orbitofrontal cortex, the "administrative" part of the brain). And we can describe the somatic effects of certain emotional states, though they may be ambiguous and only resolved by reference to contextual cues.[96] Sexual arousal may feel different from the way having a good idea feels. Ultimately, though, at the essential level both reactions are the product and salience of neural firing. Emotions, or fast thinking,[97] work more expeditiously because they generally operate best when they operate at the visceral level: do not take a closer look at that snake to decide whether it is dangerous! Rely on your revulsion of snakes and move away, quickly, before you have time to think about it. In fact, of course, when you recoil from the snake you have thought about it, for all intents and purposes. Your recoil may have been unnecessary, but in the event false positives are a better result than false negatives.

What is important, then, is that Litton's broader theory helps us break down, or at least reappraise, the neat emotional–rational dichotomy that would inform the more specific argument that relies on psychopaths' emotional incapacity. Both emotional and rational deficiencies are cognitive; to the extent that legal doctrine or normative theories rely on substantial differences between the two, that case must be made, and not just by describing distinguishable affective reactions. A good deal of the instrumental–noninstrumental tension, though, would seem to rely on just that type of simplistic physical distinction. Morse's (and Litton's) argument was ultimately based on capacity, *neural* capacity: The psychopath can no more see the morality of a situation than the color-blind motorist can distinguish red from green, except perhaps on the basis of the lights' positions. It would, therefore, be immoral, and the doctrine fails as a normative matter, when psychopaths are held criminally responsible; they simply do not have the capacity to avoid criminality, at least in the extreme cases. And it does not matter whether we label that capacity emotional or rational. It is all neural.

Morse recognized, too, that there are degrees of psychopathy and would continue to impose criminal liability on those with "less severe conditions who retain residual moral capacity."[98] Although this would seem to present something of an empirical rather than conceptual problem, Morse should be applauded for recognizing the predicament his conclusion poses for the doctrine: Once we acknowledge that some people, those we can describe accurately as psychopaths, lack moral capacity (at least the same moral capacity as the rest of us), we cannot as a moral matter punish them for their actions; their criminal action is excused because they lacked the capacity to act morally. Where we draw the line, how we decide that one psychopath has sufficient moral capacity to be subject to punishment and that the next does not, remains problematic. Perhaps we could do what we seem prone to do in other areas of the criminal law: resolve uncertainty by adjusting, maybe discounting, the sentence imposed. If we are not sure whether your psychopathy so compromised your moral functioning to preclude retribution, then suspend part of your sentence. That type of reasoning always seems to have a Solomonic quality to it, but is understandable even if it does lack a fundamental coherence and integrity.

The most provocative part of Morse's thesis here, though, suggests a pervasive critique of the moral fit between criminal doctrine and human agency. Morse anticipated the broader responsibility problem, in at least an atypical setting: "A potential objection concerns people whose acculturation, rather than biological or psychological abnormalities, may deprive them of particularized rather than general moral concern."[99] And Morse concluded that "this is a difficult problem for responsibility theory."[100] Indeed.

Responsibility

Once we see reason, at least in some contexts, to break down the lines between emotional and rational cognitive function, and, with Morse, appreciate that acculturation (writ broadly as nurture) may affect empathy and so feelings of guilt, then we have begun to see the concept of normative responsibility fray, at least at the edges. Recall that for the instrumentalist, all the responsibility necessary is responsibility in the causal sense.[101] You could conclude that an individual is responsible for a criminal act to the same extent, but not to a greater extent than you could conclude that your car is responsible for a criminal act. Both the individual and the car are caused causes, insofar as normative responsibility is concerned: which is to say, not at all. It is noninstrumental theory that needs moral responsibility to support desert to support retribution. If retribution is incoherent (or, at least, nothing more than dressed-up revenge), then criminal punishment can ignore it and focus solely on instrumental ends. That would not solve all the empirical measurement problems, but it would clear away a good deal of what obscures.

The metastudy published by Adrian Raine and discussed at length in the previous chapter supports a move in that direction. Recall that the studies reported by Raine reach important empirical conclusions about the dynamic interrelationship between nature and nurture that *may* result (in a statistically measurable way) in greater propensities to violent behavior.

About Responsibility

For Moore, Morse, and other apologists for the doctrinal and normative status quo of the extant criminal law, responsibility, a conception built on folk psychology, is indispensable. If the criminal doctrine does not take moral as opposed to mere causal responsibility seriously, indeed, as indispensable to the moral object of the criminal law, then the criminal law would be normatively incoherent. For the materialist, the criminal doctrine's insistence upon moral responsibility is akin to extracting gold from lead, a remnant of burning witches and punishing epileptics for being possessed by demons. So for the materialist, it makes no more sense to talk of the moral responsibility of the most or even moderately heinous criminal than it does to talk about the normative responsibility, desert, or culpability of your car.

It is clear, though, that the criminal law relies on folk psychology, and perhaps relies on no constellation of folk psychology conceptions more than it relies on moral responsibility. The doctrine builds responsibility from belief, desire, and intention: fundamental folk psychology realities. So we could make sense of the doctrine only if we take moral responsibility and the constituent beliefs, desires, and intentions seriously. The materialist would acknowledge that the extant criminal law doctrine does depend on beliefs, desires, and intentions and their provision of normative responsibility's foundation.

Therefore, when Morse confronted deprivation and desert,[102] he had to start by demonstrating the criminal doctrine's reliance on folk psychology generally and normative responsibility specifically and crucially: "We could decide morally and legally to abolish notions of individual responsibility and to replace them with group responsibility or *no responsibility at all*, but this would require an argument that goes far beyond the implications of [genetic and social] deprivation in the moral and legal world we inhabit."[103] In that conclusion, Morse was absolutely and completely correct: The materialist perspective changes everything, including the law's understanding of the implications of deprivation. When responsibility can have only a causal but not a moral sense (because the sense of "moral" itself comes into question), then the effect of deprivation on responsibility would seem subordinate to the point that the responsibility assumed by folk psychology is as insubstantial as the insights of alchemy. That is, deprivation does not assume a role in appraisals of responsibility because responsibility is a chimera in anything but the causal sense. So although, of course, deprivation would undermine normative responsibility, so would everything else in the monist, deterministic world understood by materialism.

But Morse must take responsibility seriously because he has recognized that the criminal law doctrine takes it seriously and because, too, he rejects materialism in favor of a less mechanistic and compatibilist, even dualistic worldview: "Law, unlike mechanistic explanation or the conflicted stance of the social sciences, views human action as almost entirely reason-governed."[104] And Morse has seemed to cash out his folk psychology in terms that resonate with Moore's reliance on emotion(s) as the determinant(s) of justice: "It is one thing to say that behavior breached a moral expectation. This is an example of objective description that follows from a moral norm and facts about the world. It is another to hold the agent morally responsible for that behavior, which involves a complex of emotions and their expression that have the force of a judgment. When we hold people morally responsible, we are experiencing the moral reactive emotions and expressing them appropriately."[105] So Morse's analysis and philosophy depends upon a moral realism[106] founded on emotion, the emotion of guilt specifically. Because we feel the emotion, we know that we are punishing people appropriately. This is not quite the guilt-emotion-as-basis-of-retribution argument, but it amounts to the same thing. The analysis relies on reactive emotions' verisimilitude generally to found metaphysical normative truth, just as Moore's theory of retribution depends on the guilt emotion's being specifically veridical. But, as explained in the preceding argument, though the emotion might travel with desert, it does not establish desert, and it certainly does not do the work of establishing normative desert.

Although Morse's reliance on reactive emotion provided him reason to excuse psychopaths, who at least have an impaired ability to feel guilt because they lack the affect that would provide the moral cue, it ultimately fails him because it obscures the folk psychology nature of a responsibility analysis that would warrant retributionary punishment. Morse, though, concluded that we need responsibility, that the criminal law would be incoherent without it.

Reason for Responsibility?

According to Morse, we need responsibility in order to make sense of the normative commitments of the criminal law; without responsibility, we would be limited to a system of state-imposed sanctions that merely prevent wrongdoing, a wholly instrumentalist system. And the criminal law could not function if reduced to such terms. But when Morse defended that proposition, when he tried to undermine the instrumentalist critique, he relied on empirical limitations to develop a conceptual argument: Although deprivation may undermine desert, we cannot know what deprivation *in fact* resulted in behavior that should provoke in us the emotional reaction sufficient to ground desert; the lines are too hard to draw, and, in any event, we do not excuse all crime that could be attributed, in whole or part, to factors beyond the control of the actor: The best predictor of violent criminal behavior is the Y chromosome, but your violence is not excused because you are male.[107]

When he made that argument, the *reductio ad absurdum* of the instrumentalist's deterministic argument, Morse did more than rely on a rhetorical flourish: He invited confusion about the nature of excuse as a defense. Keep in mind that excuse, as an affirmative defense, only makes sense in a criminal justice system founded on retributive principles; only in such a system does normative desert make a difference. So consider how the instrumentalist would react to the argument that a particular defendant's criminal behavior should be excused because the actor suffered from a neural anomaly that made him unusually and unpredictably violent. It would seem that the instrumentalist, unconcerned with responsibility, retribution, or desert, would decide that the defendant's violent proclivities are reason to limit the likelihood that the defendant would have opportunities to harm others in the future. So the particular defendant would be incarcerated *for only so long as necessary to protect society* from that defendant and only under conditions that would not enhance the dangerousness of that defendant for that defendant's particular circumstances. The instrumentalist would not excuse the crime or the defendant; that idea could gain no purchase in the analysis.

Morse responded to the vagueness of an excuse analysis by pointing out, quite correctly, that we can never know (or, at least not in the present state of the science) the extent to which the defendant in question could have acted other than he did: "There is no test or instrumentation to resolve questions accurately about the strength of desire and the ability to resist."[108] Yes, very true, and a generation ago, we did not know that psychopathy was the result of deficient emotional, moral affect (perhaps malformation of the amygdala). That was an empirical limitation of the state of the neuroscience, not an insurmountable conceptual limitation with inviolable normative significance. For the same reasons that we could excuse psychopaths—because we know, can see on an fMRI scan and discern by testing—we could excuse others when their actions are caused by circumstances beyond their control. But even that vindication of an instrumentalism that could expand excuse analysis is problematic.

It is problematic because excuse analysis itself is problematic, an inquiry designed to respond to a problem created only because the criminal law misunderstands human agency, or understands human agency only as well as it could a couple or few hundred years ago. For the same reasons that we can now appreciate the failure of folk psychology generally—it misleads us into making decisions that ultimately undermine the object of the criminal law—we can begin to see that relegating broader consideration of the forces that frame human agency in terms of a conception built on folk psychology, responsibility, will ultimately undermine the normativity of criminal law doctrine. When we excuse Morse's psychopath, we do not let him back into the general population; we limit his movement so that he cannot harm others. And when we know that he can do no more harm, because we have found a way to treat the condition or to protect society at large without imposing on his freedom as substantially, we would do so. The fact that we do not now know whether or even when we will find a cure for the condition does not mean that we should rely on retributionary principles to punish the psychopath. Morse's analysis of psychopathy makes clear that he would not do that either, but his understanding of deprivation demonstrates that he does not appreciate the ramifications of his own analysis.

The Immorality of Moral Responsibility

The premise of noninstrumental moral theory, deontology specifically, is that human agents have moral responsibility for their actions and can be subject to blame and feelings of guilt too as a result. Such noninstrumentalism does not just make room for retribution; it justifies it. Retribution need not flow from noninstrumentalist moral theory,[109] though the rationalization of retribution in the criminal law depends on a compatibilism that accommodates something enough like free will to make the imposition of punishment on the basis of desert moral. But what if moral responsibility were invalid, as an empirical matter? What if moral responsibility itself were immoral?

There are surely benefits in the cost-benefit sense to the moral responsibility system, perhaps even benefits that would provide grounds to support the moral responsibility system if it were fictional.[110] Believing that we have moral responsibility (and, even better, getting others to believe that they have it) serves a worthwhile instrumental purpose: We could rely on guilt to structure behavior in efficacious ways. So moral responsibility might well be, in that view, the type of thing that we would have to invent even if it did not exist. But whether it would in fact be that type of good thing could reduce to an empirical question: Do we gain more by imposing punishment, of the retributionary sort, than we lose by ignoring that all our actions are the product of forces beyond our control because there is no independent "I" that is "in control" of the entity that is "me"?

Moral responsibility depends on the reality of beliefs, desires, and intents that in turn rely on the supernatural, on our being autonomous gods who can cause without being caused. That is the view that neuroscientific insights challenge. It is a view nourished by our affective systems and by societal constructions (often religious systems, even those that, curiously, espouse predestination). And that sense of being an uncaused cause does more than nourish our sense of demidivinity; it provides a framework within which social cooperation can thrive, and social constraints can too. There is a predictable and adaptive synergy between the affective heuristics that found or at least seem to reveal moral responsibility and the successful interpersonal strategies that are conducive to human thriving. But heuristics are by nature rough and at the margins unreliable. They also can outlive their utility. Emotional reactions that were crucial to survival on the savannah about twenty-five thousand years ago would actually undermine human thriving today: In-group biases consistent with survival then fail us today, and we have no trouble concluding that what seemed moral then would be immoral now.[111] Once we appreciate that morality is just the rationalization of emotional reaction, we also can see clearly that morality is a social construct, an often but not invariably useful construct at that.

The Rationalization of Emotion

Our emotional reactions are uniquely salient; by definition we feel them in a way that we do not quite feel an epiphany of rationality. And the emotions, for that, seem to elude rationalization, seem to communicate, even reify, something that is otherwise inscrutable. I suspect that the case could be made (though I need not make it here) that a good deal, perhaps all of, deontology boils down to the rationalization of emotional reaction. Our feelings, affect, have an ostensible verisimilitude that cold reason cannot approach, and the efforts of noninstrumentalist theory, to a good extent, are efforts to reason about emotions, to explain their role in explaining, even justifying, moral practices. That is why Moore founded the deontological notions of retribution and desert on the guilt emotion, though then curiously constrains guilt's confirmation of the rectitude of desert by reference to guilt that is appropriate. Rationalization of the irrational can take us only so far before we must fall back on real rational argument, apparently.

Waller presents perhaps the best elaboration of the naturalistic thesis undermining moral responsibility. He recognized the relationship between emotion and moral responsibility: "Commitment to moral responsibility is based on visceral emotional reactions and locked in place by a far-reaching theoretical system."[112] That system is deontology, or at least noninstrumentalism. Further: "the unshakable certainty of many philosophers concerning moral responsibility must have some emotional source independent of rational argument."[113] And although he appreciated the "visceral and universal reaction: the basic retributive impulse, the desire to strike back when we are harmed,"[114] he saw the now *mal*adaptive consequences of a reaction that served our forbearers well before the rise of civilization and contemporary police possibilities and practices: "[I]t is clear that what may have been useful at one stage of development is now maladaptive.... [E]ven if one grants that at earlier stages of development, retributive practices were of some benefit (in terms of either group or individual selective pressures), it may be that in our present state they are maladaptive (just as human aggressive tendencies have become severely problematic in an era of handguns, not to mention nuclear weapons), as well as being morally and rationally unjustified and unfair."[115] Moral responsibility worked well, for a time: a time in which the alternatives were "lynch mobs and personal vendettas."[116]

There is much in Waller's book that challenges noninstrumentalism, belief systems that depend on ratcheting up affect to found the supernatural. For present purposes, though, it suffices to appreciate that his thesis and argument provide a perfectly and simply natural argument against moral responsibility, a moral system that by assuming the existence of a homunculus-like "I" within each of us embraces a capacitarian conception of control that justifies, even promotes, cruelty based on a desert that does not exist.

It may seem an oxymoron to suggest that moral responsibility could somehow be immoral. Waller's book makes clear just how it is certainly so: Insofar as moral responsibility assumes a human agency to support retributive punishment and that human agency is fallacious, a figment of our atavistic imagination that makes us comfortable with, in a sense rationalizes, our basest inclinations, the system premised on that fallacious conception of human agency adds insult to injury or, worse, injury to injury by punishing people noninstrumentally for their actions as though they were the author of them. But we, none of us, are in any moral way responsible for what we are: We are all, at any given moment and with regard to any action or decision, the product of the forces (genetic, environmental, economic, social, etc.) that have worked on the stuff that is "we."[117] It would be impossible for us to take control and act in any other way than we have.

So a system of social control, of reward and punishment based on the fallacious understanding of human agency that does not come to terms with the determined creatures that we are, but that instead relies on what might from time to time be emotionally satisfying reactions to the fiction that worked well enough before we could understand that it is a fiction (an understanding accommodated and, indeed, confirmed by recent neuroscientific insights)[118] is bound to assign blame, and reward too, on insubstantial bases. Such a system is, of course, immoral, unjust by reference to any coherent value system. And it also is profoundly inefficacious: "Belief in moral responsibility blocks us from looking closely at the causes, and thus hamstrings efforts to develop and enhance healthy take-charge responsibility."[119]

Now this is not to suggest that the current moral responsibility system, even with its reliance on retribution, necessarily fails abysmally all the time. Insofar as the result of a finding of moral guilt will be incapacitation, it should not surprise that the product of a retributionary calculus—removal of the criminal from society during the years when she is most likely to offend—will often accomplish pretty much what an instrumental approach would. There is nothing in the instrumental perspective that relieves a criminal of liability for his actions, or that undermines the unpleasant response of the state that is the consequence of the state's need to protect its citizens.[120] Excuse in the criminal law does not result in release of the criminal, but it may result in a response that is more considerate of the causes of the harm and the continuing threat the criminal presents.[121] When the threat is no longer present (i.e., when the tumor that caused the antisocial behavior has been successfully excised) there is just no residual work for punishment, through retributionary principles, to do; no issue of moral responsibility remains. But a system based on moral responsibility must found punishment on retribution: There would be a debt unpaid were the individual or group not made to face the consequences of their actions, at least in the world of Moore and Morse.

Conclusion

This chapter has begun the assault on conceptions of moral responsibility and demonstrated the incoherence and ultimate inefficacy of legal doctrine and theory premised on an inauthentic sense of human agency, the inauthentic sense noninstrumentalism provides and requires. The four following chapters build on the observations offered here both to fine-tune and expand the argument that the primary legal doctrine, all of it, is misconceived, not just because it misses the mark, but because it is aimed at the wrong things. In the next two chapters, we shall see that tort doctrine depends on something like a wrong, a discernable injury, even as the theory strains to distinguish the tort wrong from the criminal law wrong. Tort, in its treatment of wrong, falls right between criminal law and contract law, as it should in this book's inquiry and argument too.

CHAPTER FOUR

Neuroscience and Tort Law Doctrine

Introduction

Tort law fills a gap in legal doctrine between criminal law, governing the direct power of the state to regulate harm,[1] and contract law, governing consensual relations. The tort law generally concerns nonconsensual civil relationships. You do not choose your tortfeasor; you are his victim. So there is no basis to infer consent. The book's concern in this chapter is with the unintentional torts that cause injury, which is a subcategory of the general category torts, but it is the important core of the tort law that neuroscience challenges; to the extent that the tort law concerns intentional actions that result in injury, the tort and criminal law would overlap in their relevance to neuroscience. Unintentional torts involve a different state of brain: negligence.

Tort liability is premised on the human agent's responsibility for behavior that results in harm to others. The agent is not subject to liability for harm she causes as long as she acted reasonably, in a manner consistent with ordinary principles of care.[2] So at the intersection of tort and neuroscience should be a conception of reasonableness informed by empirical reality. Human agents are constrained to be no more reasonable, no more prudent and careful, than human agents are generally prone to be. There will, then, be harm for which no one is liable, insofar as the reasonable person is imperfect. Neuroscience can describe the contours of that space between reasonable and flawless behavior because neuroscience tells us what it means to be human both as a general matter (for humankind generally) and at the individual level (the capacities of particular parties plaintiff and defendant).

This chapter first considers the tort standard of care. The reasonable person standard is not a formless vessel; it both reflects and instantiates certain moral conclusions about our responsibilities toward one another. That normative flexibility is a by-product of uncertainties about human agency that neuroscience may address. So we might anticipate resistance to neuroscientific insights that contract the room for play in the joints of the tort doctrine. It is one thing to reach a particular normative conclusion—liability or no liability—by reference to moral conceptions uninformed or underinformed by empirical evidence; it is another to reach that same conclusion in the face of empirical evidence that discloses the absolute incapacity of the parties to act reasonably, notwithstanding doctrinal insistence that we assume capacity or ignore incapacity in fact.

This chapter then considers how neuroscience may provide a framework to discover answers to questions redolent of proximate causation, particularly insofar as plaintiff's contributory or comparative fault may be deemed to have caused the plaintiff's injury in whole or part and so eliminated or limited the defendant's liability therefor. Such affirmative defenses are, ultimately, questions of proximate cause: Who was the cause of the injury suffered by the plaintiff, and to what extent was the injury the product of complementary causes? Those questions take on unique significance in the context of causes of action premised on injuries that are a product of processes hidden in the workings of the brain (e.g., addiction), as well as injuries to the brain itself (e.g., chronic traumatic encephalopathy, or CTE). Can we *not* be the proximate cause of what happens in our brains? Or is the possibility that there is a "we" separate from our brains so that "we" could be the proximate cause of what happens in our brains a dualistic fantasy?

The chapter then turns to application of neuroscientific insights to inform analysis of issues concerning the compensability as well as monetization of injuries. Perhaps the two most significant potential contributions of neuroscience to the tort law involve translation of the mental into the physical. First, neuroscience demonstrates that what we have described as mental differences (perhaps in terms of maturity or capacity) are ultimately physical differences, the consequence of some chemical or electrical or microscopic structural feature. That would as well be pertinent to distinctions the tort law draws between mature and immature actors (between competent adults and children of a certain age, or even adults of a certain advanced age). Second, neuroscience may reveal the organic bases of both pain and emotional harm and may reveal how all injury is ultimately and fundamentally physical.

The sources of US tort doctrine are as numerous as the number of jurisdictions in the United States and perhaps even as numerous as the number of trial courts in those jurisdictions. For purposes of the instant study, it is not necessary that the doctrine be captured and presented in precise detail; bold general statements will do. The most accessible and maybe the best contemporary source of such statements is the Restatement of the Law (Third), Torts, promulgated in pertinent part by the American Law Institute in 2010 (hereinafter often the Third Restatement). Though some may quibble with the correctness of, or even with the normative choices reflected in, the Third Restatement, the presentation of the tort doctrine is sufficiently accurate for present purposes.

The Standard of Care

[Mrs. Erma Veith], while returning home after taking her husband to work, saw a white light on the back of a car ahead of her. She followed this light for three or four blocks.

The psychiatrist testified Mrs. Veith told him she was driving on a road when she believed that God was taking ahold of the steering wheel and was directing her car. She saw the truck coming and stepped on the gas in order to become air-borne because she knew she could fly because Batman does it. To her surprise she was not air-borne before striking the truck but after the impact she was flying.

Mrs. Veith did not remember anything else except landing in a field, lying on the side of the road and people talking. She recalled awaking in the hospital.

Actually, Mrs. Veith's car continued west on highway 19 for about a mile. The road was straight for this distance and then made a gradual turn to the right. At this turn her car left the road in a straight line, negotiated a deep ditch and came to rest in a cornfield. When a traffic officer came to the car to investigate the accident, he found Mrs. Veith sitting behind the wheel looking off into space. He could not get a statement of any kind from her. She was taken to the [hospital] and later transferred to the psychiatric ward of [another hospital].

The psychiatrist testified Erma Veith was suffering from "schizophrenic reaction, paranoid type, acute." He stated that from the time Mrs. Veith commenced following the car with the white light and ending with the stopping of her vehicle in the cornfield, she was not able to operate the vehicle with her conscious mind and that she had no knowledge or forewarning that such illness or disability would likely occur.[3]

If Mrs. Veith in fact had no forewarning of her hallucination, should she be held liable in negligence for harm caused by her episode? That is the kind of question this part of the chapter engages, from a perspective informed by neuroscientific insights. It is necessary to consider the apposite doctrine: How might neuroscience change the way we understand this part of tort?

Mental versus Physical Disability

The "reasonable person" conception fixes the tort standard of care. The Third Restatement provides that a "person acts negligently if the person does not exercise reasonable care under all the circumstances."[4] That formulation does not substantively change the formulation of the Second Restatement.[5] The Third Restatement, also like the Second, bifurcates physical and mental deficiency: Though physical disability may be taken into account when considering the reasonableness of an actor's behavior, "An actor's mental or emotional disability is not considered in determining whether conduct is negligent[.]"[6] The Third Restatement categorically provides that "A child less than five years of age is incapable of negligence."[7] The fact that the law takes into account age in determining responsibility but not mental capacity is curious and suggests that the reason for not recognizing mental limitations generally may have more to do with administrative convenience than normative object: There is no obvious moral difference between a five-year-old with the mental capacity of a five-year-old and an adult who, in at least some settings, behaves as would a five-year-old on account of a mental deficiency. The role of administrability is confirmed in the Third Restatement's citation of authority to the effect that the "problem of verifiability" explains "the common law's unwillingness to take mental disability into account."[8]

Now there are policy reasons that support distinction between physical and mental capacity. Those justifications largely, though not exclusively, relate to problems of proof: the general desirability of objective measures (difficulty of knowing whether the actor "did or did not do his best");[9] distinguishing mental deficiency from some other cause of negligent action; problems of attributing tortious behavior to mental impairment ("even a schizophrenic may drive well"); access to mental incompetent's insurance coverage; and the interest in providing those who care for the mentally infirm the incentive to monitor their behavior.[10] Some of those justifications might be undermined as neuroscience blurs the line between mental and physical impairment (i.e., as we develop enhanced ability to correlate chemical, electrical, or structural brain anomaly with particular behavior) and other justifications would not (e.g., desirability of access to insurance and incentives to monitor). Neuroscience may expose normative conflicts obscured by the extant doctrine: When we cannot identify and isolate the source of a mental deficiency as confidently as we can the source of a physical deficiency (schizophrenia is so far invisible on a brain image but a broken leg is obvious on an X-ray), we are prone to discount what we cannot see, and so it makes good practical sense, and perhaps good normative sense too, to rely on the second-best option, such as considerations of insurance and incentives to monitor. That conclusion is reinforced by skepticism about assertions of latent mental conditions as opposed to patent physical conditions. Neuroscience, though, is making the latent patent just as MRI has made visible soft tissue injuries that are invisible to X-ray.

But administrability may be a fragile reed on which to premise the mental–physical dichotomy when neuroscience increasingly undermines the bases of that distinction. It is true that the law has reason to favor objective measures: If A plus B ineluctably leads to C, then we do not have to waste much time or expend too many resources determining the consequences of A plus B. If your vision is so impaired that you could not see the car coming toward you, we know that you could not get out of the car's way. It may be more difficult, more expensive, to connect what we consider to be a wholly mental impairment to a physical consequence. But when the cost of doing the mathematics, determining whether A (mental deficiency) plus B (context) leads to C (injury) is lower, we might be more inclined to fine-tune our analysis to ensure that we expend those fewer resources to get a better, more correct result (more correct by reference to some normative object). As neuroscientific advances make the mental capacity of tortfeasors and victims more accessible, when we can with more confidence identify the causes of mentally aberrant behavior— more certainly connect A to B—we may be willing to inquire into mental competence just as we do physical competence. The point is not that neuroscience should or will change the tort law tomorrow; the point is that bases of distinction that now make good sense will need to be reevaluated as neuroscience breaches the notional mental–physical divide. The Third Restatement recognizes that "Courts have only recently begun to encounter cases in which the person's mental disability has a clear organic cause."[11] We may anticipate that as neuroscience demonstrates more and more certainly the organic bases of all psychiatric disorders, the extant distinction between mental and physical capacity will dissolve.

The tenuous, and maybe tentative, nature of the balance struck by the doctrine is revealed in tort's apparently inconsistent treatment of normatively similar, if not identical, questions. For example, courts occasionally distinguish whether an actor is the plaintiff or defendant in determining the reasonableness of that actor's behavior. That is, the court would recognize the general rule and not excuse behavior because the defendant was suffering under a mental deficiency but would, or might, excuse behavior that would otherwise be contributorily negligent when the plaintiff lacked mental capacity to some extent.[12] Courts observe other distinctions that seem to indicate an ambivalence about the physical–mental dichotomy: For example, although the emotional illness of an adult is not to be taken into account, the emotional illness of a child is pertinent to the liability determination.[13] But what is the legally or morally significant difference between a child suffering emotional illness and an adult suffering from Alzheimer's disease? More to the point, even if we are able to posit a basis of distinction, is that basis sufficient as a matter of the normative object of the legal doctrine?

Age and the Standard of Care

The lines drawn in the current law begin to fade as neuroscience advances. Prediction of this tendency is confirmed, perhaps unwittingly, by a relatively recent comment of Judge Richard Posner: "The court has learned from brain science that teenagers are immature! But we knew that. The problem with using it as a basis for distinguishing between murderers of different ages is that many adult murderers have problems with their brains, too. Why is it not cruel and unusual to sentence them to life in prison?"[14]

Why indeed? We rely on age because it is accessible, an administratively easy way to draw distinctions that also, coincidentally, conform to another moral emotion: compassion for the vulnerable, especially the young. But that conclusion presents the question: When neuroscience reveals more reliable moral markers, the constituents of mental makeup that are the causes of antisocial, including negligent, behavior, should we not take those bases of limited capacity into account just as we now consider age?

Indeed, even the tort doctrine's consideration of age may need to be recalibrated as neuroscience reveals more about the development of the adolescent brain. Fundamental to a system that is founded upon administrative ease is the necessity of drawing certain lines. Rather than accounting for the cognitive abilities of each individual tortfeasor, tort doctrine is concerned only with chronological age. Recall that the Third Restatement categorically provides that children under five years of age are incapable of negligence. A minority of states follow the rule of sevens: Children under seven are incapable of negligence; children between seven and fourteen are presumptively incapable of negligence, but the court will allow evidence to rebut that presumption; and children over fourteen are presumptively capable of negligence, but evidence can rebut the presumption.[15] With a notable exception,[16] children are held to the same standard as their reasonable peers.[17] Once eighteen, however, there is a duty to act as any other reasonable adult under like circumstances.[18]

The science certainly already reveals that what is reasonable for the adolescent does not necessarily mirror what is prudent for the adult. For the adolescent, interaction with peers—even benign and unobtrusive interaction—can stimulate the brain to engage in risky behavior. In one recent study,[19] adolescent and adult participants were asked to complete a computer-simulated driving course. The simple act of notifying the adolescent drivers that two of their peers were monitoring the simulation from a remote location caused the adolescents to engage in riskier driving.[20] Relative to adult participants, adolescents showed significantly greater activation of the ventral striatum and orbitofrontal cortex (two regions known to affect decision making) when notified that their peers were monitoring their progress.[21]

The divergent neural processes between adult and adolescent brains have been confirmed.[22] The ability to voluntarily suppress behavior, for instance, comes from the integration of multiple brain regions, and that integration is enhanced through synaptic pruning and myelination during childhood and adolescence.[23] Even beyond adolescence, the brain continues to develop. An increase in white matter and decrease in gray matter has been noted in subjects up to twenty-five years of age,[24] and myelination can continue into one's thirties.[25] Neural development then does not end at eighteen, or even twenty-one, for that matter. The joke among researchers is that car rental companies, which refuse rentals to those under twenty-five, have had it right all along.[26]

What Advances in Neuroscience May Reveal

Once we are able to identify the organic cause of certain aberrant mental states, we may well be better able to connect the dots between that mental state and a negligent action. Not only will the barrier between the mental and physical shift, we also may be able to draw better distinctions among mental states. That is, we may discover that the cause of the accident was not, for example, the defendant's inability to process certain types of data in certain settings on account of some neurological aberration but was instead a consequence of the defendant's lower, though not abnormally low, intelligence level. That is a line we cannot draw now, and so the tort law does not (and should not try to) draw it. We might imagine that, as we develop the ability to trace more certainly the sources of some mental deficiencies, we also would have a better sense of how they may be isolated and treated and better reason in imposing liability on those who do not take steps to avoid or reduce the consequences of such chemical, electrical, or structural anomalies. When we know that a particular behavior is likely the product of posttraumatic stress disorder (PTSD), we may better serve the goals of tort law if we require the victim of the disorder to take the medication or undergo the procedure that would minimize the risk his affliction might present to others (just as we would have good reason to impose liability on the person who suffers an avoidable epileptic seizure while driving without having taken the medication that would have avoided the seizure).

Neuroscientific insights in time will surely respond to many quandaries presented by the current tort doctrine. But neuroscience can now confirm only that a wide range of cognitive anomalies *may* be the product of real but practically invisible organic causes.[27] Further, we can see, sometimes quite clearly, neurological anomalies that look as though they should impair mental function but that do not do so, at least not in the way pertinent to the question in issue.[28] Functional MRI scans, then, may reveal something about the mental capacity (or at least mental state) of a subject that could be pertinent to liability. Yet, even while images of the brain become more vivid, indeed, more revealing, the science may be some distance from knowing exactly what it is the scans reveal.

This leaves the tort law and brain science in a difficult place: The legal doctrine bases a distinction on evidentiary limitations, and on evidentiary limitations that there is reason to believe will be overcome or at least be more often surmountable as the science advances. The distinction is not conceptual. There is nothing about mental deficiencies, as such, that should not excuse in the same way that physical deficiencies would. If the defendant is not able to appreciate a state of affairs because of some mental condition, that is not obviously different from the same defendant's not being able to respond to a wholly physical challenge because of some physical impairment.

Even as we learn more about the mental makeup of tortfeasors, the problem of appreciating the consequences of that better understanding in relation to the normative foundation of tort will remain. Today we might distinguish the mentally lazy from the mentally less competent. But why would we do so? The answer must lie in the normative object we attribute to the tort law. If we are all just the product of our unique physical and mental (also necessarily, at some level, physical) characteristics, what do we hope to accomplish by the imposition of tort liability? That is the question that should inform reevaluation of the tort law standard of care issues in light of neuroscientific advances. It is, fundamentally, the question to be treated in chapter 5.

Questions about the relationship between the physical and the mental also pertain to the proximate causation issue in tort, and that is the focus of the next section of this chapter. As we shall see, conceptions of dualism may once again intrude.

Proximate Causation

On December 1, 2012, Jovan Belcher, a linebacker for the Kansas City Chiefs of the National Football League (the NFL), "murdered his baby's mother, Kasandra Perkins. Then Belcher drove to the team's facility next to Arrowhead Stadium and took his own life in front of head coach Romeo Crennel and General Manager Scott Pioli. Before committing suicide, Belcher 'thanked' Scott Pioli and asked him as well as team owner Clark Hunt to care for his infant daughter, Zoey."[29]

A little more than a year later, Belcher's mother filed a lawsuit against the NFL: The lawsuit described Belcher as a "loving father, son, teammate and advocate for victims of domestic violence" who ended up suffering "severe and persistent headaches [postconcussion syndrome], depression, mood swings, explosivity, suicidal ideations, irresistible and insane impulses" and "neurologic dysfunction such as [chronic traumatic encephalopathy]."[30] What or who was the proximate cause of Belcher's rampage and suicide? Was he alone the cause? Or did the NFL and earlier sponsors of his participation in the sport also cause the neurological dysfunction? Were the murder and suicide the product of multiple causes, some of which were within Belcher's control and others not? Neuroscience may not yet provide the certain answers but may already suggest the means to frame the difficult causation questions.

Before a defendant can be liable to a tort plaintiff, the defendant's actions must be the proximate cause of the plaintiff's injury.[31] The proximate cause inquiry also may matter when the plaintiff's actions or inactions are in some way related to her injury. If the plaintiff contributes to the circumstances giving rise to the harm the plaintiff suffered, most jurisdictions will compare[32] the negligence of the plaintiff and defendant in determining the damages for which the defendant would be liable. If the jury determines that the plaintiff is equally at fault—contributed 50 percent to his injury—then the plaintiff would generally[33] be able to recover only one half of his out-of-pocket expenses and pain and suffering damages from the defendant. The calculus in any particular case may be opaque, but the idea is not difficult to grasp. Neuroscience might affect the proximate causation inquiry by shedding light on whether the plaintiff could have taken steps that would have reduced or eliminated altogether the consequences of the defendant's actions. There is a particularly salient context in which this cooperation of neuroscience and proximate causation principles may affect the tort doctrine.

Addiction

Engle v. Liggett Group, Inc.[34] was a large class action brought against a cigarette manufacturer for deaths and injuries allegedly caused by the plaintiffs' use of the defendant's product for many years. The plaintiffs relied on several theories, including the simple negligence of the manufacturer. A crucial issue in the litigation was whether the plaintiffs were in whole or in part comparatively negligent for the harm that the cigarettes caused them. Cigarette addiction at least impairs and may, in some cases, overwhelm the smoker's ability to refrain from smoking. In *U.S. v. Phillip Morris USA, Inc.*, the court accepted "the extraordinary hold that nicotine has on the human nervous system and the fact that such hold stems from nicotine's pharmacological properties."[35] The court examined how smoking tobacco provides the "fastest rate of absorption and highest blood levels of nicotine," noting that nicotine reaches the brain in a matter of fifteen to twenty seconds.[36] The court further explained how nicotine possesses a structure similar to that of acetylcholine, a neurotransmitter.[37] Nicotine binds with acetylcholine receptors, artificially stimulates the acetylcholine system, releases a number of hormones, and affects mood and behavior.[38] The court found that the artificial stimulation of the acetylcholine system produced desirable effects (stimulation, alertness, and stress and anxiety relief) and undesirable effects as the brain becomes tolerant of or dependent upon the effects of smoking ("irritability, lethargy, restlessness, sleeplessness, anxiety, depression, hunger, and weight gain").[39]

The court concluded that the understanding of these properties "supports the now overwhelming consensus in the scientific and medical community that cigarette smoking is an addictive behavior and that nicotine is the component in cigarettes that causes and sustains the addiction."[40]

Nicotine acts as an agonist to acetylcholine receptors within the brain, activating the ubiquitous acetylcholine-modulated pathways. One of these pathways, the classic reward pathway in the Ventral Tegmental Area, utilizes glutamate–GABA (a common neurotransmitter)–dopamine–acetylcholine interactions to create the sensations of dependence and withdrawal.[41] This reward pathway has three characteristics that present a unique proximate cause problem: euphoric effect after use, severe withdrawal symptoms, and cue-triggered recidivism. The essential effect of a drug's manipulation of the reward system is the extra dopamine left in the synaptic cleft. With extra dopamine stimulating the postsynaptic neuron, the brain perceives more reward and the euphoric effect follows. The body responds by instigating the expression of proteins that dampen reward pathways. The addict compensates by using more of the drug, and the vicious feedback loop begins. Of course, to escape this dependence, an addict must overcome painful withdrawal symptoms created by the manipulation of glutamate–GABA–dopamine–acetylcholine interactions. Even if the addict successfully abstains from use of the drug, alterations in molecular signaling, genetic expression, and synaptic connections remain, leaving the addict vulnerable to cue-triggered recidivism.[42] When addicts use their drug of choice, they automatically associate certain environmental stimuli with the use.[43] Reencountering such cues triggers the associations previously made and increases the likelihood of relapse.[44]

Evidence that cigarette manufacturers could and did manipulate the nicotine content of cigarettes confirms that those manufacturers were taking advantage of the interaction between a chemical property of their product and their customer's brain function.[45] It is easy to sell more of a product (and at a higher price) if your customers cannot resist using it or can resist using it only at great physical and emotional cost.

The profound damage that cigarette smoking does is the product of habitual use over many years. Evidence of cigarettes' deleterious health effects would result in many if not most adults giving up the habit were it not for the addictive qualities of the product. But addiction may be a very personal thing, and not all are prone to addiction to the same extent or able to overcome an addiction once established. Addiction is a brain state, and brains' susceptibility to addiction may be plotted on a curve like other physical circumstances and conditions. So even though cigarette manufacturers could manipulate the nicotine content of their products in order to increase the likelihood and durability of their customers' addiction, cigarettes could not be specially designed for each individual, with their nicotine content tailored to ensure that the cigarette is sufficiently addictive to hook all users. Some people will become sufficiently addicted; others will not. It may be that there are exogenous contributors to addiction for which the cigarette manufacturers could not control, such as the social, economic, and emotional environments of particular users. Whether Jones and Smith (or either or neither of them) become addicted to cigarette smoking and the degree of their addiction may be a function of variables both within and outside the control of the product manufacturers.

It would always seem, particularly to those not addicted to a malign product, that Jones or Smith could quit, if she chose to do so. And that conclusion does not change just because it is difficult to quit; indeed, it may not change no matter how difficult it is to quit, particularly in a world where the dangers of cigarette smoking are notorious and smoking cessation programs and products are ubiquitous. So the cigarette manufacturers' response to individual plaintiff smokers, once the primary tort liability of the manufacturer has been established, would be that the plaintiff's recovery is either barred altogether or at least diminished.

Florida is a comparative negligence state, so in *Phillip Morris USA, Inc. v. Douglas*,[46] the defendant cigarette manufacturers interposed the smoker's failure to cease use of the product that caused their illnesses and perhaps death as a defense to liability. The court described the dynamic: "individual plaintiffs do not simply walk into court, state that they are entitled to the benefit of the [general class action findings],[47] prove their damages, and walk away with a judgment against the [defendants]. Instead, to gain the benefit of [those] findings in the first instance, individual plaintiffs must prove membership in the [Engle] class [which often] hinges on the contested issue of whether the plaintiff smoked cigarettes because of addiction or for some other reason (like the reasons of stress relief, enjoyment of cigarettes, and weight control)."[48] If it is *addiction* to cigarettes that causes the lethal (or, at least, injurious) level of use of the defendant's product and if addiction is a condition over which the plaintiff has some control, then the plaintiff's recovery would appropriately be reduced as a result of the extent of use caused by the addiction. But if the plaintiff is not in control of his addiction, if the defendant's action *caused* the plaintiff's dependence on the product that damages the plaintiff's health, then is it appropriate to reduce the plaintiff's recovery on account of a variable that the defendant controlled and exploited? A good deal depends on your understanding of addiction and control; considerations redolent of dualism would seem to pertain here.

The Dualism of Fault?

David Wallace is a litigator who challenged the *Engle* premise (that a plaintiff who proves addiction may recover from cigarette manufacturers) and has drawn on arguments that put in issue the neuroscientific conception of human agency.[49] In a brief essay, he concluded that neuroscientific and psychiatric insights are at odds with the legal concepts of human agency and responsibility. Whether we are responsible from the perspective of neuroscience, he concluded, is largely irrelevant to whether we are legally responsible: "The plaintiffs [in *Engle*] are essentially arguing brain causation, which is facially absurd. Smoking behavior is not the work of a homunculus in the brain or neuronal circumstances. '[W]e can't get the macro story from the micro story' (Ref. 6, p. 135). Brains do not smoke cigarettes; acting people do, and the whole human organism is involved. For the same reason, brains are not subject to responsibility attribution; acting people are. Law is about personhood, not biophysical function."[50] That is not an unfamiliar argument and Wallace's presentation of it is no more than typically thoughtful of the dualistic premises. Knowingly or otherwise, he relied on the same dualism as some neuroscientists, philosophers, and legal theorists before him: "Brains don't kill people; people kill people";[51] recall that Bennett and Hacker also concluded that the brain and the human being are distinct entities;[52] and Pardo and Patterson reached the same conclusion though they instead juxtaposed the brain and the person, not a substantial difference.[53] That is dualism, though those who would urge separation between the brain and the "killer" or "human being" or "person" maintain that they are not dualists.[54]

But their protests are not convincing. Pardo and Patterson described a middle path between materialism and dualism by proposing that mind is merely the aggregate of a person's intellectual abilities, making it a function of the physical brain.[55] Because mind is now a collection of properties rather than a separate *ousia*, they can claim to have avoided the dualistic premise of nonphysical substance. They crossed the ontological divide, however, by positing that the person that thinks and feels is not the brain that performs those concomitant functions (firing neural synapses, etc.).[56] Personhood is the nonphysical entity linked to the now materially grounded mind. But on what grounds can this person be said to exist if it is a nonphysical entity? Only by recourse to nonphysical substance, the defining characteristic of dualism. Hence Walter Glannon pointed out that Pardo and Patterson come "dangerously close to . . . substance dualism."[57] Their method is linguistically distinct but functionally identical to Cartesian dualism; they substitute person and brain for mind and body, but dualism by any other name is still dualism.

Wallace's argument was premised on the same dualistic conception, but he was quite right about one thing: It is just that dualistic perspective that animates the law, and the tort law in particular as it relates to comparative liability built on the plaintiff's contributing to the causation of his injury. "The nub of the matter in *Engle* is what addiction means for purposes of legal cause or causal responsibility, not the pharmacological effects of nicotine on the brain or how smoking behavior comes about as a matter of microlevel biophysical function. Regardless of addiction status, smokers are otherwise reasonable legal persons for all legal purposes, from contracts, to torts, to advanced health care directives, to informed consent. The real question in *Engle* is who is in charge for responsibility or accountability purposes, the brain or the person."[58] That analysis captures well the disposition of the current tort doctrine, which relies on the very dualism that he described: the brain *and* the person as though the two are separate, distinct in some way pertinent to the normative calculus. Although the brain might become addicted, in some clinical sense, the full person retains the ability to overcome that compulsion. In fact, although Wallace's object is to criticize the *Engle* conclusion that cigarette manufacturers may be liable for the consequences of their customers' addiction, the Solomonic nature of the *Engle* calculus is entirely consistent with the dualism Wallace championed and the doctrine assumes: In *Phillip Morris USA, Inc. v. Douglas*, the plaintiff's recovery was reduced by 50 percent to reflect the plaintiff's fault in surrendering to nicotine addiction.

Wallace relied extensively on an article by Dr. Harold Kalant[59] to support his conclusions about the dualism of addiction (and so too the dualism of the apposite comparative negligence doctrine). Kalant was responding to a paper published in 1997 by Alan Leshner, former director of the National Institute on Drug Abuse in the United States.[60] Leshner had concluded that "addiction is a brain disease . . . tied to changes in brain structure and function."[61] In the paper, Leshner explained that when addiction happens "A metaphorical switch in the brain seems to be thrown as a result of prolonged drug use." At the time that "switch is thrown, the individual moves into a state of addiction."[62] Wallace noted Leshner's conclusion that a medical concept or status does not correspond to a legal consideration. Describing addiction as a disease would seem to wring from it any sense of the victim's fault: You are no more at fault for your smoking, or drug, or alcohol addiction than you would be for contracting Parkinson's disease.

Wallace's object was to detach the medical state of addiction from any legal conception of the term. That is, although medicine may have one conception of a condition or pathology for medical purposes, insofar as the medical purposes are not the same as (or at least not coextensive with) the legal purpose of reaching that medical conclusion, it would be error to conflate the two: to attribute legal significance to the medical conclusion. Addiction is defined in the *Diagnostic and Statistical Manual of Mental Disorders* to enable healthcare professionals and researchers " 'to diagnose, communicate about, study, and treat people with various mental disorders,' not to answer legal questions."[63]

Much of Kalant's analysis upon which Wallace relied described the co-operation of neuronal change effected or at least accommodated by addictive substances, such as nicotine, with exogenous social and environmental factors. Indeed, the very same intracellular adaptive changes caused by the introduction of potentially addictive substances happen to cells all the time: "They are basic parts of the processes of learning, memory, and forgetting";[64] "accumulation of the transcription factor delta FosB and suppression of c-fos, which has been seen after chronic exposure to cocaine or amphetamine, has also been found after exposure to natural reinforcers such as sucrose or sex."[65] Further on, Kalant asked "whether there is anything unique in the addictiveness of drugs as opposed to natural reinforcers."[66]

We can assume, for the sake of the present analysis and argument, that the answer to Kalant's hypothetical question is "no"; there is nothing unique about addictiveness to drugs. Drugs are just another means to create or enhance the neuronal effects of other common pleasurable experiences that we consider benign. Some individuals, though, may find it easier and more effective to create those effects by administering drugs than by eating too much or having too much sex. So from the materialistic perspective, the perspective that dualism (acknowledged or not) challenges, the ultimate explanation for addiction to nicotine may not look or be all that different (if different at all) from the explanation for behaviors in which our species is certainly predisposed, if not hardwired, to engage.

Indeed, even were we able to identify the precise and predictable neural pathways that addictive substances exploit, we would still not have an explanation for why two people, apparently similarly situated, respond differently to addictive substances: Not everyone who uses nicotine becomes addicted to it and certainly not to the same extent. Further, some people can "kick the habit" more easily than others. The reasons for those differences could, certainly, be genetic, but that path to materialism may not advance the inquiry too far either:

> There are now numerous examples of environmental factors controlling the expression of genes, so that an individual with a given genetic make-up may be vulnerable to induction of addiction or relapse under some circumstances and not under others. For example, both clinical observations and experimental models have demonstrated that addicted individuals, after undergoing successful extinction of heavy drinking behavior, experience a greater risk of relapse when exposed to stress.[67] They presumably have the same genetic make-up at all times, but various genes related to vulnerability appear to be switched on when they are under stress and not in its absence.[68]

Individuals who are more prone to addictive behavior in stressful settings may not always be able to control (in any sense of the term) whether they find themselves in a stressful situation. Surely those who distribute such substances could not control their customers' exposure to stress. Certainly, as between the two, potential plaintiff-consumer and potential defendant-manufacturer, it is not at all clear that the manufacturer would be in as good a position as the consumer to enable the consumer to avoid stress or appreciate the customer's idiosyncratic need to avoid stress. So from that perspective, to the extent that nicotine addiction, for example, is the product of a confluence of forces—neuronal, genetic, and environmental—the comparative/contributory negligence calculus might bar the plaintiff's recovery. Even if we conclude that the addiction and the negative health effects that result were a product of cooperative actions of the manufacturer and consumer, we could at best guess at what their relative levels of fault might be.

As between the manufacturer of a negligently designed or manufactured product and the consumer of that product, who *is* in the better position to avoid the inevitable losses caused by the product? One view, held by Wallace, is that insofar as individual consumers are responsible agents, the manufacturers of the cigarettes that harm them should not be responsible for the harm caused by the cigarettes because the choice to use the product is the consumer's. Notwithstanding the addictive qualities of the product, and notwithstanding the fact that we can understand something about how addiction works in the brain, the decision to smoke is more than just the product of brain chemistry, electricity, or structure: There is something more than the brain, that is, the person, that decides whether to smoke, and reducing the decision to a matter of brain function denies the substance of the choice.

That dualistic argument is conceptual: The plaintiff is more than his brain; the plaintiff is a person and the relative culpability of the plaintiff is a function of that personhood that goes beyond the physical structures and processes in the brain. It relies, though, on Kalant's description of how addiction happens to the brain; but Kalant did not suggest that there is more than the brain. Kalant's conclusions were ultimately materialistic: We cannot ignore the physical components of addiction; the problem with equating simple neuronal function to addiction is that such an equation is empirically inaccurate. In order to understand how and when certain physical processes take place in the brain or culminate in addiction, we would have to take account of *all* the variables that would ultimately affect brain function. It is not nicotine alone that causes addiction to cigarettes; it is nicotine in combination with other environmental factors (endogenous and exogenous to the brain) that will result in addiction. Kalant's conclusion would not support Wallace's philosophy, though it may agree with his conclusion *in some cases*.

That distinction is crucial. If Wallace were correct in his *conceptual* conclusion, then cigarette manufacturers should not be liable, at all, to their customers whose use of cigarettes causes illness or death. But if Wallace misunderstood Kalant, then manufacturers could be liable to smokers affected by the manufacturers' product. Kalant did not conclude that the brain is not involved in addiction; he concluded only that we need to understand the broader context in which the brain functions in order to understand the endogenous brain characteristics that result in addiction. Kalant's point was an empirical one, and Wallace misunderstood that. Wallace mistakenly used an empirical observation to make a conceptual argument.

A Basis for Liability?

Once we understand that addiction is the consequence of factors endogenous and exogenous to brain function, then we see a basis for manufacturer liability.[69] A cigarette manufacturer does not need to know or even confidently predict that a *particular* individual will have the combination of neural characteristics (the combination of chemical, electrical, and structural features) that would predispose that person to nicotine addiction. And the cigarette manufacturer does not need to know that a *particular* individual will be exposed to the environmental stresses and influences that would cooperate with a certain neural predisposition to cause addictive behavior. All that the manufacturer needs to know, in order to take advantage of the commercial benefits of its customers' addiction to its products, is that *some* relatively determinate number of customers will have the right combination of neural propensities and encounter just the right environmental circumstances to become hooked. It is not the *who* that matters; it is the *how many*.

Just as the individual who shoots negligently into a crowded hall does not know whom his bullets will hit, the cigarette manufacturer does not know, does not need to know, who among the consuming population will become addicted to its product. Of course, cigarette manufacturers may make efforts to reach those most vulnerable to smoking in the first place: the young, the less well educated, and the more easily impressionable.[70] If you reach members of any of those groups at a time when they have less ability to do the cognitive and social mathematics, they are more likely to develop a taste (as determined by neuronal and environmental factors) for the high that nicotine provides, and then, when later somewhat removed from that initial neuronal and environmental circumstance, they will have a more difficult time overcoming the physical dependence on the drug that developed from that earlier, less mature time.[71]

It would also be true that at the time cigarette manufacturers sell their product they would know roughly how many people will suffer serious health impairment or even death as a result of smoking.[72] So it would seem that cigarette manufacturers know both how many people will smoke and the consequences of their doing so. Is that enough to impose liability irrespective of individual smokers' comparative fault, or would doing so ignore the difference between the brain and the person? Is that a legal question somehow distinct from the biological question?

Emerging Contexts

Neuroscience may inform that inquiry, and so affect the tort doctrine by causing us to reconsider if not revise conceptions of comparative fault.

But although the contours of the inquiry may be most clearly presented in the case of nicotine addiction (a rare if not unique case of a legal product that would seem, almost by definition, to be defective),[73] it may as well inform other contexts in which conceptions of comparative fault pertain. Consider the liability of sports organizations and leagues for injuries that are endemic to the sport, such as head injuries in football, hockey, rugby, or even soccer. It is clear that when those who will spend a significant percentage of their lives playing those games first start to play them in an organized setting, they may not be in the best position to make mature decisions about their own neural well-being.[74]

An obstacle to any litigation that might be brought against equipment manufacturers or sports organizations or leagues is proximate causation: How would the plaintiff prove who is at fault for an injury that is compounded over years, maybe decades, at a potentially wide range of times during which brain development might be impaired or brain injury might result from repetitive concussions or brain trauma short of concussion caused by persistent blows to the head?[75] And even if we were able to fix the time during which we were most certain brain injury had occurred, would the plaintiff bear some responsibility for continuing an activity that, we would imagine, the plaintiff or plaintiff's parents were aware caused at least some distress? As the risks of CTE become better known, participants in activities that present CTE risks would be in a better position to avoid the behavior that causes the condition.

Until neuroscience answers more of the questions about the condition, it is unlikely that the tort law will be able to draw the lines necessary to allocate responsibility and liability. It is not difficult to imagine that the causes of CTE may be numerous, and that not everyone who participates in contact activities assumes the same risk of debilitating injury. Besides, it may be the case that injuries caused at the earliest stages of participation, when the brain is less developed, may lead to the most profound long-term consequences.[76] CTE is a progressive degenerative disease and so may work much the way nicotine addiction works: The more injured (or addicted) the person's brain becomes, the less able the person may be to control the behavior that causes further and greater injury.[77] Although the parameters and dimensions of CTE are still the subject of study and discovery, the condition has been associated with depression, suicidal tendencies, aggressive behavior, cognitive impairment, and impairment of motor control.[78] It is not difficult to imagine that some of those symptoms might make it difficult to control the behavior that would lead to further brain injury.

In the case of both addiction and CTE, then, our understanding of the affected neural processes is incomplete. Further, and most significantly, we do not understand, yet, the relationship between chemical, electrical, and structural dimensions of brain function and exogenous environmental factors that might affect the liability calculus. But it does seem clear that division of the plaintiff into brain and whole person does not advance the inquiry. Neuroscience can tell us more about how brain and environment interact and, also, can make clear the actuarial certainties upon which liability assessments might be founded.

Finally, neuroscience also may result in *less* liability: An object of learning more about brain function and about conditions that impair brain function is that we empower plaintiffs to take steps to limit or avoid injuries, even injuries that in the first instance were caused by someone else. So, for example, if there were a certain treatment for some form of addiction,[79] we would expect a plaintiff to take advantage of that rather than contract serious cardiopulmonary disease and try to impose liability on the manufacturer of the addictive substance.[80] We are already seeing the effect that CTE risk has on the participation of children and young adults in activities that present concussion risks.[81]

The role that neuroscience might play in proximate causation analyses demonstrates the fine line between doctrinal and evidentiary questions. Those questions also merge in tort law's determination of what constitutes a compensable injury. Just as tort relies on proximate causation to resolve liability questions and neuroscientific insights may refine causation analyses, tort distinguishes injuries that are compensable from those that are not by means that neuroscience may elucidate. The important distinctions are drawn both by the doctrine and the evidence pertinent to the application of the doctrine.

Compensable Injury

The tort law distinguishes physical from mental or emotional (terms often used interchangeably) injuries in determining what is and what is not compensable. Again, administrative feasibility—the accessibility of the proof that would support consideration of things we cannot see, or at least not see as well as others—seems to be determinative. Some injuries are just more obvious than others: The compound fracture of a broken limb is more obvious, easier to confirm without sophisticated medical equipment, than is the brain alteration that gives rise to PTSD. The tort law does not value broken limbs more highly than it does emotional injury, but it can see the physical nature of the broken limb better. Even if the fracture is not compound, an X-ray will reveal what the naked eye would not. And now an MRI will reveal soft tissue injuries that were invisible a relatively short time ago.[82]

The materialist, though, knows that all states of mind, including PTSD and profound emotional injury, have physical referents; there is no purely mental or emotional injury that does not have a physical correlate. Depression is every bit as much a physical injury or condition as a lacerated spleen.[83] Indeed, as we better understand the physical causes of what we have deemed to be wholly mental or emotional conditions, the familiar doctrinal bases of distinction may disintegrate, compromising or confounding the normative foundation of extant tort doctrine. It is one thing to limit recovery for emotional harm when you cannot see it, or at least objectively verify it; it is altogether another to cling to a particular basis of distinction that is inconsiderate of the science. Brain scans may enable us to see the scars left by emotional trauma.[84] We already have rudimentary means to see the signature of pain, a development that will certainly affect the operation of the doctrine and should probably effect change in the doctrine itself. Once administrative hurdles are overcome, once we can be as certain of emotional injury as we are now of some physical injuries, doctrinal barriers to recovery for the negligent infliction of emotional distress should fall. Tort doctrine that makes sense as a second-best solution to problems of proof is no longer an acceptable solution when the problems of proof are overcome. And well before they are certainly overcome, we may expect that developments in neuroscience will begin to undermine moral choices that made sense when we knew and could see much less.

Nevertheless, tort doctrine, for the time being at least, does recognize a fundamental distinction between physical and nonphysical injury and treats the two separately and differently. The doctrine also distinguishes between intentionally and negligently inflicted mental or emotional harm. Administrative convenience and accessibility of proof might explain treating emotional harm differently from how we treat physical harm, but it is not immediately clear that emotional harm is manifest any more clearly just because it was intentionally inflicted. Tort doctrine does not say that *A* is liable for breaking *B*'s arm on purpose but not if *A* inflicts the same or even greater damage negligently. Although *A* might be answerable for punitive damages in addition to compensatory damages for intentional or even reckless[85] behavior, *A* would be liable to *B* in some measure for either negligent or intentional actions that physically harm *B*. But that is not so in the case of emotional distress: There are barriers to recovery for the negligent infliction of emotional distress that do not apply if the plaintiff can prove the intentional infliction of emotional distress. To the extent that the disparate treatment of physical and emotional injury proceeds from considerations of access to proof of injury, we may still be able to make sense of the distinction, though its coherence may not be immediately obvious.

Rationalizing the Distinction

The tort doctrine in this area has not changed substantially over time and for the most part the Third Restatement of Torts continues the approach of the Second Restatement and seems to rely on the same reasoning. The Second Restatement rule on the negligent infliction of emotional distress was definitive in its terms: "If the actor's conduct is negligent in creating an unreasonable risk of causing either bodily harm or emotional disturbance to another, and it results in such emotional disturbance alone, without bodily harm or other compensable damage, the actor is not liable for such emotional disturbance."[86] The comments to the provision explain that

> The reasons for the distinction, as they usually have been stated by the courts, have been three. One is that emotional disturbance which is not so severe and serious as to have physical consequences is normally in the realm of the trivial, and so falls within the maxim that the law does not concern itself with trifles. It is likely to be so temporary, so evanescent, and so relatively harmless and unimportant, that the task of compensating for it would unduly burden the courts and the defendants. The second is that in the absence of the guarantee of genuineness provided by resulting bodily harm, such emotional disturbance may be too easily feigned, depending, as it must, very largely upon the subjective testimony of the plaintiff; and that to allow recovery for it might open too wide a door for false claimants who have suffered no real harm at all. The third is that where the defendant has been merely negligent, without any element of intent to do harm, his fault is not so great that he should be required to make good a purely mental disturbance.[87]

The first of those three reasons is unassailable: As long as the emotional harm is minimal, the tort law should take no more account of it than it would a superficial physical injury. But it will be important to keep in mind that as our understanding of the brain improves, we may be in a better position to distinguish a severe but transitory headache from an insult to the brain that could, in the longer term, impair cognitive function. Indeed, until we know what we are looking for, even CTE may not be readily distinguishable from moodiness or forgetfulness.[88] So as neuroscience matures, our ability to distinguish the serious from the trivial, and appraise plaintiffs' credibility, may improve as well. As a result, there could be fewer cases for which that first basis of distinction remains viable.

The second basis also concerns matters of proof: A physical injury confirms that the alleged emotional harm is substantial. That distinction too loses some validity in that we have other means, for example, brain scans, to confirm that there is a visible and objective correlate of claimed emotional injury. But here a note of caution is warranted: The brain may manifest what seems to be, at least to the untrained eye, a profound insult without neural function being impaired to the extent that the appearance would suggest. That is, the brain is plastic: If some portions of the neural network are compromised, others may compensate.[89] Further, some emotional injuries may heal, much as would a broken limb, and so the plaintiff may recover from even a profound emotional injury, certainly more likely than she would from a severed spinal cord. Until we know more about the brain and how its systems cooperate, we will not know enough about how the brain heals. So it will not be enough to confirm the organic basis of emotional harm; for the tort law to make full sense of it, we will have to know more about the brain's ability to recover from obviously physical trauma that takes an emotional toll.

The third basis of distinction would seem to be the most problematic from a more enlightened neuroscientific perspective: Why should the fact that a disturbance is purely mental compel the conclusion that the injury is less deserving of the tort law's protection? It is not difficult to think of mental or emotional injuries that are every bit as significant as mere physical injuries: Who would not trade depression for a broken arm? PTSD for a torn ligament? Insofar as there are less certainly effective ways to treat mental injury than there are to treat at least some physical injuries, it may be that even the apparently lesser mental or emotional injury is, in the fullness of time, the greater injury. As advances in neuroscience reveal the physical (chemical/electrical/structural) basis of mental and emotional injury, that will surely be prologue to treatments that respond to the organic deficit. But there is no reason to believe that there will not be lag time, perhaps great lag time, between the discovery of the neural basis of depression (in all of its forms for all those afflicted) and the development of a treatment. We can now accurately diagnose many, many more diseases than we are able to treat (at all, much less effectively). In the meantime, then, the tort doctrine has to be ready for the day when we can better see the physical source of what we consider to be a wholly nonphysical injury but have not yet developed the means to treat such an injury. So here the Second Restatement does not seem to be on very firm ground. It may be that this error will be self-correcting, or could be were the doctrine not cast in ostensibly inflexible terms.

In the meantime, Third Restatement treatment of the mental or emotional injury caused by negligent conduct continues the specious distinctions of the Second Restatement. The first of the two apposite Third Restatement sections is section 47.

Section 47. Negligent Conduct Directly Inflicting Emotional Harm on Another
An actor whose negligent conduct causes serious emotional harm to another is subject to liability to the other if the conduct:
  (a) places the other in danger of immediate bodily harm and the emotional harm results from the danger, or
  (b) occurs in the course of specified categories of activities, undertakings, or relationships in which negligent conduct is especially likely to cause serious emotional harm.[90]

The comments explain that subsection (a) applies to the situation in which there was a danger of bodily harm but the victim suffered only emotional harm.[91] And the second subsection continues the current rule, according to which certain categories of circumstances warrant the conclusion that emotional harm is likely to occur.[92] This confirms that the presumptive reluctance to compensate for emotional harm is informed by skepticism about the significance or even existence of the harm. Both subsections are based on that skepticism. The Third Restatement captures (subsumes, actually) the common law effect and zone of danger rules in section 47 and the common law bystander rule in section 48:

Section 48. Negligent Infliction of Emotional Harm Resulting from Bodily Harm to a Third Person
An actor who negligently causes sudden serious bodily injury to a third person is subject to liability for serious emotional harm caused thereby to a person who:
  (a) perceives the event contemporaneously, and
  (b) is a close family member of the person suffering bodily injury.[93]

Section 47 of the Third Restatement replaces section 436A of the Second Restatement and section 48 of the Third Restatement replaces sections 47, 48, 312, 313, and 436 of the Second Restatement.[94] So we have good reason to conclude that sections 47 and 48 of the Third Restatement are a comprehensive presentation of the state of the tort law relating to liability for negligently inflicting emotional harm. The discussion and analysis that follow focus first on section 47 of the Third Restatement, in terms that pertain generally to the dimensions of the physical–mental injury dichotomy, and then turn to section 48.

Negligent Infliction of Direct Emotional Harm

The commentary and sources cited in the reporter's notes to section 47 focus on the balance struck between compensating those who have suffered real harm and the administrability of rules that would distinguish real from insubstantial (or less than substantial) harm. Section 47 requires either that there be evidence of a coincident objectively verifiable risk of imminent bodily harm or that the emotional harm occurred in circumstances that are "especially likely to cause serious emotional harm." The substance of those categories is left for determination by the courts, but the idea is that they will involve contexts in which third parties are easily able to imagine that emotional harm would result: It is one thing to be told negligently and erroneously that your car has been struck by lightning; it is quite another to hear the same inaccurate report about your child. We all, just as a function of being human, can better imagine that serious emotional harm would result from the false report about a loved one.

It is not certainly and inviolably true that one type of misreport would necessarily cause greater emotional harm than another, but it is the case that we have no reliable means to confirm subjective descriptions of emotional harm, and a tort plaintiff might have reason not to be entirely or accurately forthcoming. The test is not one of foreseeability, a standard that might result in too much liability; the test is one of duty and so may be determined by the court as a matter of law to avoid the risk that juries would do some justice at the expense of the law's integrity. To some extent, then, the rule of section 47 is designed to cut off liability for what might even be real emotional harm. At some point the law must draw a line, and that does make sense. There is a level of emotional disquiet that comes just with being alive in a community. That level of anxiety is the ambient emotional noise that we must all suffer as a cost of enjoying the benefits of society. Further, we have different, perhaps even idiosyncratic and idiosyncratically variable, anxiety levels, and the law simply does not take into account diverse levels of emotional disquiet not within some normal range. Section 47 does not require that the emotional harm be accompanied by any physical manifestation, though. The test depends purely on circumstances that would suggest the substance of emotional harm as a consequence of the circumstances surrounding the defendant's actions: The section concerns actions of the defendant that would, in the circumstances, be likely to inflict emotional harm either because the defendant exposed the plaintiff to risk of immediate bodily harm or because the relationship between the plaintiff and defendant suggests that emotional harm would likely result from the defendant's negligence.

As a means to further limit the plaintiffs who may recover for having been put in danger of bodily harm, subsection 47(a) includes an immediacy requirement. That limitation is designed to provide more objective confirmation of the emotional harm. Emotional harm experienced by "passengers in an apparently doomed aircraft"[95] is deemed more likely real than the emotional harm suffered by those who are negligently exposed to a known carcinogen (e.g., asbestos) or, perhaps, pathogen (e.g., HIV).[96] The test, the basis of distinction, is how long it takes for the plaintiff to know that she has not contracted the disease to which the defendant exposed her. Because it might take years to confirm or disconfirm cancer caused by asbestos, the bodily harm is not immediate.[97] But because the plaintiff could know within just a few months whether exposure to HIV has communicated the disease, exposure to HIV would be immediate.[98] The apposite comment explains that the distinction really has less to do with immediacy than with raising "the potential of multiple lawsuits."[99] It is determinative that the HIV plaintiff but not the asbestos plaintiff "can determine within a known and relatively short interval whether or not the exposure actually did cause physical injury," and so the emotional harm would have been created "in a way that does not raise the potential of multiple lawsuits."[100] The multiple lawsuits the comment has in mind are the successive actions the plaintiff could bring between the time of exposure to the risk (when the emotional harm occurs) and the time that the consequent bodily harm is manifest. The same comment explains that that is "one reason" for the immediacy requirement. It may be easier to make sense of the requirement as another means to monitor the administrability of the emotional harm rule: Perhaps everyone can imagine living for some time with the fear of even a terminal illness; few of us can grasp the unimaginable horror of falling out of the sky in an airplane that was negligently designed, constructed, or flown.

The comments also make clear that plaintiffs may recover only for severe emotional harm and refer to PTSD specifically. An illustration[101] describes PTSD as "a serious emotional harm" and so distinguishable from emotional upset that admirers of a public figure might experience when the defendant's negligence caused the violent death of that celebrity. The line that the Third Restatement draws between emotional harm that is serious and emotional harm that is not serious is not elaborated. The comment explains that the court, as a matter of law, may decide that the emotional harm alleged is insufficiently serious much as any court could determine that there was no tort liability because the defendant did not owe a duty to the plaintiff.[102] The comment recognizes the seriousness of PTSD but does not premise that finding on any recognition of the disorder's physical constituents. PTSD is, apparently, serious because it is diagnosable and may be identified by competent health care professionals: There is no appreciation of degrees of PTSD or of recovery from the disorder.

The apposite reporter's note takes account of the thin-skull plaintiff issue with regard to emotional harm. We would not care if the defendant's negligent blow resulted in a particularly egregious break of the plaintiff's arm because the plaintiff's bone was unusually weak—defendant would still be liable if the defendant's action would have broken a normally constituted arm—and so it would seem to follow that we would not relieve a defendant from liability for emotional harm just because the plaintiff is more emotionally sensitive, more prone to serious emotional injury, than would be the norm. The reporter concludes that the case law, though, does distinguish application of the thin-skull rules in the case of physical injury from application of the rule in the case of emotional injury.[103] Nonetheless, the commentary notes decisions finding that once the threshold has been crossed, that is, once the plaintiff has established that the emotional harm was serious, the court would not limit the plaintiff's recovery just because the actual injury suffered by the plaintiff exceeded that which a usually susceptible plaintiff would have suffered. So the threshold is objective: The plaintiff must not have suffered the harm only because the plaintiff was hypersensitive, but once the harm crosses that threshold there is no reasonableness limitation that would limit the *extent* of harm for which the plaintiff could recover.

It seems clear that the Third Restatement's rule in section 47 on negligence directly causing emotional injury to another is designed to serve normative objects not directly related to the severity of the injury actually suffered by the plaintiff. The bias seems to be in favor of denying recovery, absent certain proof of emotional injury, and the Third Restatement's formulation relies on ostensibly objective indicia that are related to emotional injury: If it is the type of emotional pain that a third party can almost feel, then recovery would be allowed.[104] If the emotional injury alleged does not lend itself to such empathic sensibilities, then recovery is not available.

The reporter's notes do, though, acknowledge that technology has affected (and, presumably, may continue to affect) the objective threshold upon which the rules are based. The apposite commentary describes specifically the decision of the Rhode Island Supreme Court in *Perrotti v. Gonicberg*.[105] Prior case law in the state had allowed recovery for a pregnant woman who feared that her unborn child might have been injured in an automobile accident caused by the defendant's negligence. But that earlier case was decided in 1917,[106] an era before fetal heart-rate monitoring and ultrasound. By 2005, the year that *Perrotti* was decided, such technological means to appraise fetal injury were commonplace and could, in fact, confirm that the plaintiff's unborn child had not suffered an injury. Plaintiff's fears could be and had been allayed. So it is clear that neuroscientific advances (just as any scientific advances) will not necessarily result in more liability; indeed, they may as often or more often result in less liability.

Emotional Healing

The ameliorating benefits of neuroscientific insights are noteworthy. Just as we would not allow the plaintiff with a negligently broken leg to let the wound persist and fester, we would not allow the individual with a severe emotional injury to continue to suffer without taking advantage of adequate and available treatment. Indeed, developments in neuroscience might provide new channels for potential plaintiffs to reduce the debilitating effects of traumatic emotional injuries, possibly even heal them, further limiting defendants' liability. Neuroscience surely can help us better discover the location, formation, and interaction of the neural processes that form the memories we associate with emotional injury, and such better understanding may lead to better treatment. For example, one promising area where the treatment of PTSD could develop was implicated in a recent study led by Dr. Roberto Malinow, a neuroscientist at the University of California at San Diego, that identifies a physical explanation of memory, a process called long-term potentiation (LTP).[107]

In order to show that LTP and its counterpart, long-term depression (LTD), could alter memory activation, Malinow and his team conditioned a group of rats to respond to a classic fearful stimulus—an audible tone that corresponded with an electric shock.[108] Using a method called optogenetics, the researchers were able to manipulate genetically modified photosensitive neurons by delivering light pulses through fiber optics implanted in the rats' brains. Once the rats displayed a conditioned response to the electric shock, the rats were exposed to light pulses that triggered LTD, which dampens synaptic activity associated with memory formation. Rats that had undergone LTD exposure no longer responded to the fearful stimulus. The team then delivered light pulses that triggered LTP, and the conditioned response was thereby reactivated.[109] Malinow, then, was able to bidirectionally trigger and erase the rats' associative memory at will. The result suggests a process that could one day be replicated in humans who suffer from PTSD, effectively eliminating painful memories underlying emotional harm. If Malinow's procedure finds an analogue in humans, then the treatment of PTSD will serve the same functions as fetal heart-rate monitors and ultrasound in alleviating the deleterious effects of mental harm. Yet the fear-association at the base of PTSD is much more complex than the straightforward model used by Manilow. The spatial understanding of memory location and targeting specificity of the optogenetic methods would certainly have to advance significantly before present hurdles could be overcome. Even then, greater normative questions would remain.

Emotional Harm Resulting from Physical Injury to Another

Section 48 of the Third Restatement too is built around administrative limitations, particularly the obstacles to our confidently appraising the extent and even the very fact of emotional injury. The successful plaintiff must "perceive" the tort "contemporaneously" and be a "close family member" of the party suffering the "sudden serious bodily injury."[110] As in section 47, the emotional injury must be "serious." So, at the outset, the doctrine excepts those who contemporaneously witness a family member suffering sudden serious bodily injury but whose own emotional injury is not serious. That is, the doctrine does not trust the perceptual, timing, relationship, and suddenness requirements to police the tort: The plaintiff still must establish that the emotional injury suffered was serious. Indeed, one might wonder what the jury would conclude about a close family member who witnessed in real time the sudden bodily injury of a close family member and did not suffer serious emotional injury; in any event, it is not likely that the plaintiff would acknowledge such impassiveness.[111]

There is a further limitation: The cause of action is also, somewhat curiously, deemed to be derivative. So if the defendant had a valid defense against the injured family member, say, contributory negligence, at least some jurisdictions (and the Third Restatement) would deny recovery to the family member who suffered serious emotional injury as a result of witnessing the accident.[112]

Most telling is the comment's acknowledgment that the lines drawn by the black letter of the doctrine are, if not "arbitrary,"[113] at least not the product of a careful calculus balancing harm against the burden of having avoided the harm: The object of the limitations is to reduce incidents of liability because "The law of negligence has never applied the ordinary rules of foreseeability to emotional harm. . . . [A]s a matter of policy [emotional harm] is an injury whose cost the legal system should not normally shift."[114] The comment does not explain why that is so. You would imagine that if emotional injury could have as profound a negative effect on the plaintiff as demonstrable physical injury, the tort law would have no obvious reason to deny the plaintiff recovery.

The doctrine, though, seems to reach the conclusion it does for two inter-related reasons: (1) We remain skeptical of emotional injury—it is too easy to claim and too difficult to disprove; and (2) the same negligent act may, at least in theory, expose too many potential plaintiffs to emotional harm. A great many people may see (through the media) the commission of a tort that results in serious bodily injury. But it also is true that a large number of people may suffer bodily injury (serious or not, just as long as compensable under the tort law) when a poisonous gas is negligently released, or a boiler bursts, or a bridge fails, or a drug manifests fatal side effects. We would not deny those injured in such calamities recovery merely because they are part of a large group. So the number of potential plaintiffs alone would not seem to be necessarily determinative. The reason to deny recovery to all those who suffer serious emotional injury on account of a defendant's negligence would, again, seem to be premised on concerns about the substantiality or even just existence of the injury: problems of proof.

We may anticipate, then, that the tort doctrine could shift as neuroscience discovers reliable markers of emotional injury, physical correlates of emotional harm. Once we have a way to see emotional injury as clearly as we can see a broken bone or confirm a connection between the negligent release of a carcinogen and resulting cancer, there would be no reason to maintain the tort law's distinction between physical and emotional injury. And it may be that long before we can certainly confirm the accuracy of such a physical marker of emotional injury we may have reason to be more indulgent of emotional injury claims. The limitations contained in the existing law could succumb as we better understand the human emotional system, well before we can confirm that a particular chemical, electrical, or structural anomaly in the brain is the certain evidence of emotional injury.

Indeed, section 48 already indulges some sensitivity to what may be reliable indicators of emotional harm: the fact that the defendant caused serious bodily harm that the plaintiff perceived contemporaneously and that the plaintiff was a close family member of the defendant's victim. The doctrine could simply provide that defendants are liable for emotional harm negligently caused, without qualification. The restrictions built into the doctrine are designed to provide confidence in the claim of emotional harm by specifying circumstances that, we may all imagine, would certainly cause emotional harm were we in the plaintiff's position.

The Problem of Pain

Pain is an enigmatic injury that we may recognize but cannot certainly measure in others. The biological function of pain is basic: Pain is aversive.[115] Our bodies interact with the environment in various ways and sometimes those interactions are harmful. Pain tells our bodies not to do that again. It is essentially the way that we learn. Pain attaches persuasive and aversive emotional significance to deleterious actions. By engaging disparate centers of the brain, such as the limbic system,[116] pain is naturally an individualized sensation—likely a different qualitative experience for different people in different contexts. Yet its function is consistent in all people: It acts as an alarm system to the brain, eliciting repetitive punishments. Tort law has found this repetitive injury repressible.[117]

Tort operates under the "legal fiction that money damages can compensate for a victim's injury."[118] Since courts cannot numb pain or instill pleasure, they offer monetary compensation in the hope that the injured may find consolation. But challenging that strategy is the subjective nature of pain and suffering. Too much relies upon a plaintiff's own evaluation of her discomfort, with the prospect of hefty recovery encouraging exaggeration or even downright simulation. The thin-skull plaintiff doctrine, making the defendant liable for the full extent of pain and suffering, only increases the stakes in the fact-finder's pain and suffering valuation.[119] That circumstance presents both evidentiary and doctrinal challenges that neuroscience might be able to address.

The study of pain is a familiar topic in neuroscience literature. Only recently has the field formulated something akin to an objective signal for subjective perception of pain. Dr. Tor D. Wager's neural pain signature has found that fMRI data can validate self-reports of pain.[120] By selecting brain regions implicated in pain processing and assigning weights to the signals in those regions, Wager's study was able to predict the self-reports of acute thermal pain from the fMRI signal strength. Though commentators have recognized that the finding itself is limited,[121] the study suggested that there may be a common indicator of subjective appraisals of pain between individuals. We should not expect a pain fingerprint or thermometer, however, that would obviate at least some reliance on the self-report of pain.[122] We may, though, imagine that advances in the neuroimaging of pain could make self-reports more reliable and that the complementary use of self-report and imaging could provide greater confidence than reliance on either alone. It also would seem likely that developments in either method will improve the accuracy of the other and so the two cooperatively.

Kolber is certainly one of the most thoughtful commentators considering the effect that neuroscience may have on the law and has recognized that insofar as we award damages for pain and suffering, "our tort system must make inferences about the magnitude of people's pain if it is going to optimally deter future harmful behavior or correct harms that have already occurred."[123] His important 2007 article[124] was a statement of the relationship between emerging neuroscientific insights on pain and the tort law. Writing before the most recent discoveries that might bear on the objectification of pain, Kolber recognized the consequences for tort that "Pain responses are 'significantly influenced by psychological context, the meaning of the pain to the individual, the patient's cultural background, and the individual's beliefs and coping resources.'[125] Emotional states like anxiety and depression also 'dramatically influence[]' pain perception.[126] Thus, 'severity of pain does not bear a simple relationship to the degree of tissue damage.'"[127]

Kolber alluded to the fundamental tension between tort compensation and pain. Plaintiffs receive a single figure that purports to represent the countless inputs that create pain sensation: the efficacy of the plaintiff's Aδ fibers in transmitting instantaneous pain information to the brain, the potentiation of the C fibers sending long-lasting pain sensations to the brain, the sensitivity of the pain circuits that flow up the spinal cord and project to the somatosensory cortex, and every other gateway, pathway, projection, and input meshing together to make that awful painful feeling. The collusion of so many processes make the arithmetic behind a calculation of damages burdensome—even with ample neuroscientific understanding. That is, the more we understand pain, the more we may appreciate that it defies rather than accommodates objectification, even if we discover some reliably observable physical correlate. This issue is similar to the problem of tracing the efficient causes of addiction and allocating tort responsibility therefor.

Yet the neural pain signature offers hope of a one-to-one metric for pain perception and compensation—a common currency of redress. On the one hand, that might mitigate the evidentiary uncertainty confounding key doctrinal questions such as how much pain and suffering is worth, how resilient we expect victims to be, how much liability defendants must bear, and so forth. On the other hand, advances in neuroscience might confirm that the status quo, the apparent second-best solution, is actually *the* best solution because of the nature of pain. But if we come to that conclusion—and it is too soon to do so—then it is neuroscience that will have provided the answer, the confirmation of pain's inscrutable idiosyncrasy.

Neuroscience then may ease the burden that tort law presently places on the fact-finder to gauge the truthfulness of a plaintiff's suffering and then assign a valuation. Like the brain imagery that substantiates emotional harm, neuroscience might insert a degree of objectivity to a pain and suffering analysis. Neuroscience validates pain and suffering in terms of its existence, not in terms of its redressability. If neuroscience can discover malingering plaintiffs, the law might then more accurately compensate.[128] Exposing feigning plaintiffs is merely one example of neuroscience's evidentiary potential and the ways it could influence the doctrine's moral choices.

Ultimate Inscrutability?

Even if we could certainly determine the fact and severity of mental harm and pain, it may still be the case that the doctrine would not change. There just might be something about nonphysical harm that could, in the estimation of triers of fact and developers of the doctrine, warrant different treatment for emotional harm than we afford physical harm. We react differently to emotional harm than we do to physical harm, but it is not clear that differences in those reactions are just the consequence of the different forms that those harms take: Both the broken arm and the chemical or electrical or structural anomalies that are manifest in depression are physical; remember, from the neuroscientific perspective presented here, everything is physical. In fact, though, the disruption of life caused by PTSD, for example, may be far greater than that caused by a broken limb or even loss of a limb. The prognosis for recovery may be much worse for the victim suffering emotional harm than it would be for the victim suffering physical injury. Before the tort doctrine changes, our sensibilities may have to change.

Neuroscience may have its greatest effect on the doctrine by changing attitudes of judges and juries about harm that is not obviously physical: When (if ever) neuroscience can confirm mental injury and pain in objectively accessible ways, when (if ever) we can see depression (as we can PTSD) on a brain scan, we may be less likely to discount the emotional injury. We may wince when we witness an accident that results in severe physical injury; we do not yet wince, at least not in the same way, when we witness the onset of depression (though if the precipitating event is graphic, we may better appreciate the fact of the injury). For now, though, it is likely that most of us who have not suffered from depression or PTSD would have more difficulty feeling that emotional pain than we would have feeling the physical pain caused, say, when a football quarterback's lower leg is bent in a freakishly abnormal way and we see his sock turn bright red. And we likely would have many opportunities to witness that injury, in slow motion, and in high definition. It is difficult to see pain or emotional injury in the same way. Perhaps that notion that we all must just persevere through emotional injury will persist. We would not dream of saying to the person who lost the use of his legs as a result of someone's negligence to just tough it out, but there may remain something that causes us to imagine that the person suffering from traumatic stress should just "snap out of it." Biases developed over the course of time, maybe tens or hundreds of thousands of years, will not likely be overcome very easily, or quickly.

Conclusion

The examples that support the foregoing survey of the effect that neuroscientific insights may have on the tort law in the years to come necessarily

paint with a broad brush and treat summarily many issues that would warrant further attention in a study that focused solely on the tort law. The
object of this chapter, though, is to suggest how neuroscience may recast
the normative commitments of the tort doctrine and to demonstrate as
well how ostensibly evidentiary concerns affect the doctrine itself.

The next chapter builds on that examination of exemplary tort doctrine to engage the dominant normative perspectives that endeavor to describe and explain what it is that the tort law does, if anything, to further
moral ends. As we will see, crucial to those inquiries is finding a place for
tort between the criminal law and the contract law. Negligence liability
must discover and isolate a crucial disruption of relations between and
among human agents; someone must have done something that is wrong
in some way for the tort law to intervene, but not wrong in the way the
criminal law proscribes. The doctrine, as the foregoing survey has suggested, often seems confused about what that something must be. Theory
dependent on Aristotelian and Kantian premises is, we shall see, no less
confused (or at least confusing). And, as was the case in the appraisal of
criminal law theory, conceptions of moral responsibility are pertinent: indeed, pervasive.

CHAPTER FIVE

Neuroscience and Tort Law Theory

Introduction

Noninstrumental normative theories of criminal law focus on retribution; noninstrumental normative theories of the tort law focus on corrective justice rather than retribution. The difference is significant because corrective justice would support the imposition of liability when retributionary theories would not, just as (at least analogously) tort would impose liability where the criminal law would not. Now that could certainly be more than analogy, if we contemplate that corrective justice does not require moral blameworthiness or at least the same moral blameworthiness that the imposition of retributionary punishment would require. That is, we might say that someone is merely negligent, that his state of mind could be described in folk psychological terms as not having desired to bring about the harm to the victim in the same way that he would have had to desire that result to satisfy the criminal law *mens rea* requirement. In rough terms, that captures the legal distinction between negligence, or inadvertence leading to harm, and the intent to do harm.

Negligence is the less culpable mental state, and it also is the mental state that presents the lesser threat to public welfare. We may well imagine that those who intend to do harm will cause more of it than those who have no such intent but are just not as careful as similarly situated others would be. That would provide an instrumental basis of distinction and could, once we take account of human agency, support some noninstrumental weight as well. Indeed, such a distinction would work well with instrumental theories of tort designed to explain the doctrine in terms of accident avoidance[1] and cost spreading.[2] It would, for instance, provide all the normative theory necessary to support important aspects of the strict products liability law.[3]

Tort law can actually function without reference to negligence or fault at all. In the case of strict liability for products containing manufacturing defects, the successful plaintiff need show no negligence on the part of the defendant-seller: It is enough that the product that harmed the plaintiff was not in fact manufactured as the defendant designed it to be, and it is no defense that the seller used all reasonable care or even extraordinary care in the manufacture of the product.[4] Tort, then, can function without moral responsibility at all. So it certainly does not require the same moral responsibility as is required in retribution-based accounts of the criminal law. But liability without fault, without moral responsibility would rest on instrumental premises, and not all theorists find market-based or efficiency accounts sufficient or even convincing in part.

This book does not focus its critique on the empirical challenges presented by instrumentalist theories: Such problems are great and may not be certainly solvable.[5] But even if we cannot do the mathematics to be certain that we have accurately fixed the variables to encourage just the efficient level of activity, by some measure of efficiency, we may agree to settle somewhere near the best we can do now and hope to do better in the future as, for example, we better understand incentives and can control monitoring costs. Those efforts almost certainly will be advanced as the neuroscience advances, but the extent to which they have is wholly an empirical question for instrumentalists.

The focus of this chapter, consistent with the primary object of this book, is to consider the impact of neuroscientific insights on noninstrumental theories of tort. Insofar as the dominant contemporary noninstrumental theory of tort is corrective justice, a concept distinct and readily distinguishable from retributive justice, the work to do here is beyond that done in chapter 3. Again, central to the task is fixing the conception of human agency formulated by the apposite noninstrumentalist theory, here corrective justice, and comparing that conception with human agency as revealed by neuroscience: Is the human agent contemplated or assumed by noninstrumental normative theory authentic in light of what neuroscience tells us it means to be human?

This chapter endeavors to formulate the understanding of corrective justice that informs extant noninstrumental tort theory, to do the same for its elaboration in civil recourse theory, and to see whether tort doctrine consistent with those conceptions can withstand analysis in terms of the human agency that naturalistic, neuroscientific observations reveal. It would likely be impossible to survey *all* the phases of corrective justice in this single chapter or perhaps even any single volume. But it would be possible to formulate the central conception of human agency upon which any theory of corrective justice must depend.

Focus on particularly prominent treatments of corrective justice in tort will most effectively describe the essential challenge that neuroscientific insights present to the theory's interpretation, or justification, of tort doctrine. The work of Ernest Weinrib[6] and Jules Coleman[7] merits particular focus.[8] The chapter begins with a description of corrective justice that distinguishes it from retributive or distributive justice and then considers the relational focus of corrective justice supporting the normative theory. The fit, such as it is, between corrective justice and civil recourse theory bears consideration as well. From those premises the materialist critique may proceed in terms accommodated by the emerging neuroscience.

A caveat: Chapter 3 engaged directly the normative theoretical arguments in favor of noninstrumentalism, particularly retributive justifications of punishment premised on moral responsibility. Questions about moral responsibility with regard to retribution could confront noninstrumental theory head-on and demonstrate the dualistic premises of such perspectives. Insofar as neuroscientific, materialistic insights undermine noninstrumental, largely deontological, arguments, the critique of extant criminal law doctrine and normative theory that would support it presents the moral conflict of law and neuroscience starkly. The argument is ongoing and chapter 3 joined it. But the moral conflict of neuroscience and noninstrumental normative theory of tort is not yet so well developed. The prevailing noninstrumental tort theory, corrective justice (and its clarification, correction, or elaboration in civil recourse theory), has not similarly been appreciated in opposition to the refutation of human agents' moral responsibility that the materialism of neuroscience would assert. This chapter, then, more begins a train of thought than joins an ongoing conversation. The approach here to the leading noninstrumental tort theory is necessarily more oblique. The object of corrective justice and civil recourse is, at least in part, to distinguish responsibility in tort from responsibility in criminal law. So the basis of moral responsibility that matters to tort is different from that supporting retribution in the criminal law. This chapter considers whether the moral responsibility that the dominant noninstrumental tort theories claim for tort remains, nonetheless, vulnerable to critique in light of the materialism that neuroscience supports. The conclusion is that corrective justice retains and relies upon a conception of human agents' moral responsibility that is infirm—actually, incoherent. The path, though, is not as clearly illuminated as it is when criminal law theory (and, later, contract theory) is exposed to the light of neuroscience's materialism.

Weinrib's Corrective Justice

Aristotle is the source of corrective justice, as, indeed, he is the source of much noninstrumental theory. In the *Nicomachean Ethics*,[9] Aristotle distinguished distributive from corrective justice: Distributive justice would be concerned with the allocation of resources among the members of a group;[10] principles of corrective justice would govern transactions, consensual or nonconsensual (likely poles on a continuum in the private law), between two actors.[11] A simple illustration of a context in which distributive justice principles operate would be the provision of a taxation system or system of so-called entitlements. Corrective justice, in contrast, has a relational focus: When *A* harms *B*, either in person or property, corrective justice principles would inform resolution of the imbalance created by that harm.[12] An instrumentalist theory of tort, one founded on efficiency or the market, would rely on distributive principles: As between transactional groups—for example, buyers and sellers of defective products, or pedestrians and negligent drivers, or railroads that operate steam locomotives and the farmers whose crops are burned by the occasional spark that flies from the locomotive's smokestack onto the farmers' fields—which group should the loss fall upon in order to create welfare (or limit waste)? Although the relationship between distributive and corrective justice may be problematic (is corrective justice a theory of justice at all if its only object is to restore the initial just distribution?),[13] the distinction between the two suffices for present purposes: to illustrate the relational focus of the currently dominant noninstrumental normative theory of tort, corrective justice.

Palsgraf *and Corrective Justice*

Certainly those who would found tort on corrective justice could simply posit the requisite relation from an interpretation of the doctrine: insofar as tort affords *A* the right to recover from *B* when *B* harms *A*, corrective justice principles must be the source of that right.[14] The commentators, though, have gone beyond the bald doctrine and have enlisted Justice Benjamin Cardozo's seminal opinion in *Palsgraf v. Long Island Railroad*

Co.[15] Succinctly, Mrs. Palsgraf was injured when a scale on a commuter train platform fell over. The scale fell as a result of the explosion of fireworks that were contained in a package that another commuter was carrying when a railroad employee, attempting to help the owner of the package onto the train, pushed the package's owner, causing the package to fall and the fireworks to explode.[16] Starting from the premise that the railroad employee's action was negligent, the issue concerned the operation of the proximate causation requirement.[17]

Negligent defendants are liable, responsible in the tort sense, for the consequences proximately caused by those actions. If *A* drives negligently down a residential street and *B*, wholly oblivious to that act, trips over the rug in his living room and is injured as a result, it is clear that *A*, though negligent, will not be liable to *B* for *B*'s loss. It is the relation between (or, rather, among) *A*, and *B*, and the injury-causing event that is the concern of the tort law: It is not sufficient that *A* is negligent and *B* coincidentally suffers an injury; tort law only intercedes when there is a connection, a *sufficiently* direct or proximate connection between *A*'s action and *B*'s injury. That is what proximate causation requires, and tort liability requires proximate causation. Cardozo's precise language enlightens: "The conduct of the defendant's guard, if a wrong in its relation to the holder of the package, was not a wrong in its relation to the plaintiff, standing far away. Relatively to her it was not negligence at all. Nothing in the situation gave notice that the falling package had in it the potency of peril to persons thus removed. Negligence is not actionable unless it involves the invasion of a legally protected interest, the violation of a right. 'Proof of negligence in the air, so to speak, will not do.' "[18] Weinrib explained Cardozo's reasoning in terms that resonate with the corrective justice interpretation of tort doctrine by contrasting Cardozo's majority opinion in *Palsgraf* with Andrews's dissenting opinion. Cardozo's opinion was about duty and the relational basis of duty in the tort law that tracks, it would seem, the duty contemplated by corrective justice:

> Cardozo's majority judgment treats wrongdoing and the resulting injury as intrinsically unified; Justice Andrews' dissenting opinion treats them as disconnected requirements that can be independently satisfied. . . . Cardozo held that because the defendant's negligence was not a wrong relative to her, the plaintiff could not recover. Andrews, dissenting, held that the duty to avoid creating unreasonable risks is owed to the world at large, not merely to the person who might be expected to be harmed. . . . Cardozo's majority opinion emphasizes the relational quality of negligence.[19]

As Cardozo most pithily (and famously) observed in the same opinion: "The risk reasonably to be perceived defines the duty to be obeyed, and risk imports relation; it is risk to another or to others within the range of apprehension."[20] Leaving aside the obvious question "why," what there is about relation that connects negligent action and injury in a normatively significant way, the point for present purposes is that an interpretation of tort in terms of Cardozo's relational theory can do some positive work: It explains why defendants pay damages to plaintiffs rather than, say, to the state, and why the source (and limit) of the plaintiff's recovery is that payment rather than some insurance fund for those who are the victims of misadventure.[21] It is the breach of an existing duty defined by relation that supports the plaintiff's right to recover from the defendant and the defendant's obligation to the plaintiff: "because the plaintiff's right is the ground of the duty that the defendant breached, the parties are intrinsically united in a single juridical relationship."[22]

If Andrews's construction of the duty requirement were correct, Weinrib reasoned, it would be incoherent to measure the plaintiff's right and measure of recovery and the defendant's obligation and liability by the same calculus: It would, that is, be serendipitous in the extreme were the measure of plaintiff's injury the same as the measure of the deterrence necessary to discourage the particular defendants and similarly situated actors in the future from injurious activities. But Weinrib found a role for corrective justice: "[Andrews's] failure to integrate injury and wrongdoing brings into question the appropriateness of entitling this plaintiff to damages from this defendant. Andrews treats injury as singling out the plaintiff, and wrongdoing as singling out the defendant. The difficulty lies in finding a basis for joining these two parties, out of all those who suffer injury or commit negligence, in one lawsuit that makes this particular defendant liable for this particular plaintiff's injuries."[23] Corrective justice provides that basis. It is the relation between plaintiff and defendant that connects the moral dots in a way Andrews's understanding could not.

Weinrib enlisted corrective justice, and its focus on relation, to provide a noninstrumental alternative to instrumental, economic, normative theories of tort. It could be that Cardozo had such a relational construct in mind, though it does not seem particularly important why it should make much difference whether *he* did or did not (just as likely, perhaps, may have been Cardozo's desire to limit the reach of negligence liability generally, for broader social or economic reasons).[24] At the end of the day, it may be just as likely that Andrews's view was more correct if more correct means better at predicting the future. Certainly in a world where strict product liability makes particular plaintiffs foreseeable to particular defendants only rarely, the fabric of the relation Cardozo had in mind and Weinrib relied upon to rationalize all of tort does seem a bit strained. In any event, it may be just as likely that tort does not admit any type of corrective justice explanation: It may be the way it is just on account of a series of historical accidents.[25] But that may be an inconsequential quibble: The point remains that corrective justice principles, refined, have provided the foundation of the dominant noninstrumental theory of tort.

Weinrib based his corrective justice vindication of tort on Kant's construction of the original Aristotelian idea: "I situate corrective justice within Kant's philosophy of right. For Kant as for Aristotle, corrective justice is the justificatory structure that pertains to the immediate interaction of doer and sufferer. Kant, however, differs from Aristotle in presenting corrective justice not as an isolated category but as part of a ramified legal philosophy. His treatment therefore enables us to see the place of corrective justice within its family of associated concepts."[26] Weinrib's interpretation of tort doctrine could be right: Tort may focus on relation in the way understood by Cardozo and vindicated by Kant's reading of Aristotle. Weinrib's theory offers both a positive and a normative explanation of the doctrine. Cursory consideration of the role Kantian deontological theory plays in that theory will support a critique from neuroscientific premises.

Corrective Justice and Kant

Weinrib acknowledged his dependence on notions that the neuroscientific critique of folk psychology and vindication of materialism render incoherent: "Kant locates the conceptual roots of corrective justice in the *free purposiveness* of *self-determining* activity. He thereby connects corrective justice to his obscure but powerful analysis of the process of willing. The equality of corrective justice turns out . . . to be the equality of free wills in their impingements on one another. In the Kantian view, such equality is normative because it reflects the normativeness intrinsic to all *self-determining* activity."[27] Corrective justice depends on Kant's "free purposiveness," "self-determining," and "free wills." The focus, in fact the animating principle, of institutions such as tort is the idea, the reality, of "self-determining agents":[28]

"Legality, when conceived in Kantian terms as an idea of reason, is the articulated unity applicable to the external relationships of freely willing beings."[29] Weinrib described Kant's free will in libertarian (not in the political sense) terms and quoted Kant to that effect: "By 'the practical,' I mean everything that is possible through freedom. . . . A will is purely *animal* (*arbitrium brutum*), which cannot be determined save through sensuous impulses, that is, *pathologically*. A will which can be determined independently of sensuous impulses, and therefore through motives which are represented only by reason, is entitled *freewill* (*arbitrium liberum*), and everything which is bound up with this will, whether as ground or as consequence, is entitled *practical*."[30] So the premise of Kant's idea of reason and, in turn, the basis of Weinrib's corrective justice is the "conception of the will as free."[31] And Weinrib fixed that foundation in terms that reject the determinism that monism establishes: "Freedom of the will is for Kant [and, we may assume, Weinrib] what most sharply distinguishes purposiveness activity from the passivity of a sequence of efficient causes."[32] In fact, the terms of Weinrib's elaboration are redolent of the supernatural, or at least of homunculi: "For purposive action to be *free*, it must have the capacity to abstract from the immediacy of inclination, to reflect upon the content of the mental representation, and spontaneously to substitute one representation for another. . . . [T]he purposive being—although *affected* by inclination, which can suggest a content for action—is not *determined* by inclination and is therefore not in the coercive grip of any particular representation or object of desire."[33] Though some room is left for inclination, the will is still free; and here it becomes a bit difficult. Weinrib bifurcated Kant's conception of free will: "free choice . . . as independence from determination by sensuous impulse, and practical reason . . . as the determining ground of purposive activity."

Now the first constituent, "independence from determination by sensuous impulse," is easily understandable, though fallacious. The second constituent is more opaque. According to Weinrib, it seems essentially contractarian as universally valid maxims: "the determining ground of free activity is not the *content* of any particular purpose . . . but the very *form* of purposiveness as a causality of concepts."[34] Weinrib then offered as "the most general expression of this formalism," the categorical imperative: "'Act upon a maxim that can also hold as a universal law,' which entails at a minimum that one's reason for acting be capable of being conceived in universal terms without contradiction."[35] I will leave critique of the categorical imperative and Kantian morality generally to others;[36] the object here is more modest: to formulate the foundation of corrective justice in the Kantian theory upon which Weinrib relied.

Weinrib further described the two aspects of free will as interrelated: "Free choice and practical reason can both be defined in terms of each other: free choice is the capacity for determination by practical reason rather than by inclination, and practical reason is free choice determining itself as a causality of concepts."[37] So, it would seem, if either of those two premises is infirm, reliance on Kant's conception of freedom would fail, or at least be ill placed. It might still, though, provide a positive account of legal doctrine, at least as applied (and as Weinrib would apply it) to legal institutions, such as tort.

Weinrib's account of Kant's "practical reason is free choice determining itself as a causality of concepts" seems obscure, but the fault may be with Kant rather than Weinrib. Kant, of course, had something of an excuse: He was writing in the eighteenth century, about three hundred years before fMRI and similar technological innovations. For Kant, the brain was a black box in ways that it would not (need to) be for Weinrib. We can only imagine where (or even whether) deontology would be today had Kant not proceeded from essentially supernatural premises that would have seemed more plausible when bloodletting did too. (Bloodletting fell out of fashion in the late nineteenth century, but it was in use for about two thousand years before then.)[38]

The portion of Kant on corrective justice Weinrib utilized to make sense of the legal institution of tort depends on free will, of the libertarian strain. That is, Kant's theory is correct, in the sense of apposite to human agents, only if human agents have free will. So if Kant was wrong about that, then Weinrib's invocation of Kant to make the normative argument in favor of corrective justice would necessarily fail, though his reliance on Kant to posit the corrective justice positive theory of tort could remain plausible. There is no reason to believe that extant tort doctrine understands human agency any better than did Kant.

Now it is difficult (and of limited utility) to suggest that any unitary theory, positive or interpretive, provides *the* accurate depiction of tort or any other area of the law. But there is, nonetheless, much to commend any interpretation that makes sense of the cases, particularly leading cases such as *Palsgraf*. That commendation is even more clearly deserved when the interpretation makes sense of the fundamental apparent incongruities of the doctrine: specifically why defendants' liability is measured by harm to their victims and why negligent actors are liable only if there is an injured victim. Corrective justice, wherever it comes from and however plausible its foundation, does offer an interpretation of tort duty that can tell a coherent story about the doctrine. That, of course, does not necessarily commend the doctrine, but it does provide a gloss that could at least fill the gaps when instrumental interpretations fail.

Coleman also wrote in the corrective justice tradition and focused our attention on the nature of the action the defendant must have taken in order for tort principles animated by corrective justice to support (in the normative sense) the imposition of liability. It is worthwhile, then, to take account of Coleman's contribution so that we may get closer to the ultimately crucial moral responsibility paradox of the doctrine.

Coleman's Corrective Justice

Jules Coleman recognized that instrumental theories, specifically efficiency and market theories (not the same thing),[39] explain a good deal but not all of tort. It is when, perhaps within interstices, instrumental theory fails that noninstrumental, moral theory intervenes to provide rules of decision. Markets do well when conditions are favorable to market-based transactions; morality matters, is dispositive, when markets fail: "Morality constrains the extent to which individuals are free to act on the basis of utility-maximizing reasons. . . . It follows from the fact that morality is irrational under the conditions of competitive equilibria that moral constraints are rational only under conditions of market failure. . . . Morality, like law and political association more generally, is a solution to the problem of market failure."[40] For Coleman, then, economic analysis can make sense of a good deal of the tort law; indeed, economic analysis justifies strict liability: "The theory of strict liability is a response to both the failure of the retributive theory of fault liability and to the success of the economic analysis of it."[41] It is strict liability that demonstrates "that liability and recovery in torts do not always depend on fault."[42] But strict liability is not all of tort; it is just the part that is amenable to interpretation in terms of economic analysis; strict liability works where (and because) markets work.

Outside that market context, though, fault is the foundation of tort, and it is here that Coleman confirmed the operation of the "moral" principles Weinrib discovered in Kant: "Fault is central both to the institution of tort law and, in my view, to its ultimate moral defensibility. The principle of justice that grounds the fault rule is corrective justice."[43] If fault is not part of, indeed, is not the very foundation of tort, then tort is normatively incoherent. What supplies that coherence, then, is corrective justice, as construed by Coleman, and the role of fault in that construction.

Coleman's corrective justice seems, at least for present purposes, to be Weinrib's corrective justice,[44] and Coleman reported that Weinrib convinced him that relational responsibility is at the heart of that conception: "[C]orrective justice provides particular persons with reasons for acting, and it is that fact about it that distinguishes corrective [justice] from distributive justice. Without the imposition of duty or *responsibility*, corrective justice is, at best, reducible to one or another form of distributive justice. . . . [C]orrective justice . . . specifies a framework of rights and *responsibilities* between individuals. In the relational view, it is the wrong . . . that must be annulled, that specifies the content of the relevant duty."[45] It is the wrong in a relational sense that triggers the right under corrective justice. So Coleman's corrective justice needs (at least a sense of) wrong to found liability; it is a moral imperative of corrective justice every bit as important as the relational focus of the tort law. It is wrong and relation that must be coincident for the imposition of tort liability. And Coleman's "mixed conception"[46] relies on a sense of responsibility related to the wrongdoer's agency: "the duty to repair those wrongful losses is grounded not in the fact that they are the result of wrongdoing, but in the fact that the losses are the injurer's *responsibility*, the result of his agency."[47] We need to make sense of both the nature of wrongfulness and responsibility in Coleman's rendition of corrective justice.

Wrongfulness

Now *wrong* seems to both denote and connote a moral failing, something like moral responsibility, which chapter 3 of this book exposed as inapposite in the case of criminal responsibility of determined human agents (which is to say *all* human agents). So if Coleman's sense of wrong in fact tracked the moral responsibility sense, his normative theory of tort doctrine would fail in being based on an inauthentic conception of human agency (though his positive account of tort doctrine in such terms may well be as accurate as any attempt to depict tort could be). But if he understood wrong and wrongfulness in *a*moral terms, then he may have found a sense of wrong that can coexist, even cooperate, with the authentic sense of human agency neuroscientific principles would vindicate.

Coleman, Hershovitz, and Mendlow, writing in the *Stanford Encyclopedia of Philosophy*, explained corrective justice in such amoral terms, or, at least terms that do not rely on moral desert:

> Corrective justice theory—the most influential non-economic perspective on tort law—understands tort law as embodying a system of first- and second-order

duties. First order duties prohibit conduct (e.g., assault, battery, and defamation) or inflicting an injury (either full stop or negligently). (Some theorists believe that corrective justice has nothing to say about the character of these norms; others think that it helps define their scope and content.) Second order duties in torts are duties of repair. These duties arise upon the breach of first-order duties. That second-order duties so arise follows from the principle of corrective justice, which (in its most influential form) says that an individual has a duty to repair the wrongful losses that his conduct causes. For a loss to be wrongful in the relevant sense, it need not be one for which the wrongdoer is morally to blame. It need only be a loss incident to the violation of the victim's right—a right correlative to the wrongdoer's first-order duty.[48]

But can conceptions of corrective justice, such as those of Weinrib and Coleman, based, as they apparently are, on Kantian notions, really capture amorality that correctly reflects the materialism of naturalistic normative approaches vindicated by neuroscientific insights? It would seem to be something of a challenge.

The presentation of Weinrib's thesis above, recall, relied on free will. Weinrib construed Kant as actually integrating a conception of the human agent in terms of free will "as independence of determination by sensuous impulse."[49] Further: "When purposive activity is free, the purposive being is linked to its particular purpose by a rational operation and not by the imposition of sensuous impulse."[50] There is, of course, ultimately no essential material difference between rational operation, the operation of neurons, and sensuous impulse, the operation of neurons. One is not normatively different from the other; or, if they are, nothing in the Kantian theory utilized by Weinrib demonstrates the difference. The materialism neuroscience confirms leaves no room for such independence: We *are* the products of what Weinrib, after Kant, described as sensuous impulse. There's no other mysterious stuff; it is all sensuous impulse insofar as sensuous means what it only can mean: the *material* that defines and determines who we are and how we act. There is no room for moral blame in that accurate account of human agency, and so any normative theory that depends on the immaterial, the nonsensuous, is fantasy, supernatural (pretty much by definition).

And the whole normative point of positing free will, at least as it relates to the law, is to fix the contours and limits of moral responsibility. If there is no moral responsibility, there is no need for free will; just as if there is no free will, there can be no moral responsibility. Indeed, following

Weinrib, we can see how both can be defined in terms of each other: A morally responsible agent is an agent that has free will, one whose actions (including thoughts and all cognitive processes) are not determined (in the incompatible deterministic sense), and in order for an agent to have moral responsibility, the agent must have free will (in the libertarian, or perhaps compatibilist, sense).[51] Weinrib concluded that, per Kant, "The meaning of normativeness is precisely the determination of free choice in accordance with its own nature."[52] That certainly seems right, but it assumes a free choice that is not consistent with human agency, and so any theory built on that fiction would be infirm. Finally, Weinrib posited his (and Kant's) corrective justice precisely in terms of free will: "Freedom of the will, the integration of free choice and practical reason, is the principle that unites the various aspects of the practical idea of reason into a network of conceptual independencies."[53] Would there be, though, a way to build corrective justice without invoking free will, as Coleman's *Stanford Encyclopedia* entry suggests?

Coleman distinguished "fault in the doing" from "fault in the doer" and concluded that tort relies only on the former: "the standard of fault in torts is that of fault in the doing. The question is whether corrective justice implicates a similar notion of fault or wrongdoing. It does."[54] Culpability is not part of the calculus: There can be wrongful action in the absence of moral culpability, of blame.[55] So it certainly seems as though Coleman would be on the right track, or at least on a track that would not depend upon the Kantian free will posited by Weinrib. But the path is not as clear as it might be and seems confounded when Coleman tried to elaborate on wrongness that is not morally culpable: "The sort of wrongdoing required by corrective justice is objective. It does not require moral culpability or blame. It requires only that the conduct in question fail to comply with the relevant or appropriate norms of conduct. In the typical case, this failure is the failure to exercise reasonable care; it is a failure to abide by governing community norms. Because wrongdoing is objective in this sense, culpability-deflating excuses are normally irrelevant to the demands of corrective justice."[56] The problem with moral responsibility in an incompatibilist, determinist conception of human agency is that there is no sense in which "I" am responsible *in a moral sense* for what I do for either (or both) of two reasons: (1) there is no "I"; there is only a locus of apparently coordinated (at least coincident) activity; and (2) even if there is an "I," that locus of coordinated activity or something else (more), that entity, that locus, is merely the product of forces acting upon it that it does not control in any moral sense. There must be moral control before there can be moral responsibility. (Keep in mind that that reasoning in no way undermines instrumentalist responses to actions that result in injury; there still can be fault and damages, as long as such serve only instrumental purposes: That would not be immoral.)

It may be the case that one fails to act in a manner consistent with community norms and that we could decide that the law does impose liability therefor on actors irrespective of their lack of moral responsibility. Nothing obvious in the reference to community norms necessarily implicates moral responsibility. My car can violate norms of automobiledom when the steering fails (part of what it means to be a functioning car) without any moral opprobrium attaching. But does Coleman's sense of wrong nonetheless cross the line and posit a sense of corrective justice that relies on moral responsibility in ways that fabricate a human agency that does not exist (other than in Aristotelian and Kantian imaginations)? To answer that question, we need to appreciate what Coleman's theory understands responsibility to be.

Responsibility

There are two aspects to moral responsibility: First, the term contemplates a normative failing and so assumes an agent who can make moral choices. Second, the term distinguishes mere causal responsibility, in the way the cue ball is responsible for the eight ball's going in the corner pocket when the cue ball strikes it (after having been struck by the cue stick, which was propelled by the player, who . . .). So we need a sense of connection between wrong and responsibility that is (as Coleman termed it) objective, a responsibility that "does not require moral culpability or blame."[57] Coleman applied that description of *objective* to the "sort of *wrongdoing* required by corrective justice"[58] in order for it to be objective. It just stands to reason that if morality is removed from the sense of wrongdoing, it could not be reinserted in the responsibility calculus without undermining the objective nature of liability in corrective justice. So Coleman's responsibility must be similarly objective, insofar as what he means by objective is amoral.

It is important to recognize that the reason for tort theorists to distinguish responsibility in the objective, merely causal, sense from moral responsibility is to make clear that tort does not require the same moral responsibility that retribution, in the criminal context, would require. Coleman et al. explained:

Many theorists believe that a principle of retributive justice—say, that the blameworthy deserve to suffer—does a good job of interpreting and justifying criminal law. Yet most theorists think that such a principle does a rather poor job of interpreting and justifying tort law (except, perhaps, for the part of tort law concerned with punitive damages). First, tort liability does not communicate condemnation, since (as explained above) a defendant can be liable in tort even though he did nothing blameworthy. Second, the duty of repair in tort is treated as a debt of repayment, in that it can be paid by third parties—and not just when the creditor (the plaintiff) has authorized repayment. By contrast, "debts" incurred as a result of criminal mischief can never be paid by third parties. You cannot serve my prison sentence. Third, a person cannot guard against liability to criminal sanction by purchasing insurance. Yet it is common to purchase insurance to guard against the burdens of tort liability. Indeed, in some areas of life (e.g., driving), purchasing third-party insurance is mandatory.[59]

But is that objective responsibility amoral in the sense that would make it coherent from the materialistic perspective (and so consistent with neuroscientific insights) or just a different form of moral responsibility from that supporting retribution? It is clear that tort does not depend upon retributionary principles for the reasons Coleman et al. rehearsed. That alone, though, would not wrench all morality from the sense of responsibility corrective justice theory does incorporate.

Coleman explained his sense of the requisite responsibility in terms of three cooperative (or at least coincident) elements:

> In the case in which an injurer is responsible for another's loss as a result of his wrongdoing, his responsibility depends on the truth of the following proposition: the victim's loss is his fault. The sentence, "*P*'s loss is *D*'s fault," is true only if the following set of propositions is true:
> 1. *D* is at fault.
> 2. That aspect of *D*'s conduct that is at fault is causally connected in the appropriate way to *P*'s loss,
> 3. *P*'s loss falls within the scope of the risks that make that aspect of *D*'s conduct at fault.
> A loss cannot be someone's fault unless his conduct is in some way at fault.[60]

But what constitutes "at fault"?—mere causal connection in the physical sense? In what way can a determined agent be "at fault"? And what to make of the qualifier "in the appropriate way"? Those operators appear to be doing some normative work, but without elaboration, they are no more than place savers. Perhaps that was Coleman's intent.

It would seem, from the sense of fault and wrongdoing that emerges from Coleman's theory, that corrective justice necessarily implicates a non-instrumental normative theory. But Coleman claimed it does not: "Corrective justice depends on a substantive theory of wrong or wrongdoing, but its point or purpose does not. In that sense, it can remain an independent principle, not swallowed up by the norms it sustains or protects."[61] (So that sense of wrong could be fixed by instrumentalism: an efficiency or market-based theory?)[62] Yet then Coleman went to lengths to explain why corrective justice *needs* a moral foundation: "Ought we to conclude that corrective justice is compatible with any underlying theory of wrongdoing? I think not. Its moral independence assured, the question before us is whether corrective justice itself constrains the norms it can sustain. I argue that it does, and in important and interesting ways."[63] For Coleman, corrective justice is based on morality: "corrective justice imposes moral reasons for acting."[64] We get a sense of what these moral reasons could not be; they could not be purely instrumental: "In order for the principle of corrective justice to apply to the underlying system of rights sustained by its application the rights must be such that they are worthy of protection against infringement by the actions of others, even if they would not be protected against infringement by state action designed to replace them with the set of entitlements which could be defended as required by the best theory of distributive justice."[65] Because the rights are defensible irrespective of the best theory of distributive justice considerations, they are moral in a noninstrumental sense: "within limits, the rights in place [and vindicated by corrective justice] may help to sustain an institution that generally improves individual well-being and social stability, and that does so in ways that encourage individual fulfillment, initiative and self-respect."[66] The calculus of "individual fulfillment, initiative and self-respect" may be obscure, but it does suggest that the operative sense of morality is noninstrumental, or only instrumental in the strictest sense, a sense that would seem to defy objectification. It is not immediately clear how those three constituents—individual fulfillment, initiative, and self-respect—could be objectified across human agents, or whether we would even want them to be.

For present purposes, though, it is enough to demonstrate that Coleman's description of the corrective justice thesis apparently tried to avoid reliance on responsibility in any more than the causal sense: It would suffice that the defendant's action (or inaction) was the physical cause of the wrong. But the underlying wrong would have to be a *moral* wrong: the defendant's violation of the plaintiff's rights to individual fulfillment, initiative, and self-respect. The defendant's behavior need not be blameworthy, but it must infringe those rights (or something like them, we imagine). In that way we might find moral responsibility in Coleman's corrective justice. It is just not clear.

Keep in mind: It was not the object of either of the commentators considered so far to remove the normative focus of tort from matters of moral blameworthiness for the sake of developing a coherent normative argument of what tort doctrine should be if the law were to correctly recognize that human agents, as tortfeasors and their victims, are not morally responsible. Weinrib based his corrective justice interpretation of tort in terms of Kantian free will and Coleman's corrective justice too depended on a morality inherent in it: recognition of rights to individual fulfillment, initiative, and self-respect, seemingly noninstrumental objects that the materialism revealed by neuroscience and captured by determinism cannot corroborate. Coleman's perspective, though, may make less room for the type of moral responsibility dependent on control that neuroscience would challenge. But his analysis is not a model of clarity in that regard; perhaps we should not expect it to be, his focus being as it is.

It is worthwhile to consider now the civil recourse theory refinement (or partial repudiation) of corrective justice, which strains even more mightily than Coleman to remove moral responsibility from tort. And presentation of the civil recourse response to Coleman may also cast light on the role of moral responsibility in Coleman's corrective justice theory.

Civil Recourse

Civil recourse theory endeavors to provide a noninstrumental interpretive and normative theory of tort: It both explains what tort is actually up to and describes how that object is consistent with a normative, albeit not *necessarily* moral, vision. Civil recourse theory is either a corrective to responsibility-based normative theories of tort[67] or the source of a gloss that may be imposed on corrective justice theory to better explain what can be correct about a "corrective justice" theory when tort, according to advocates of civil recourse theory, is not really so much about correction or justice.[68] The fit between corrective justice and civil recourse is manifest in the two perspectives' appreciation of the relational nature of tort liability and its elucidation in *Palsgraf*. Zipursky has explained Cardozo's opinion in terms that suggest the contours of civil recourse:

> The central point of Chief Judge Cardozo's *Palsgraf* opinion is that a defendant's failure to use due care must have been a breach of the duty of due care owed to the plaintiff; the breach and duty elements of the negligence claim must fit together in the right way. The opinion infers this requirement from the broader principle that a plaintiff may not sue in tort for a wrong to another, which itself flows from the idea that a tort claim is fundamentally a private right of action to redress a wrong to oneself. Chief Judge Cardozo utterly rejected the sort of private attorney general conception of tort law that has become prevalent in contemporary tort thinking. So long as scholars and students reading his *Palsgraf* opinion resist his private-law mindset, they are doomed to misunderstand what the opinion actually says.[69]

Zipursky then emphasized that tort "is fundamentally a private right of action to redress a wrong to oneself," and so it depends on a concept of wrongness. But wrongness need not connote any blameworthy action or failure to act; wrong for civil recourse theory can be the same type of wrong that could support at least Coleman's corrective justice: violation of a plaintiff's right, whether that violation was the result of culpable behavior in any sense whatsoever, that is, whether or not the defendant was morally responsible.

In an article focusing on the sense of wrong upon which civil recourse relies, Goldberg and Zipursky distinguished the necessarily moral wrong that is the subject of criminal sanction from the amoral wrong that supports tort:[70] Tort does not require the type of wrong that supports the imposition of blame; tort contemplates a legal wrong but *not necessarily* a moral wrong.[71] So "The fact that an act falls under an authoritative legal directive that characterizes it as a legal wrong does not entail that such an act, in the circumstances it actually occurred, warrants categorization as morally wrongful."[72]

The authors' object was to distinguish the relational wrong-based account of torts provided by a civil recourse theory elaboration of corrective justice from noninstrumentalist loss-based constructions, such as those offered by Perry[73] and, they argued, by Coleman in his *Risks and Wrongs*.[74] If "wrong" in civil recourse invoked moral violations, then the distinction between the scope of criminal law and tort law would be compromised.[75] And if the basis of tort were loss rather than wrong, civil recourse theory

would be dubious because the relational basis of tort would be undermined: There just would have been no reason why Cardozo was right
and Andrews incorrect in *Palsgraf* if the object of the tort law were loss
allocation. Recall that Weinrib too located (or, at least, discovered) the
normative foundation of corrective justice in tort in Cardozo's opinion, a
conclusion confirmed by Zipursky's reliance on the same majority opinion
to found civil recourse.

Much depends on conceptions of blame, and we can better understand the role of moral responsibility in a normative theory of tort such as
civil recourse when we come to terms with the theory's understanding of
blameworthiness.

Blame, Blameworthiness, and Resentment

Closer inspection confirms that the civil recourse interpretation of tort
invokes blameworthiness, so may, then, depend on a conception of human agency that neuroscience reveals as inauthentic. In order to base tort
on wrong rather than loss, civil recourse must explain the normative significance of the injury. From Goldberg and Zipursky's civil recourse perspective, wrong just works better than loss as long as we understand "wrong"
in the right way. Certainly the negligent action that results in harm to
the plaintiff may be, from all accounts, indistinguishable from the same
defendant's negligent action that results in no harm. The distinction, so
far as tort is concerned, focuses on the loss (or lack of one); the wrong
seems to be the same. In both events the defendant may have been driving negligently: In one event the defendant's car struck a child and in the
other it did not. The difference was a matter of luck, nothing more. So in
order to explain why civil recourse (and, presumably, corrective justice)—
focused as they are on wrong rather than loss—would support the imposition of liability in one case and not the other, Goldberg and Zipursky
had to discover and demonstrate a normative difference between the two
settings: Why should the law care about the one wrong accompanied by
loss and not the other wrong without loss? Their answer? The two wrongs
are not the same. And how are they different? Blameworthiness. This is
tricky, and so reproduction of their argument at length is appropriate,
even necessary:

> [I]n tort law it is particularly clear that a defendant's vulnerability to an action
> by the plaintiff should turn on whether the defendant actually injured the plain-
> tiff, for the injury is intrinsic to the wronging of which the plaintiff complains.

From the plaintiff's perspective, it is not correct to say that there just happens to have been a conjunction of her loss and wrongful conduct by the defendant: In her eyes the defendant's wrong is mistreating her or interfering with some aspect of her well-being (or failing to protect or assist her in ways that would have prevented her from suffering a certain kind of setback). More importantly, the court's obligation to provide an avenue of civil recourse against the defendant hinges on the defendant having wronged the plaintiff in a manner that renders her a victim entitled to respond to the wrongdoer.

Those puzzling over the relevance of harm to degree of responsibility have always recognized the fact that the victim of the harm will be naturally disposed to feel differently towards the wrongdoer when the wrongful conduct ripens into harm than she would if no harm ensues. What they have wondered about is why it should make a difference to how the wrongdoer's acts are categorized and evaluated from a more objective perspective. As to this question, tort theory helps moral theory. One of the things we are asking about when we evaluate someone's conduct is what acts he has done. And there is no ground for insisting that a classification that abstracts from results carves at the normative joints—often, in fact, the opposite may be true. A morally significant aspect of what an actor has done is whether his acts—described in a result-inclusive way—are ones that another person could fairly demand that he be held accountable for. That is, even assuming that the increased resentment felt by the victim is not itself to be converted into an attribution of greater blameworthiness to the author of the injurious act, tort law helps us to see a distinct but related point. The question of whether a defendant's blameworthiness is greater is a question of whether the degree and nature of the resentment (not improperly) felt by a victim of the result-inclusive wronging is greater than that of a victim of a harmless wronging. A heightened degree of blameworthiness does not necessarily entail an increased level of wrongfulness, but it may reflect an increase in the level of blame by others to which a third party (like the state) would regard the wrongdoer as properly vulnerable. To say that an actor could reasonably be resented to a greater degree is not to say that there was some respect in which the actor's conduct ought to be deemed more wrongful; on the other hand, increased blameworthiness in the sense of increased grounds for resentment may indeed be an attribute of the actor's actions, not simply a reification or projection of the spontaneous or natural reactions of others.[76]

If the distinction Goldberg and Zipursky suggested does not maintain, then their civil recourse interpretation of tort fails, both as a positive and normative matter. So does the distinction hold up? The perspective vindicated by neuroscientific insights may inform the analysis.

Goldberg and Zipursky's discovery of something of normative significance in the negligence that does result in an injury that is not present
when the same negligence does not result in an injury is, perhaps, elusive.
Goldberg and Zipursky grounded civil recourse, and so their elaboration
of corrective justice in tort, on the wrong rather than on the loss that results from the wrong. Indeed, they went to some lengths to distinguish
their perspective from Coleman's, particularly because Coleman focused
so much of his attention on loss:

> Coleman himself goes out of his way to insist that tort law is fundamentally
> about losses, not wrongs. In his mind, tort law distinguishes itself from criminal
> law precisely on this score. If law is going to respond to wrongs qua wrongs, he
> says, it should be in the business of punishment. Tort law, by contrast, shifts
> losses. It is true that, for Coleman, as for Perry, Epstein, and Ripstein, the deter-
> mination of when a loss is to be shifted hinges on the identification of grounds
> for holding the defendant morally responsible. Still, the type of responsibility
> that generates tort liability is the moral responsibility for a loss one has caused,
> rather than responsibility for having committed a wrong, a point that Stephen
> Perry makes clear by invoking Honoré's notion of "outcome-responsibility"
> and distinguishing it from responsibility for actions. Instantiating the principle
> of corrective justice, tort law specifies that an actor's having caused a loss to an-
> other, and having done so by means of conduct that falls short of an applicable
> moral standard, is a sufficient reason for deeming the loss to be the defendant's
> moral responsibility and not the plaintiff's. The fundamental question to which
> the principle of corrective justice provides an answer is thus: Whose mess is it?[77]

Yet according to Goldberg and Zipursky, Coleman did expound a theory
of corrective justice attentive to the prominence of wrongness: "The duties one has in corrective justice arise as a result of wrong or wrongdoing,
not as a result of wrongful loss."[78] But his regard for loss, in the portion
of *Risks and Wrongs* with which Goldberg and Zipursky took issue, suggested, according to them, the shortcoming of a corrective justice perspective not informed by the insights of civil recourse theory. Goldberg and
Zipursky focused instead more prominently on the wrong[79] rather than
the loss, and loss itself was denied some normative significance, at least
initially. But then at the point in the argument when Goldberg and Zipursky have to find something to add to the wrong in order to distinguish negligence that causes injury from negligence that does not, they afforded loss
normative significance: Wrong alone could not do the normative heavy
lifting.

The distinction Goldberg and Zipursky fashioned focuses on the resentment felt by the victim, the extent to which the victim could blame the wrongdoer: And so blameworthiness is a constituent of the wrong itself, and not an insignificant constituent at that. It is blameworthiness ("in the sense of increased grounds for resentment") that distinguishes actionable wrongs from wrongs that are not actionable: But actual loss, apparently, has nothing to do with it. And, most pertinent for purposes of neuro-scientific inquiry, the determinant of blameworthiness is the emotional reaction of the victim: "increased grounds for resentment may indeed be an attribute of the actor's actions, not simply a reification or projection of the spontaneous or natural reactions of others."[80] That does seem to be alchemy: Somehow, what would be a reaction (an affective reaction) of the victim (or, rather, an affective reaction imputed to similarly situated victims generally) becomes the keystone, the support that distinguishes civil recourse from loss allocation theories of tort. But that fails.

It fails as a positive matter because there is no indication in the doctrine that the blameworthiness of the defendant is determinative of the viability of a tort action. Cardozo's opinion in *Palsgraf*, upon which corrective justice and civil recourse theories rely, did not excuse the Long Island Railroad from tort liability because its employee was less blameworthy, if blameworthiness was a function of the victim's resentment toward the employee's actions. If blameworthiness, as measured by degree of resentment felt by the victim, is a function of whether the victim suffered an injury (else how could she be a victim?), then Mrs. Palsgraf suffered an injury and so would have every reason to feel that resentment (or, at least, no obvious reason not to). There is nothing, then, in *Palsgraf* to suggest that blameworthiness, or appropriate resentment, was a constituent of the calculus. Mrs. Palsgraf was a victim of the railroad in the same sense in which the person *not* struck by a negligent driver would be a victim of that negligent driver. But Mrs. Palsgraf did not obviously therefor have no cause to *blame* the railroad; she did, after all, sue the railroad and pursued her action to the New York Court of Appeals. Her suit failed because there was not the appropriate *relation* between the actor's action and the injury. It would seem curious that blameworthiness could matter to support focus on wrongs rather than loss but not to determine the relation necessary to support tort liability. For, after all, the issue when a negligent action does not result in an injury is, at bottom, a relation issue: The negligent action was not related to the harm suffered *because there was no harm suffered*, no harm for the negligent action to be related to. So the positive case for the distinction Goldberg and Zipursky would draw is suspect, at best. It certainly compromises their perspective's reliance on *Palsgraf* as a seminal case.

Further, the focus on resentment resulting from blameworthiness would seem to invoke something like emotional harm, though it is (admittedly) difficult to be sure. Surely there are only two types of loss, physical and emotional, that could support tort liability, unless blameworthiness giving rise to resentment founds a third type of harm, perhaps a psychic harm (whatever that might be). It does seem as though Goldberg and Zipursky had in mind something at least more akin to emotional harm: "the victim of the harm will be naturally disposed to *feel* differently towards the wrongdoer."[81] Now it could be that that feeling is an indicator of something that had normative significance rather than something of normative significance itself: We "feel" someone is more blameworthy (has wronged us more?) only when they have wronged us in the normatively significant way; they have not wronged us in the normatively significant way only because we feel they are more blameworthy. (Or something like that.)

If, though, the distinction Goldberg and Zipursky offered is premised on some distinction in the affective reaction of the victim, then it would run headlong into tort law's persistent suspicion, or at least trepidation, insofar as pure emotional injury is concerned. The general rule is that the negligent infliction of emotional distress is not actionable. It is only actionable in certain exceptional cases in which we have good reason to appreciate the substantiality (and, indeed, actual existence) of the emotional injury.[82] So it would be curious if the foundation of a normative theory of tort relied upon a distinction that the doctrine generally rejects, or so it would seem.

Goldberg and Zipursky did specifically engage the tort distinction between physical and emotional harm and found it to be entirely consistent with civil recourse theory: "No one can dispute that tort law as it presently stands has adopted and maintained limited duty rules for negligently caused emotional distress—rules that, among other things, relieve employers from any general duty to take care against causing employees emotional distress."[83] Their argument was that the tort law's distinction between physical and emotional harm makes sense, or, at least, can be vindicated by civil recourse principles. So there must be something about greater blameworthiness born of resentment that is not inconsistent with that distinction, as a normative matter, else the law's distinction between the physical and emotional would be suspect, contrary to Goldberg and Zipursky's opinion on that matter.

But those portions of Goldberg and Zipursky's argument are not dependent on reasoning that the materialistic perspective would necessarily undermine; they are just curious potential non sequiturs in their analysis. Indeed, it might be that once neuroscience confirms that there is no such thing as nonphysical injury—even emotional injuries have a physical correlate[84]—the relation between wrong and loss posited by corrective justice and civil recourse theories will be clarified: Either the resentment-blameworthiness normative symbiosis would be cashed out as Goldberg and Zipursky suggested, or we would have reason to question their suggestion of some kind of quasi-emotional harm (loss).

Moral Responsibility and Civil Recourse

I suspect that Goldberg and Zipursky's argument ultimately is intended to be normative (or at least tends toward the normative pole of interpretation). Neuroscientific insights and the materialism that they vindicate make clear that moral responsibility, at least for determined human agents (and all human agents are determined in the incompatibilist sense), is a chimera: For the reasons outlined by commentators such as Bruce Waller,[85] Galen Strawson,[86] and Neil Levy,[87] there is no such thing as moral responsibility of human agents in anything like the libertarian free will sense assumed by Kant and theorists who would extend (or, at least, depend upon) his deontology to support a corrective justice and civil recourse normative justification for tort.

If A acts negligently and injures B, A is in no helpful sense morally responsible for the predicate negligent action. There is no essential A who could have that ultimate uncaused cause divinelike moral responsibility. Just as in the criminal context, though, we do not let the tortfeasor "get away" with his actions: The imposition of damages does play a deterrent role, both specifically and generally. That deterrent effect is watered down a bit, perhaps even considerably in some cases, by the availability of insurance, but insurers can and do encourage the type of loss mitigation strategies that deterrence of risky behavior accomplishes generally. Indeed, the prevalence of no-fault regimes even confirms that statutory modifications of the tort law can better focus civil liability and its adjudication on those losses of a sufficient magnitude to justify the cost.[88]

More significant for the instant inquiry, though, is the fact that civil recourse makes a very modest normative claim, and perhaps no unique moral claim at all. Its proponents (at least one of the leading proponents) go to some lengths to distinguish civil recourse from corrective justice.[89] The distinctions drawn concern the basis of the two perspectives' descriptive power. In fact, Zipursky maintained, civil recourse posits no guiding theory of morality:

> Now, the tort theorist need not develop a foolproof theory of morality or moral metaphysics, for whether we are dealing with the "true" morality, or even whether there is such a thing, is not really the point. The point is that our tort law—whether it can ultimately justify doing so or not—embodies the kind of moral principles and rationales just articulated. . . . The reasons for requiring wrongdoers to provide compensation are reasons entrenched in our system; they are reasons about the duties of repair owed by the wrongdoers (tortfeasors) to those whom they have injured. Moreover, when these duties of repair are dispatched—when the defendants pay the plaintiffs—a sort of justice is done. The wrongful injury is rectified when the defendant carries out the obligation to compensate the plaintiff. In this sense, courts applying the common law of torts see to it that corrective justice is done—that the defendant-injurer provides the compensation owed to the plaintiff-victim, thereby correcting the improper disturbance created, so far as possible.[90]

So civil recourse would not be a moral theory as such, though it does instantiate some benefit that could be normative, at least in an instrumental sense:

> Can a relational conception of duty be given a rational reconstruction within an intelligible framework of moral principle? We begin with the mundane observation that being a moral person involves, in part, constraining one's conduct in light of certain aspects of the well-being of others.
>
> Having a sense of duty is critical to being a moral person because it involves a recognition of the importance of acting in light of others' well-being. The existence of duties of care to others—parent's duty to his child, or a physician's duty to her patients—causes individuals to focus on certain aspects of the well-being of others. This enables individuals to prioritize certain aspects of their conduct. It also enables them to sustain and develop an internalized normative pull towards a certain set of actions. This is the feeling of being obligated in certain ways to those others—a parent's internal orientation to fulfill his duties to his child, or a physician's recognition of the necessity of doing what her patients' well-being requires. . . . Duties of care are a subset of relational duties more generally. They are relatively open-ended duties that take a wide variety

of shapes and forms depending on context and relationship. Because we recognize a wide range of duties of care in our society, each of us prioritizes certain needs of others and certain required courses of conduct in some manner, and each of us is motivated by this sort of pull to action. Duties of care enable us as actors to select courses of conduct for ourselves that are consistent with important aspects of others' well-being. They also enable us to sustain friendships, family relations, professional relationships, business contacts, employment relationships, and so on. That is not only because we could not remain in such settings if we failed to conform our conduct to the relevant norms—though that is largely true. It is, more deeply, because the creation of these relationships goes hand-in-hand with the cultivation of an internalized motivation and disposition to focus on another's interests.[91]

A human agent, however, is not morally responsible for "acting [or not acting] in light of others' well being." It would seem clear, then, that the imposition of liability, responsibility, without a coherent basis would undermine rather than serve general well-being. And to the extent that conceptions of tort liability, and the contours thereof, vindicate the imposition of moral responsibility when the nature of human agency does not support such a conclusion, the tort law fails as a normative matter. Now that does not mean that there is no role for tort to play; it will, certainly, be necessary to allocate loss in such a way as to encourage well-being, if that is possible. There is, though, no reason to believe that generalized duties of care, particularly those that rely on objective standards, would encourage well-being. Indeed, to the extent that legal doctrine allocates loss by reference to standards that serve administrative convenience while ignoring normatively significant idiosyncrasy, the law will fail to vindicate well-being.

The Dangerous Vacuity of Civil Recourse Theory

Civil recourse theory, then, may actually subvert morality by providing an apology for the normatively infirm status quo: The tort doctrine fails, in important ways, to allocate loss on a normatively defensible basis because, for example, it posits distinctions between emotional and physical injury,[92] draws lines that ignore normatively significance differences in capacity,[93] and allocates fault to victims by operation of contributory and comparative negligence principles while ignoring the actuarial nature of mass tortfeasor wrongdoing (and that certainly borders on the immoral, at least from the instrumental perspective).[94]

The normative naïveté of civil recourse theory was noted, albeit summarily, by Coleman et al.:

> Despite its explanatory power, civil recourse theory is vulnerable to a potentially serious objection—or else it seems to leave tort law vulnerable to such an objection. Because civil recourse theory offers little guidance as to what sort of redress is appropriate, the theory depicts tort law primarily as an institution that enables one person to harm another with the aid of the state's coercive power. Tort law may well be such an institution, of course. But if it is, it may be deeply flawed—indeed, it may be unjust. This problem can be posed in the form of a dilemma. Either the principle of civil recourse is grounded in a principle of justice or it is not. If the principle of civil recourse is grounded in a principle of justice, then civil recourse theory threatens to collapse into a kind of a justice-based theory. If the principle of civil recourse is not so grounded, then the principle apparently does no more than license one party to inflict an evil on another. If that is what the principle does, we might reasonably wonder whether it can justify or even make coherent sense of an entire body of law.[95]

Granted, that is not the objection that proceeds from the materialistic perspective that neuroscientific insights would support, but it is of a piece, and confirms that civil recourse theory *does indeed* offer a moral argument even by not expressly (or, at least, primarily) endeavoring to do so. If civil recourse theory would champion the extant doctrine and find moral justification for it, then civil recourse must confront critiques of the extant doctrine that demonstrate the doctrine's normative deficiencies. Coleman et al. pointed out that by failing to provide a limitation on (or, we might say, morally coherent justification of) "one person['s right] to harm another with the aid of the state's coercive power" civil recourse theory, or perhaps all of tort (if civil recourse theory *is* the accurate interpretation of the doctrine) is normatively infirm, if not incoherent. Any justice-based theory of liability would have to take account of human agency in a non-idealized fashion in order to do the right moral work the right way. So if tort doctrine's conception of human agency is wrong, as was suggested in chapter 4, then any apology for the doctrine fails either as a positive or normative matter. It would seem that the moral heavy lifting that civil recourse theory could do would require something of moral substance between bad and blameworthy, but that something may be suspect from the perspective that neuroscientific insights would inform. It is worthwhile, then, to take account of the morality that civil recourse would recognize.

Goldberg and Zipursky's (as of this writing) most recent elaboration offers a glimpse.

Goldberg and Zipursky have continued their civil recourse interpretation of tort law[96] by focusing on Benjamin Cardozo's opinion in *MacPherson v. Buick Motor Company*.[97] The opinion is canonical, both for what Cardozo said about the relation prerequisite to the imposition of tort liability (and its source) and for the weight the decision has assumed in the tort doctrine: It is construed by some—erroneously, Goldberg and Zipursky have concluded—as a crucial constituent of the fabric of strict products liability doctrine. For present purposes, though, the Goldberg and Zipursky piece is important for what it said about the moral valence of tort, a normative perspective that the authors seemed to eschew in much of their earlier work on civil recourse theory considered so far. Indeed, the argument of this installment seems to represent a primarily normative rather than merely positive interpretation of the doctrine.

Recall that *MacPherson* was a car case. The plaintiff was injured in an accident when the wheel of the new Buick he had purchased from an intermediate retail dealer broke.[98] The issue was presented in terms of privity: Insofar as Buick had sold the car to its retail dealer, rather than directly to the plaintiff, Buick argued that there was no privity between it and the plaintiff and so no available negligence action. Cardozo found a sufficient relation between MacPherson and Buick, notwithstanding the lack of contractual privity, to found the action. His reasons for doing so seem to be based on considerations of public policy: What sense would it make to limit Buick's liability for negligence to the retail car dealer, the one entity in the distributive chain that we could say with most confidence would *not* regularly be driving the car and exposed to personal injury on account of its failure? Subsequent courts and commentators have emphasized the public policy basis of *MacPherson*[99] and have used that focus to support an instrumentalist reading of the opinion.[100] It is, though, Goldberg and Zipursky admonished, crucial to keep in mind that Cardozo enlisted public policy to reach his conclusions on the *relationship* issue, and it is *relation* that is the focus of civil recourse theory, a *non*instrumental theory.

Goldberg and Zipursky's concern seemed to be with readings of *MacPherson* that would permit judges and juries to do social engineering along instrumental lines informed by policy objectives. Such reading would, Goldberg and Zipursky feared, deny the fundamental relational nature of negligence law. Reading *MacPherson* as the establishment of a policy-oriented nonrelational tort law would undermine the good sense Cardozo made in *Palsgraf*, good sense that, recall, animates corrective justice and civil recourse theory too. Such a misreading of *MacPherson* would result in vindication of instrumental normative theory that the corrective justice and civil recourse perspectives could not abide.

> Gilmore, Posner and others have felt compelled to offer implausible interpretations of *MacPherson* because they supposed (correctly) that Cardozo was a great judge, and further supposed (incorrectly) that one cannot be a great judge without being an instrumentalist about adjudication. The latter supposition is but one expression of the familiar, though deeply confused, thought that, since law is a human creation that serves human purposes, it can only be applied and analyzed instrumentally. It should go without saying that practices and institutions created by humans often serve human purposes indirectly.[101]

Their conclusion focused on Cardozo's attribution of responsibility, in what would seem more than just the causal sense: "Quite evidently, Cardozo's line of attack on the privity rule come[s] from within a moralized understanding of negligence as a legal wrong. The principle he puts forward is not about where costs are best placed to provide compensation or achieve deterrence, but about who is really *responsible* for an injury."[102] Goldberg and Zipursky have in mind a sense of wrong and responsibility that determinism questions: "it is implausible to read *MacPherson* as somehow downplaying the connections between negligence liability and ordinary notions of *wrongdoing and responsibility*."[103] So "ordinary notions of wrongdoing and responsibility," the very notions a materialist perspective vindicated by neuroscientific insights would challenge, is in fact the basis of negligence liability, at least if Goldberg and Zipursky's positive account of the doctrine is accurate. And it may well be. Indeed, for purposes of the instant inquiry—critique of the noninstrumental morality of tort doctrine—keep in mind that if Goldberg and Zipursky's positive account is right, the tort doctrine is indeed incoherent because it relies on an inauthentic conception of human agency, a conception that requires a sense of moral responsibility that is inconsistent with the nature of human agency.

Goldberg and Zipursky's ultimate concern was that if *MacPherson* were read as supporting an instrumental policy perspective, the imposition of negligence liability would be divorced from its moral underpinnings. It would then surrender to instrumentalist reasoning and in turn be subject to the type of social engineering, generally done by legislatures,[104] that Goldberg and Zipursky found to be particularly pernicious. There is, they maintained, room for moral and pragmatic thinking to make sense of the negligence law in the modern world, even as it pertains to the negligence liability of corporate entities: "[O]ne might be able to fashion a justification for strict retailer liability on the ground that a commercial seller's injuring of another person through the sale of a defective product is a distinct, strictly defined *wrong*—one that is difficult to avoid committing, sooner or later—within the family of legal wrongs that constitutes tort law."[105] That conclusion is curious: The authors have described strict liability in terms of a strictly defined wrong as "within the *family of wrongs* that constitute tort." So within that family must there be wrongs that are not strict, ones that it is not "difficult to avoid committing," ones to which moral responsibility would bind? There seems to be inconsistency between development of the idea of wrong here and development of the same idea in their earlier writing.

At the end of the day, you can define *wrong* as anything that causes harm, but it would seem to wring all the normative force out of the idea if a wrong is the kind of thing that is inevitable *no matter how careful you are*. Indeed, that would seem to be the problem with conceptions of wrong in tort generally, particularly in light of the vacuity of the idea of moral responsibility in the case of human actors. That is a conclusion that would be difficult for corrective justice theorists too, as a recent contribution by Heidi Hurd made clear.

There Is No Wrong without Moral Responsibility

Ultimately, then, corrective justice theorists such as Weinrib, and Coleman, and even Goldberg and Zipursky, who posit the civil recourse alternative to corrective justice, seem to require, or at least contemplate, that tort does moral work, imposes liability by reference to blame, on account of blameworthiness. The foregoing exposition has made clear that corrective justice and perhaps even civil recourse perspectives contemplate that the subject of tort law is the type of moral responsibility that insights supported by neuroscientific depictions of human agency cannot abide: Hard determinism leaves no room for moral responsibility. There is a recent, and particularly cogent, argument from within the corrective justice tradition that, without appreciating the hard determinism that neuroscience compels, has recognized the moral failure of the extant tort doctrine as well as the deficiencies of corrective justice. And we can here follow that argument to demonstrate the fundamental moral failure of the tort doctrine, a failure that the hard determinism vindicated by (or, at least, accommodated by) neuroscience confirms.

Heidi Hurd, an avowed corrective justice theorist, recently argued, quite convincingly, that the moral foundation of negligence liability is infirm.[106] Her thesis, in a nutshell, is that there is nothing immoral about inadvertence or bad character, and so negligence liability premised on either is effectively strict liability. And, because corrective justice can explain the basis of negligence liability but not strict liability in terms of the moral blameworthiness of the defendant, corrective justice (and, for that matter, all deontologically based interpretations of the doctrine) necessarily fail. Her conclusion was that if we cannot discover a moral basis for fault (negligence) liability, if the doctrine really imposes strict liability, then tort must be reconceived: Corrective justice theorists "must denounce negligence liability in tort law and work towards the adoption of doctrinal requirements that genuinely map civil liability onto conditions of moral blameworthiness—by, for example, requiring that defendants be at least reckless (if not knowing or purposive) with regard to the harms of their actions."[107] That suggestion, though it ignores the realities of human agency (recklessness is no more the product of moral responsibility than is negligence), makes clear that any positive (or, for that matter, normative) account of the doctrine based on blame fails: The "claim is that, however 'average the man,' and however much 'ordinary intelligence and prudence' he has, he is without moral blame if the harm he causes another is a product of genuine inadvertence (however unreasonable) to the riskiness of his behavior."[108] If Hurd is right, and she is, extant tort doctrine fails, even from the corrective justice perspective, and for reasons that a neuroscientifically sound sense of human agency can confirm.

Hurd's point was that control must precede blame. She cited Westen: "To assess an actor's individual blameworthiness by idealized standards that make no allowance for traits over which he has no control is to risk blaming the blameless."[109] And inadvertence, Hurd concluded, is not something over which human agents have control. She posited several hypotheticals to support her thesis, but one is particularly resonant, and regrettably familiar: the generally attentive, devoted, loving parent who inadvertently leaves her infant child strapped into a car seat in a sun-drenched parking lot for several hours and returns to the car to find that the child has died.[110] The mother may have thought that she had dropped the child off at day care. She may have been distracted by an argument with her husband, a cancer diagnosis, or even world events that should not have so effectively distracted her. Fact is, she was sufficiently distracted to inadvertently leave the infant in the car seat. And tragedy followed. Does it matter to the moral calculus that the mother failed to take precautions that could have avoided the consequences of such inadvertence? No: "[I]n speaking of negligence as risk-inadvertence, I intend to be covering both cases in which defendants had no inkling that their conduct could produce harms of the sort that materialized, and cases in which defendants identified the appropriate types of risks associated with their conduct, but failed to advert to the factors that would enable them to accurately assess the discounted value of those risks and to determine their (un)justifiability in light of accurate measures of the costs of relevant risk-reducing precautions."[111] What is lacking in either form of inadvertence is choice, and choice is prerequisite to the imposition of moral blame, and so, necessarily moral and legal consequence premised on moral blame. So if the legal regime, tort, is premised on moral blame, it could not coherently impose legal consequences where there is no moral blame. Insofar as corrective justice and civil recourse theories rely on moral blame, they could not justify the imposition of negligence liability premised on inadvertence. (The imposition of liability on instrumental bases inconsiderate of choice and moral blame would, of course, not be similarly incoherent.)[112]

Hurd's exploration of the immorality of moral blame (and legal liability) premised on inadvertence is convincing, but the portion of her argument most pertinent to the inquiry and argument of this book concerns the morality of imposing tort, specifically negligence, liability on the basis of deficient character: That is, if you were negligent because you were drawn that way, is it moral to hold you liable for being negligent? She concluded that it is not, but in the course of developing that argument she said much that pertains to the conception of human agency that would emerge from the hard determinism that neuroscience compels. Although Hurd concluded that inadvertence is not the proper subject of moral blame, she must strain a bit more to establish that bad character also is not. Her argument is enlightening, even if it fell just a bit short of the mark. We can appreciate the effect of neuroscience on the normative foundations of tort if we follow her argument with nodding acquiescence as far as we can, and then diverge from her path when we must.

It is when Hurd shifted her focus to bad character that the prescience of her critique was revealed: The *capacity* to do other than one has done is the crucial determinant of moral responsibility, and capacity is a function of what may be (at least colloquially) referred to as character. Consider the case of the coward, or one who fails to act heroically when heroism would seem particularly laudable. Now cowardice, not in the pejorative sense (it is important to recognize that one person's cowardice may be the next person's prudence), is a matter of character. Hurd relied on the example of the young translator in *Saving Private Ryan*, who "cowers in the staircase, unable to prompt action that comports with his own knowledge that he should respond to the desperate cries of his doomed mate."[113] From the remove provided by cinematic setting, our emotional response is focused: "we are as morally contemptuous of his paralysis as we are morally appalled by one who does evil intentionally."[114] Hurd then accurately traced the source of our emotional reaction:

> I suspect . . . that our response to many cases of weakness of will is a reflection of our condemnation of the character of the person who succumbs to such temptation. We use their akrasia as evidence of their possession of character traits on which we pass independent judgment—sloth, gluttony, greed, narcissism, cowardice, etc. We take akratic actors' failure to act in rational ways towards ostensibly desired and worthy ends to be a reflection of unworthy dispositions. They do not do what they (know they) should do because they are not the sort of persons they should be. So while the cowering translator in *Saving Private Ryan* might truly have believed that his legs failed to take him where his will commanded him to go, we, in the audience, believe that he revealed and indulged a disposition for cowardice.[115]

It is not much of a leap to appreciate that our emotional reaction would be adaptive: Surrounding ourselves with those who sacrifice their welfare for others (including ourselves) would better assure our safety and even ultimate survival. Indeed, evolutionary pressures may account for a good deal of what makes certain men particularly attractive to women.[116] And that fact alone would explain the attractiveness of bravery, apart from Hurd's extension of our judgment to matters of sloth, gluttony, greed, and narcissism. (I suspect there are similarly adaptive reasons we find such characteristics unattractive too, at least insofar as there are adaptive reasons for pretty much all our dispositions, even acknowledging the risk of indulging "just so stories.")[117] Hurd then put the question of ultimate significance to her thesis: "then the question becomes whether we can blame people for their characters (ourselves included). . . . Can we morally condemn a person for possessing unfortunate character traits?"[118]

At this point, because Hurd is an avowed compatibilist, we might expect that she would track the compatibilist party line and conclude that, although we are determined creatures, there remains sufficient (trace amounts of?) free will to impose blameworthiness and moral responsibility, sufficient free will to support the imposition of negligence liability. And in fact, that seemed to be where she was headed:

> Were one to dismiss the notion of super- and suberogatory actions[119] so as to escape the implication that our practices of praise and blame reveal a concern not just for people's actions, but for their character, one would be forced to give up other even-more-core concepts. As Neera Badhwar has argued, what it means to love another or to be a friend to another is to stand in a relationship in which one ought to supererogate, and in which one is properly condemned for one's failure to do so. Inasmuch as friendship and love necessitate supererogation, and inasmuch as we blame our friends for failing to be good friends when they do not do more for us than we are owed, we cannot make sense of our most vital relationships without being committed to the view that the character assessments that give moral content to these practices are legitimate. Similarly, we cannot make sense of "a favor that is owed" without such a category. When we think worse of others for failing to reciprocate what were themselves gratuitous deeds we are drawing on the notion that people can be morally criticized for failures that are within their rights.[120]

There will certainly be discomfiture when we are forced to accept the fact that there is no moral responsibility, that we cannot take credit or blame in any morally meaningful sense for what we are or become. But that is not in any way an argument for the reality of moral responsibility; it is, at best, a reason why the sense of moral responsibility feels so real, indeed, so desperate.[121] But Hurd doubled down.

Misapplying Occam's Razor

Not only must we believe we are morally responsible agents, the fact that we believe we are may be bootstrapped to confirm that we are, in fact, moral agents. Let her explain:

> To take stock of our position, then, we are left with the raw fact that our moral practices, as reflected in both daily experience and in centuries of cultural cre-

ations, exhibit a remarkable preoccupation with assessing character. If we are going to avoid the naturalistic fallacy of deriving an "ought" from an "is," we need to treat this as mere evidence of the truth of the matter believed. Belief is often (although not always) evidence of the truth of the thing believed, and if it is otherwise here, then it is hard to imagine how we could sustain the understanding we have of ourselves, others, and our unique cultural history, for it would mean that we would have no defense of pursuits that give meaning to our lives. Whether one characterizes this argument for the blameworthiness of bad character as a transcendental one, or whether one simply takes it to exhibit the sort of "best explanation strategy" demanded by the use of Occam's razor, the bottom line is that the most parsimonious explanation of our moral practices is that bad character is blameworthy and good character is praiseworthy.[122]

Reliance on the parsimony argument from Occam's razor, though, is dubious: surely the simpler, more parsimonious explanation of our moral practices is mechanical, the explanation vindicated by neuroscience and confirmed by evolutionary theory. That is simpler than the morass Aristotelian and Kantian deontology would leave us. We can explain the moral reactions Hurd would sanctify in very mechanical terms, ultimately at the neuronal level. Hurd started from the wrong premise (a veridical moral human agent), and her faulty premise deceived her into a misapplication of Occam's razor. The mistake is, of course, not hers alone; it is the foundational mistake of compatibilism.

Wegner's magisterial *The Illusion of Conscious Will*[123] revealed the contingency of our own perceptions of will and the illusory nature of the claim that will is substantial, a necessary tenet of free will and its accommodation in compatibilism. And he explained why the illusion is so powerful, and so useful, indeed indispensable to human thriving: "the experience of conscious will that is created . . . need not be a mere epiphenomenon. Rather than a ghost in the machine, the experience of conscious will is a feeling that helps us to appreciate and remember our authorship of the things our minds and bodies do."[124] But the illusion is an illusion; consciousness is not, *reliably*, the cause of action: "The processes of mind that produce the experience of will may be quite distinct from the processes of mind that produce the action itself."[125] Further, and crucially, compatibilists may fundamentally misunderstand the nature of human agency, assuming in it a divinity for which it is difficult to account: "Pointing to the will as a force in a person that causes the person's action is the same kind of explanation as saying that God has caused an event."[126] Only God would, could be an uncaused cause; mere human agents could not be. Wegner cited Hume for a similar idea, the inference we may draw "from the constant relation between intention and action."[127] Wegner found Hume prescient in *A Treatise of Human Nature*, where Hume observed that

> Some have asserted . . . that we feel an energy, or power, in our mind. . . . But to convince us how fallacious this reasoning is, we need only consider . . . that the will being consider'd as a cause, has no more a discoverable connexion with its effects, than any material cause has with its proper effect. . . . In short, the actions of the mind are, in this respect, the same with those of matter. We perceive only their constant conjunction; nor can we ever reason beyond it. No internal impression has an apparent energy, more than external objects have.[128]

Illusions can be efficacious; they can encourage adaptive activity. But they are of limited efficacy, and we need to discard them when they undermine (by supporting retribution and the imposition of negligence liability) rather than serve human thriving. Now while the metaphysics here may go beyond the limited sense that Wegner needed, the point is fundamental: The notions of will and intention are mental processes, not mysterious in their content. They are manifestations of ultimately physical processes. So the simplest, Occam's-razor-like explanation for our experience is not what Hurd concluded it is; the simplest explanation is ultimately mechanical and so not beyond understanding in causal terms. Character, then, is no more than a generalization across a run of mechanical events, like describing a car as fast because it can translate four hundred horsepower into acceleration to sixty miles per hour in less than four seconds. In that way, good character and fast are alike and reducible to mechanical causes.

Wegner described the illusion of conscious will although Hurd concluded that character is substantial, but keep in mind that, for Hurd, there could be no character if there is no will. So if will is illusory, character is chimerical, all our feelings about its efficacy notwithstanding. Indeed, will itself is nothing but a feeling: "The experience of will is merely a feeling that occurs to a person. It is to action as the experience of pain is to the bodily changes that result from painful stimulation, or as the experience of emotion is to the bodily changes associated with emotion."[129] Character, built on will, is just compound feeling: It is neither the product of uncaused causes nor the efficient cause of anything. Character is just not sufficiently fundamental to do the work Hurd (and the compatibilists) would have it do. Character may, for what it is worth, be the sum total of will, but "this feeling [of will] is not the same as an empirically verifiable occurrence of mental causation."[130]

Occam's razor would support Hurd's (and the general compatibilists') perspective in just the same way as it would support the conclusion that the magician in fact sawed the lady in half or caused her to levitate several feet above the ground, or that a rabbit emerged from the magician's hat after cake ingredients were poured into the hat. Illusions may be quite powerful, but they are illusions. Wegner's study made clear that our conscious will is in fact an illusion, and so not the stuff of which something like supernatural character could be built. There is no question that the illusion is powerful, but keep in mind that "people once held tight to the Ptolemaic idea that the sun revolves around the earth, in part because this notion fit their larger religious conception of the central place of the earth in God's universe. Conscious will fits a larger conception in exactly this way—our understanding of *causal agents*."[131] In sum, "The experience of conscious will feels like being a causal agent."[132] For present purposes, though, the fact remains that Hurd concluded, notwithstanding the shaky compatibilism, that bad character could *not* support negligence liability. That is, she reached the same solution the materialist perspective would reach. How could a compatibilist get there?

Compatibilism and Character

Hurd started down the slippery slope of compatibilism when she acknowledged that though we do have (will) power over our character, that power is not absolute: "while character can be affected by willed actions [uncaused causes?], and is therefore not immutable, one's ability to affect it (either by raw choice or through what one hopes will be character-altering experiences) is clearly imperfect and unpredictable."[133] So, because our power of will to determine character is imperfect (though why is that? Are there degrees of perfect from one person to the next? If so, what determines those abilities, other than nature and nurture?), Hurd argued that while the criminal law could not punish bad character (and it *claims* not to),[134] corrective justice theorists such as herself could allow that tort law based on corrective justice principles, which are concerned only with restoring a moral balance, shifts the loss caused by negligent behavior from the injured plaintiff to the defendant whose bad (or, at least, imperfect) character was the cause of the loss.

According to Hurd, the corrective justice imposition of liability is doctrinally and morally based on blameworthiness, and at this point Hurd separated faulty character from blameworthiness: "it does not follow from the fact that poor character is blameworthy that the inadvertence caused by such character is thereby blameworthy."[135] She then demonstrated, convincingly, that extant tort doctrine does not support a "bad character" basis to impose liability.[136] That portion of her argument would pretty profoundly undermine positive corrective justice theories of negligence liability based on bad character.

For present purposes, though, it is Hurd's engagement of the moral challenges to normative corrective justice theories premised on bad character that is most helpful and ultimately revealing of the fragility and failure of her and compatibilists' general perspective on moral responsibility and tort law. Her conclusion was based on the distinction between deontic (duty-based) and aretaic (character-based) blame:

> When an actor's poor character exhibits itself in a moment of inadvertence that
> ultimately results in injury to another, the deontic rights of that injured party
> have indeed been violated. A deontic wrong has been done. But the culpability
> with which that rights violation has occurred is of an aretaic sort, not a deontic
> sort. In the defendant's mind was not a depiction of the harmful act, complete
> with its wrong-making characteristics. Rather, in his mind (perhaps not even
> fully consciously) were dispositional desires, emotions, or beliefs that inclined
> him towards his (harmful) action and that blinded him, or distracted him, or
> otherwise diverted him from carefully contemplating the implications of that
> action for others.[137]

Hurd concluded that while the inadvertent, the merely negligent tortfeasor may have had imperfect character that caused the loss to the plaintiff, no moral wrong has been committed: "The defendant did a wrong—a deontic wrong. He did it culpably—but only aretaically so. He did not possess a mental state that can be described as blameworthy in the usual sense of that word—a sense standardly reserved for talk of deontic wrongs and deontic culpability."[138] We could quibble with the deontic–aretaic distinction Hurd posited (more a matter of degree than kind? Both ultimately functions of chemical, electrical, and structural incidents of the brain?), but that would be inapposite, for present purposes. And she is so frustratingly close to getting this right!

Now if the deontic–aretaic distinction Hurd drew intimates something about the dangerousness of the tortfeasor, that might matter. An instrumental approach could take that into account, so long as the distinction could be empirically confirmed. It is enough to concede Hurd's point, and to wonder whether she really is as much of a corrective justice compatibilist as she claims to be. It would appear, though, that Hurd's condemnation of tort doctrine, its imposition of liability for inadvertence (as the consequence even of bad character) comes down to her impatience with tort's responding to behavior that seems more the subject of criminal law, and only when the criminal law's blameworthiness prerequisites have been satisfied. Only then would there seem to be the deontic harm to which even civil liability may attach in something other than a strict liability system. Tort law can redress the wrong, the bad, without recourse to blame.

Neil Levy has concluded, and demonstrated, that the moral responsibility of human agents is an illusion, and a potentially pernicious one.[139] His point, essentially the same as those of Waller[140] and Galen Strawson,[141] among others,[142] is that there is normative space between the attribution of wrongdoing (or fault) and the imposition of blame. In responding to Scanlon's "quality of will"[143] argument that those who act badly are blameworthy, Levy drew a normative line between "badness and blameworthiness."[144] "[W]e need to ask why Scanlon, and those who follow him in this regard[145] think that for an agent to act in a manner that reflects their indifference to others[146] just *is* to be morally blameworthy. After all, there is a natural alternative available: we can say, simply, that such an agent is *bad*, leaving open the question of whether they are also *blameworthy*. There seems to be conceptual room for such a distinction, after all, yet it is not a distinction that the quality of will theory is equipped to make."[147] Corrective justice and civil recourse theories, insofar as they both ultimately rely on Kantian free will (whether compatibilist or libertarian), deny that conceptual room and discover grounds for moral opprobrium to found the tort doctrine. And, as this book argues, those theories may well be right as a positive matter: Tort doctrine seems to deny (or ignore) that room, or to rationalize it by distinguishing actions for which the state may seek recompense (crimes) from those in which victims are entitled to recover privately (torts).

Although normative theories of tort doctrine that take the moral responsibility of human agents seriously, like those of Weinrib, Coleman, and Zipursky and Goldberg, distinguish the consequences of the imposition of tort liability from the retribution that crime exacts, such tort theories depend on desert no less (just differently, perhaps, from the way criminal punishment depends on desert). But any imposition of liability (or punishment) based on desert of human agents is fallacious: Human agents do not deserve, because human agents do not have free will and they would have to have free will before they could have the moral responsibility that would support desert. Hurd recognized that.

Conclusion

The corrective justice and civil recourse theories considered in this chapter seem to need a sense of moral (and so immoral) behavior to operate. They proceed from an assumption about the moral agency of human actors. But the moral human agency remains insufficiently examined by Weinrib, Coleman, Goldberg, and Zipursky. While Hurd recognized the challenge, she failed to meet it. The result is interpretive theory of dubious value (at best) because it fails to come to terms with its crucial object: the human agent.

Just as is the case with criminal law, the need for tort law does not evaporate once we identify the problems with the status quo. We would not loose the negligent to do what harm they will just because "they can't help it," and we would not excuse those whose behavior undermines general well-being. We would, though, rely on an authentic conception of human agency, a conception that makes no room for moral responsibility. That would leave a great deal of room for liability and compensation for harm; it just might provide us reason to reconsider the best systems for configuring a liability and compensation system. The contours of such a system is beyond the scope of this project, but it should be clear that existing lines between regulation (the state's police power) and common law tort doctrine would have to be reconsidered. The current system is expensive and subject to real moral failure, because of its disconnection with the realities of human agency. It is not clear how the costs could certainly be aligned with the benefits that could be realized, even were an authentic conception of human agency embraced by the tort law, but it is clear that we ask the wrong question—and so could never get the right answers—as long as we persist in making the fundamental moral mistakes that deontological theories based on Kantian hallucinations of free will (in either libertarian or compatibilistic terms) would ensure.

The next two chapters extend the analysis and argument developed so far to the contract law. We shall see that the fundamental constituent of contract, consent, is essentially a folk psychological idea that has lost all connection with the doctrine and with leading noninstrumental theories of the contract law. Kantian ideas matter, and confound, once again as well.

CHAPTER SIX

Neuroscience and Contract Law Doctrine

You check your mail to find yet another promotional mailing from a credit-card company, just like hundreds of others you have thrown away. But walking toward the trash can to deposit it, the large print on this one catches your eye: "Zero-percent interest on balance transfers." And this is not just another zero-percent-for-three-months offer; this card promises zero percent until the balance is paid off. Figuring that zero is less than the 10 percent you are currently paying on your credit-card balance, you fill out the application online as directed and click "agree" to the terms presented after a brief review, and shortly thereafter, your credit-card balance transfers to an account on which you pay no interest. So far, so good. While you pay down the balance on the new credit-card account—you figure you can do it in two years—you also begin using your new card to buy groceries, put gas in your car, and the like. You understand that the interest rate on purchases is not zero, but it's a modest 7 percent, still less than the 10 percent you had been paying.

All is well, until you get a bill. Then you see that your monthly payment goes to pay off the transferred balance, not your subsequent purchases. So those purchases you have made will accrue interest at a rate of 7 percent until you pay off the entire transferred balance, at least two years, and there is nothing you can do about it. Frustrated, you shove the bill in a desk drawer and forget it. Three weeks go by; your payment is late. Then you get a reminder from the credit-card company. The letter informs you that because "your minimum payment from the preceding billing period remains due and unpaid, the APR for your account will now be billed at 18 percent." When you applied for this card, you had no idea that payments would be allocated to the transferred balance before current charges would be paid off. You had no idea that one late payment could be so disastrous.[1]

B ut you did *promise*; you did *consent*.

Introduction

Though the contours of the relationship between contract and promise are subject to dispute, there is no question that promise is a constituent of contract. It is not difficult to understand contracts as legally enforceable promises. But before the promise comes the decision, and before the decision comes the perception, and, perhaps, before the perception comes the bias that underlies the perception that informs the decision that gives rise to the promise that the law of contracts would enforce.[2] Even as it appears too simplistic to capture "all that is going on," the reductive process described may adequately depict the pertinent dynamic in that neuroscience rationalizes such a progression in terms that reveal a fundamental normative calculus, a calculus that does not skip any of the crucial steps.[3] There is, though, a trade-off between what we may be able to discover and what we can discover at reasonable cost.

To date, it would *seem* that neuroscience has had less of an effect on contract law than it has had on other areas of the law. But that is not true if we expand our understanding of neuroscientific inquiry and recognize that neuroscience describes a *level* of inquiry rather than a single *form* of inquiry: Neuroscience inquires into all the bases of mental processes, *including* the operation and cooperation of chemical, electrical, and structural properties of brain function at the neuronal level. Prerequisite to, and very much a part of, the neuroscientific endeavor is the effort to discover the properties and characteristics of behavior that brain function at the neuronal level explains. So we need to know what type of decisions the brain is prone to make before we can determine what chemical, electrical, and structural brain characteristics facilitate or even dictate a particular decision. That is, before we can look for neuroscientific reasons why transactors do not read the fine print, we must first establish that people, in fact, tend not to read the fine print.[4]

Here the inquiry becomes opaque and controversial. If *every* thought is the product of brain function, then neural forces are responsible for *every* decision presaging the promise that a contract would enforce, which means neuroscience itself cannot identify those contract decisions that warrant further neuroscientific inquiry. The line cannot be drawn by neuroscience because no decisions are immune from better understanding by reference to neuronal processes. We must, then, look to contract law doctrine to fix the parameters of the inquiry, and contract law distinguishes important and unimportant decisions in terms of consent.

The scope of the consent doctrine is obtuse. The most recent comprehensive Restatement of the Law of Contracts[5] does not so much define consent as describe it as the confluence of contract, promise, bargain, and agreement.[6] Consent is something we may infer from the promise, bargain, and agreement that results in a contract. The relationship among those constituents is, at least in some measure, tautological and, in significant degree, vague. Perhaps consent is sufficiently established when we are able to decide that it would be just to hold a promisor to certain terms of his undertaking. Or perhaps sufficient consent is present even when there is no normative reason to hold the parties to all the terms incorporated in the memorialization of their undertaking: You might consent enough to be contractually bound, but not enough to be bound to all the terms upon which your counterparty would insist. Consent is not a matter of black and white; it comes in shades of gray.

Notwithstanding the mysteries and ambiguities of the consent criterion, consent seems a very good place to pursue an inquiry into the effect that neuroscience may have on the contract doctrine. The primary reason for that is the central, defining role played by consent in contract law. This chapter will first describe the consent calculus in contract, focusing particularly on the role of boilerplate in both arms'-length (business to business form agreements)[7] and consumer transactions (in which a dominant party imposes terms on a subordinate, generally less sophisticated, party through the use of forms). The consent dynamic in those two settings is not always the same, though the operation of the consent criterion in the doctrine operates as if it were. It is easier to discover what is problematic about consent than it is to respond to those problems in ways that preserve the extant doctrine.

The Consent Problem

Contract law enforces obligations that parties have consensually assumed, and that enforcement is limited by the substance of consent. The reality of contract has evolved over time such that our conception of what will suffice as consent also has evolved. Increasingly, there is a difference between the arms'-length deal that is the product of negotiation between relative equals and the bargain that results from a dominant party's imposition of terms on a subordinate party who is unaware of the consequences of such agreement. The doctrine, though, neither well distinguishes those two distinguishable cases nor describes the continuum between them. Contemporary contracting practices endeavor to adjust transactional realities in order to preserve the efficiencies of contracting on a large scale, while, at the same time, trying to accommodate doctrinal traditions.

Nowhere has the tension between requisite consent and efficiencies of scale been more evident than in the context of form contracting: the reduction of potentially great contract liability to formulae reproducible across a broad cross-section of transactions. Boilerplate—the language built into standard contracts that generally ensures the dominant party (not coincidentally, the party drafting the form contract) leverage over the subordinate party—has been the topic of much contention in contemporary contracts jurisprudence.[8] There is a sense that nothing less than the essence of contract, of legally enforceable consensual undertaking, is at stake in confronting the tensions that arise when boilerplate is imposed on those not in the position to negotiate in any meaningful way. The doctrine has yet to overcome the challenges to consent presented by contemporary transactional patterns.[9]

Just as generally subordinate parties like consumers may sign writings (forms or otherwise) that are not the product of meaningful bargaining, even parties of relatively equal bargaining power may memorialize their agreements by the use of boilerplate writings that are not, in whole or in part, the product of bargain. Meanwhile, the ever-increasing proclivity for formation of contracts over the Internet (even by computers without the intercession of sentient human agents)[10] has challenged contract conceptions dependent on earlier models and made boilerplate consent issues hard to ignore.

At the same time and perhaps since the emergence of the "law and . . ." movements, legal analysis has evolved in ways that acknowledge the law's dependence on frameworks drawn from empirical perspectives. Powered by empirical evidence particularly pertinent to contract law, our enhanced understanding of cognitive function and capacities—more accurate assessments of the nature of human agency when decisions are made and legal obligations voluntarily assumed—will revise our account of the core determinants of consensual undertakings. It is not difficult to imagine how duress undermines consent, but every bit as real, albeit latent, is the delusive effect of contemporary transactional realities that one party exploits to disadvantage the other.

And this is the final piece of the ultimate consilience on consent in contract: As more refined behavioral and cognitive insights inform our understanding of what it means to be a human agent, we will better under-stand the failure of consent as a determinant of contract liability. Indeed, empirical advances enable us to verify the artificiality of consent that commercial contracting law actually began to appreciate more than a half century ago.[11] Karl Llewellyn[12] understood that agreement was both the determinant and the measure of contract, and the Uniform Commercial Code's (U.C.C.) contract formation provisions demonstrate that under-standing. What Llewellyn intuited, contemporary behavioral economists, social psychologists, and neuroscientists now confirm. It is worthwhile to consider that intuition.

The "Battle of the Forms"

At common law, *A* and *B* may attempt to form a contract by exchanging forms. Usually the form proffered by each will be indulgent of that party's concerns, that is, the buyer's form will include warranties and the seller's form will disclaim them. The first form, sent by *A*, would constitute the offer. If *A* is the buyer, that would typically be a purchase order. As the seller, *B* will then respond by sending *A* an order acknowledgment form. The terms contained in the two forms will likely be dissonant, as in the case of conflicting warranty and warranty disclaimer. If the forms diverge, then this exchange does not constitute a contract under the common law. *B* is deemed to have rejected *A*'s offer by sending a response that differed in its terms from the offer by *A*.

It is probably the case that *some* of the terms in *B*'s form were the mir-ror image of terms in *A*'s form. The mirrored terms will normally be those describing the quantity, price, delivery date, and subject matter of the con-tract. It will be the so-called boilerplate of the two forms that will differ. But the common law of contract does not distinguish between boilerplate and negotiated terms. Divergence in any substantial respect precludes con-tract formation on the writings.

Even still, if neither *A* nor *B* notes the discrepancies and both proceed as if they are in agreement, then the contract will be formed when *B* deliv-ers and *A* accepts the goods. *B*'s order acknowledgment form will consti-tute a counteroffer on *B*'s terms and *A*'s acceptance of the goods will con-stitute acceptance of the offer on *B*'s terms. So the "last shot," the terms of *B*'s responsive form, will be the terms of their contract. The last-shot reso-lution of the mirror image problem in the context of form contracts relies on an inauthentic conception of consent. In no real way has *A* consented to the terms in *B*'s form. Indeed, *A*'s form likely included boilerplate to the effect that a contract between *A* and *B* would be formed only on the terms of *A*'s writing. But under the last-shot rule, *A* will be bound to the terms of *B*'s form, terms that *A* never read.

If we freeze the frame at this point, the normative balance is precarious. *A* did accept delivery after receiving *B*'s form. So *A* could have reviewed *B*'s terms, rejected *B*'s delivery, and avoided *B*'s terms. Saddling *A* with terms he could have avoided provides a certain rough justice. *B* has not done anything to deserve such favor, however, but for the fact that *B* sent the last shot. That said, the normative balance is no less precarious if we advance a first-shot rule and favor *A*'s terms only because *A* sent his form first. Both the last-shot rule and the first-shot rule create normatively arbitrary results because consent is irrelevant when transactors conduct business at high volume through the use of boilerplate forms, which are not read because it generally makes no sense to do so. The solution of the battle of the forms problem is not to be found in forms that do not capture the real consent of the parties.[13]

Section 2-207 of the U.C.C. was Llewellyn's effort to find the deal in a setting where there is not the type of actual consent that contract requires. Section 2-207 recognizes that a contract may be formed without a complete meeting of the minds. A contract may be the product of *sufficient* consent, even if the terms that bind the parties are not terms to which the parties have agreed in a single, negotiated writing. When parties exchange dissonant forms, Section 2-207 provides that a contract exists on the terms common to both writings combined with terms supplied by prior practices or accepted trade usage.[14] The effect of that rule is to ensure that those commercial actors who think that they are parties to a contract and act as such *are*, in fact, parties to a contract. Any other result, either no contract or a contract on terms provided by only one of the parties, would be untenable.

Consumer Consent[15]

Contractual consent today receives perhaps the most attention in consumer contracting. Treated to extensive review in secondary sources and with reasoning relied on by other courts, three cases have significantly affected our understanding: *Carnival Cruise v. Shute*; *ProCD, Inc. v. Zeidenberg*; and *Hill v. Gateway. Carnival Cruise* sets the stage.

In *Carnival Cruise v. Shute*,[16] Mrs. Shute sustained an injury when she slipped and fell on the defendant's ship. Mrs. Shute and her husband brought suit in the State of Washington where the Shutes purchased their cruise tickets through a travel agency. Received by the couple in the mail after they paid for their trip, the tickets listed terms to govern causes of action arising from the transaction, including a forum selection clause stipulating that any action brought by ticketholders would be tried in the State of Florida. The forum selection clause was in no way the product of bargain; it was presented to the purchasers after they paid for the cruise and at a time when they could no longer cancel the contract and obtain a refund.[17]

The Court's analysis focused on whether the forum selection clause was enforceable notwithstanding that the term was not the product of negotiation. There had been no bargaining over the terms; the Shutes were not sophisticated business people and so were not likely to appreciate the operation of the provision; the parties were of unequal bargaining power; the term was included in a form; and the term was sent to the Shutes after they purchased the tickets. Insofar as consent is the foundation of contract, the Court had to articulate a sense of consent that could do the work necessary to support enforcement of the clause absent (even the real opportunity for) meaningful negotiation.

In writing for the *Carnival Cruise* majority, Justice Blackmun relied on economic speculation: "[I]t stands to reason that passengers who purchase tickets containing a forum clause like that at issue in this case benefit in the form of reduced fares reflecting the savings that the cruise line enjoys by limiting the fora in which it may be sued."[18] First, that argument proved too much, suggesting that every term that reduces the dominant party's risk inures to the monetary benefit of the subordinate party. Parties are not bound by what they *should* value; parties are bound by what they value sufficiently to agree to. Second, the Court offered no empirical support for what was an empirical conclusion. Other than claiming "it stands to reason" that the Shutes paid less than they would have had there been no forum selection clause, the Court offered no proof to support that assertion.

Justice Stevens's dissent turned directly to the consent issue, questioning whether there can ever be real consent in such a transactional dynamic. Relying on Judge J. Skelly Wright's opinion in the canonic unconscionability ruling, *Williams v. Walker-Thomas Furniture Co.*,[19] Justice Stevens wrote, "when a party of little bargaining power, and hence little real choice, signs a commercially unreasonable contract with little or no knowledge of its terms, it is hardly likely that his consent, or even an objective manifestation of his consent, was ever given to all of the terms."[20] The *Carnival Cruise* dissent's invocation of the Wright opinion brought the focus back to the consent basis of contract. And Judge Easterbrook's contributions built on the consent discussion in *Carnival Cruise.*

Judge Easterbrook and the Reformation of Consent

In *ProCD, Inc. v. Zeidenberg*[21] and *Hill v. Gateway*,[22] Judge Easterbrook's opinions articulated well the tensions apparent in consent as the basis of contract in contemporary business and consumer contexts. *ProCD* treated a contract between businesses[23] and *Hill* was a prototypical consumer contracting case. Provoking considerable reaction,[24] Judge Easterbrook's opinions in the two cases understood consent in unorthodox terms that would not be familiar to careful students of the contract doctrine. Because *Hill* built on *ProCD*, it is best to consider *ProCD* first.

At issue in *ProCD, Inc. v. Zeidenberg* was the sale of a ProCD computer disk at a retail store. When Zeidenberg purchased the software, he did so intending to resell related information to businesses—in direct contravention of the license accompanying the software. Not disclosed until *after* purchase, the terms of the license appeared inside the box in which the software was packaged and again on Zeidenberg's computer screen when he installed the software. Recognizing the relevance of *Carnival Cruise* to the consent calculus, Judge Easterbrook found that Zeidenberg agreed to be bound by the terms of the license when he used the software—after he became aware of the use restrictions—and so Zeidenberg's acceptance of the terms did not occur at the time of purchase.[25]

Judge Easterbrook concluded that contract doctrine does give effect to "pay now, terms later"[26] arrangements. Such a contracting form operates in myriad common consumer transactions, for example, the purchase of insurance or of travel and theater tickets. He suggested that, by Zeidenberg's logic, no warranties in the box would pertain to the sale.[27] But nothing in contract law precludes a consumer's accepting a proposed modification of an existing contract. Section 2-207 of the U.C.C. provides that when an acceptance attempts to add new terms to a contract, the new terms "are to be construed as proposals for addition to the contract."[28] Ultimately, Judge Easterbrook relied on Section 2-204 of the U.C.C.,[29] reasoning that "A vendor, as master of the offer, may invite acceptance by conduct, and may propose limitations on the kind of conduct that constitutes acceptance. A buyer may accept by performing the acts the vendor proposes to treat as acceptance."[30] Following that rationale, the judge concluded that Zeidenberg accepted by acting in the way that ProCD provided would constitute acceptance: use of the software (which could occur only after the buyer had become aware of the use restriction).[31]

Judge Easterbrook hoped to preclude a consumer's being bound by a clause designed to exploit, noting that "Ours is not a case in which a consumer opens a package to find an insert saying 'you owe us an extra $10,000' and the seller files suit to collect. Any buyer finding such a demand can prevent formation of the contract by returning the package, as can any consumer who concludes that the terms of the license make the software worth less than the purchase price."[32] But—without any requirement that guarantees a consumer does not automatically accept an oppressive term by accident—had the fine print within the packaging provided that the buyer *did* owe the seller an extra $10,000, there is nothing in Judge Easterbrook's analysis that would preclude the seller's recovery were the seller to bring suit. If Judge Easterbrook believed that would not be the case, his sense of consent is unfathomably opaque. But he offered no elaboration.

Judge Easterbrook's opinion in *Hill v. Gateway*[33] continued the *ProCD* analysis, and extended it. The Hills called Gateway to order a personal computer. The computer arrived, and within the packaging Gateway included additional terms as part of its contract with the Hills. If the Hills did not agree to the additional terms, they could return the computer within thirty days and avoid the contract with Gateway altogether. The additional terms also specified that any dispute arising from the contract would be resolved in arbitration. The Hills retained the computer beyond the thirty day window and then brought an action in court against Gateway on account of an alleged deficiency in the computer.

Judge Easterbrook relied on *Carnival Cruise* and *ProCD* and found the Hills to be bound by the terms in the box and, so, constrained to pursue their claim in arbitration.[34] Echoing the majority's economic speculation rationale in *Carnival Cruise*, Judge Easterbrook then described the advantages of Gateway's contracting process for consumer-customers.[35] But without endangering transactional advantages enjoyed by the consumer, the employee who took the order over the phone could have told the Hills that there would be additional terms delivered with the computer and asked the Hills whether they would prefer to review the terms before placing their order. Or Gateway might have sent the terms for review and acceptance before sending the computer.[36]

There are many reasons why it is usually irrational for a buyer to read forms. First, reading every form we encounter would take time, and we will read forms only in transactions involving sufficient value to warrant the investment of time. Few consumer contracts are sufficiently valuable to justify that expenditure. Second, for reading to be worth the time required, the buyer also must understand the form and be able to make an informed judgment about its enforceability. Third, even if the consumer were to uncover an objectionable term, there is no reason to negotiate over it. All the typical consumer could do is decide not to consummate the transaction. Fourth, in failing to read such forms, nothing goes wrong enough. Consider the many goods and services you purchase each day, month, year. How many of them have resulted in litigation, or even arbitration? Rational buyers will conclude that perusal of the governing boilerplate is a game just not worth the candle. If disappointed by a transaction, buyers will avoid that seller in the future and may share their disappointment in customer reviews online. And, even if those safeguards are not as effective as buyers believe them to be, their existence reinforces the rational conclusion of buyers that what sellers put in the box is no cause for concern. Buyers believe themselves to act rationally when they do not read.

Once sellers can have confidence that buyers will think it is rational not to read, sellers can exploit that rational ignorance; indeed, not to do so would be *ir*rational. For example, credit-card companies make a good deal of money providing credit to those who are not very good at mathematics. The fact that federal legislation was necessary to curb, in part, such practices[37] confirms that credit-card customers may be tricked into believing that they are acting rationally when they are not.

Making Sense of Situation

As demonstrated by the foregoing brief survey of consent in the common law of contracts, to find that parties have consented is not a finding in itself, but rather a normative conclusion indicating that the law has discovered sufficient facts to impose liability on the resisting party. That conclusion may be based on instrumental or noninstrumental premises, but the fact remains that consent is more of a conclusion than it is an analytical tool (perhaps much like proximate causation in tort law; see chapter 4). Neuroscience may reveal that the consent conclusion—which strings together isolated facts to produce a complete constellation that leads to certain outcome-determinative inferences—is an unreliable if not altogether deficient means of performing the normative task the doctrine ought to perform. Consent is preceded by the decision to consent, but consent as a normative conclusion in contract law does not demand an examination of the decision to consent, remaining satisfied with assumptions made about that decision. In turn, neuroeconomic analysis, which begins before consent—at the decision stage—is more likely to reveal and will better serve the normative calculus by gathering more actual facts for analysis and relying on fewer assumptions.

If the decision to consent is subject to variables that may be manipulated by one party acting unconscientiously, then the normative object of the contract doctrine (whatever we decide that is) would not be well served by enforcing manifestations of agreement that effectively reward behavior that accommodates such rent-seeking. Neuroscience can reveal the incidents and operation of such unconscientious behavior, and, indeed, neuroeconomics may represent a consilience of perspectives that, in turn, reveals the impossibility (or, at least, insubstantiality) of certain conceptions of consent.

Situationism[38]

Contract law is based on a "dispositionist" perspective.[39] Dispositionism entails a conception of human actors generally, and contracting parties particularly, as rational (in a constrained sense) agents able to conduct sufficient cost-benefit analyses[40] as they contemplate transactional alternatives. But social psychology research suggests that a situationist rather than a dispositionist perspective may tell us more about transactional dynamics.[41] Situationism attributes behavior to external factors outside the rational actor's control. Indeed, it may be the case that we have been deeply captured by the dispositionist perspective because those with substantial power to inform and even determine our situation have real incentives to propagate what amounts to near religious zeal for a Marlboro Man–like dispositionism.[42]

A focus on the context of contracting—the transactors' situation—may complement descriptions of microeconomic theory, insofar as advertising is concerned, to demonstrate that the party in control of a situation—the seller of base goods and add-ons or the drafter of standard form agreements—has the incentive to actively obscure elements of the transaction that would counsel consumer caution. The market in misinformation should arise as long as no seller has an incentive to create more sophisticated consumers at the cost of decreasing the pool of myopic consumers. Further, sophisticated consumers have the same incentive to maintain the pooling effect that ensures myopic consumers' subsidization of sophisticated consumers.[43]

The Illusion of Rational Choice

The rational actor seems illusory in a world in which sophisticated people make manifestly improvident choices.[44] Those who challenge the rational choice paradigm focus on certain cognitive biases[45] that undermine our idealized conception of human rationality. Neuroscientific insights confirm those biases.[46]

Although focus on specific familiar cognitive biases has engendered considerable academic attention,[47] work by Hanson and Yosifon has comprehensively offered a rejoinder to rational choice theorists' reliance on the standard model.[48] Hanson and Yosifon compared dispositionism (the rational choice model) with situationism (behavior as the product of situational influences). They argued that the situationist paradigm offers a more robust rendering of what it means to be a human actor confronting real choices in a world more authentic than that depicted by welfare economists.[49] Hanson and Yosifon argued that we are "captured" by a dispositional self-image, yet we navigate through a situational world with situational proclivities that overwhelm our dispositional selves.[50] The advertising industry is in no small way responsible for nurturing our dispositional sense of self, and that enables "mad men" to manipulate a situation to exploit consumers.[51]

The picture of the transactor as a "preference-driven chooser," with a sense of what she wants that may be informed but (generally) not manipulated, is central to political theory,[52] microeconomic analysis of law,[53] and legal doctrine.[54] Hanson and Yosifon concluded that contract law, "For the most part . . . mirrors our basic dispositionist self-conceptions."[55] Although contract certainly makes allowances for situation, particularly in deal-policing mechanisms,[56] the general rule of contract is dispositional, emphasizing agreement, bargain, and consensual liability. Even the objective senses of contract posit a tort-like reasonable person, an actor who makes decisions based on dispositional qualities, such as "her conscious thoughts (her 'attention'), her perceptions, her memories, her intelligence, and, finally, the culmination of all those features, her judgment."[57]

Hanson and Yosifon described legal and economic theory preoccupied with the disposition of the human agent, rather than with the situational dynamic that informs behavior, as committing a "fundamental attribution error."[58] The error is understandable, of course, because our dispositionist self-conception is not imposed on us by lawyers and economists (though lawyers, economists, and others provide daily reinforcement for such dispositionism),[59] but instead is central to our self-image as autonomous actors;

we are hardwired to be captivated by dispositionism even were it not reinforced in the course of our engagement with the world.[60]

As Wegner described it, in *The Illusion of Conscious Will*, we have a propensity to ascribe our actions and reactions to the operation of our conscious will, even in cases where it is clear that our consciousness is misleading us.[61] Wegner proposed that "The experience of will, then, is the way our minds portray their operations to us, not their actual operation."[62] Hanson and Yosifon captured that idea succinctly: "Our experience of will . . . is not only an internal illusion, it is an internal illusion that is susceptible to external situational manipulation. . . . Our point . . . is that our *experience* of will—our familiar experience that our will is responsible for our conduct—is often not a reliable indicator of the actual cause of our behavior. . . . The experienced 'will,' rather than a mirror and measure of our true selves, may be another mask in the disguise of dispositionism that keeps us from seeing what really moves us."[63] If what really moves us goes unseen, then we are most susceptible to guerrilla tactics, including hidden guerrilla terms in our contractual agreements.

"Guerrilla" Terms and the Manipulation of Consent: Shrouding

How do we distinguish the guerrilla term from the less innocuous type? Gabaix and Laibson discovered several modern contract consent manipulation devices in their seminal study.[64] They identified what amounts to a market in misinformation, in which the incentives that we normally expect to police sharp practices instead reward unconscientious behavior, a market in which the weight of the contract doctrine is used to discourage competition and to exploit behavioral biases.[65] While the work of Gabaix and Laibson concerned the particular pricing schemes for loss leaders (the base good) and add-ons (necessary accouterments), their conclusions apply to contract terms generally because price and risk are directly correlated (the more risk you assume, the lower the price you pay, and conversely, the less risk you assume, the higher the price you pay).[66] When so applied, their work reveals incongruities that undermine the operation of the consent criterion in contract doctrine: "We show that informational shrouding flourishes even in highly competitive markets, even in markets with costless advertising, and *even when the shrouding generates allocational inefficiencies*."[67]

Behavioral biases persist in contract law in the form of misjudgments made by less sophisticated, myopic, or naive consumers. Of course, we all take turns acting the part of the less sophisticated consumer. A person who knows what constitutes a good deal on a computer may not know what is a good deal on a new car. And, in turn, shrouding is nothing more than exploiting behavioral biases to hide the true cost of contracting. Shrouded product attributes—such as hidden fees (e.g., overdraft fee, late payment fee), maintenance costs (e.g., oil and filter changes, inspections), and prices for necessary accessories (e.g., printer cartridges, adapters)—are those attributes not likely to be considered by a consumer in his initial purchase decision.[68] Even presumptively more sophisticated consumers, like investors buying personal investment products, generally lack an awareness of shrouded fees, such as those to be paid to mutual fund management companies.[69]

Sellers of goods and services are able to exploit consumer naïveté because competing educational advertising will *not* arise in equilibrium, that is, in a competitive market. Accordingly, Gabaix and Laibson confirmed, "In equilibrium, nobody has an incentive to deviate except the myopic consumers. But the myopes do not know any better, and often nobody has an incentive to show them the error of their ways. Educating a myopic consumer turns him into a (less profitable) sophisticated consumer."[70] Even if educational advertising would hurt a competitor's bottom line, sellers have an incentive neither to drive myopic buyers out of the market nor to alert myopic buyers to the fact that they (the myopes) subsidize sophisticated buyers. In fact, sophisticated buyers are (perhaps unwitting) coconspirators in sellers' efforts to take advantage of myopic buyers' naïveté.

Gabaix and Laibson's crucial discovery was that, contrary to earlier economists' suppositions, sellers have no incentive to make more buyers sophisticated.[71] Their findings have been replicated and reiterated in subsequent literature.[72] So we cannot simply trust the market as objectivists would have us do.[73] Indeed, sellers of goods and providers of services have an incentive to shroud additional charges and fees so that buyers will not have easy access to the true cost of their transactions and so that such sellers can maintain (or increase) the number of myopic buyers: "In a search model with only rational consumers, firms will choose to disclose all of their information if they can do so costlessly. In [a] model [with sophisticated and myopic consumers in the same market], with enough myopic consumers, shrouding is the more profitable strategy."[74] There is, then, often a very real disincentive to educate.[75]

In standard form agreements, shrouding involves the inclusion of guerrilla terms,[76] provisions in form contracts that take advantage of "rational ignorance"—the irrationality of reading terms in forms. More powerful market actors, those drafting form contracts, use guerrilla terms and contract doctrine to exploit naive consumers. Form drafters can use a kind of three-card Monty game to ensure maintenance of the pool of naive consumers: Each time consumers discover a particularly egregious term, form drafters hide the risk-shifting card by reshuffling the deck or by sleight of hand.[77] That is just effective marketing.

Recent Responses to the No-Reading Problem

So consumers do not read standard form contracts, and, in the overwhelming majority of transactional settings, to do so would be wholly irrational. Ostensibly with that transactional dynamic in mind,[78] a recent project of the American Law Institute, a Restatement of the Law, Consumer Contracts,[79] set out to reformulate the contract doctrine to account for standard form contracts between businesses and consumers. By effectively codifying Easterbrook's reasoning in *Hill v. Gateway*, however, the Consumer Contracts Restatement would exacerbate the no-reading problem of "pay now, terms later" rolling contracts by disregarding the empirical evidence that shows that consumers do not read any standard forms, whether original or additional.

According to the current iteration of the Consumer Contracts Restatement, if a consumer receives notice of additional terms before giving assent, enjoys ample opportunity to review those terms, and does not return the product or reject the terms in another manner within a reasonable amount of time, then that consumer has accepted those additional terms.[80] But if we know that consumers do not read standard forms, why should the law assume that a consumer does read additions to a standard form?[81] And, even in a counterfactual world where a consumer does read additions to standard forms, why should the law assume that a consumer would understand the additional terms presented? The restatement ignores empirical evidence, the realities of human agency in this context, and attributes legal significance to an empty gesture without offering any normative argument for doing so.

It is surprising that the two reporters and the associate reporter of the Consumer Contracts Restatement have written extensively and quite well about the lack of real consent in consumer contracts[82] and suggested that any conception of consumer contracting premised on consent fails.[83] Recently, restatement reporter Omri Ben-Shahar wrote at length about the incoherence of legally mandated disclosure requirements, pointing out that mandatory disclosures do not serve their stated purpose of benefiting consumers. As it stands, the legal regime requires certain transactors to make disclosures to consumers in order to protect consumers, but instead dominant transactors invoke the mandatory disclosure justification to incorporate oppressive boilerplate (incomprehensible to the consumer) into consumer contracts in order to vindicate the dominant party's legal position, shielding themselves from liability via the fiction of informed consent: "The reason is that mandated disclosure is ill suited to its ends. Exactly because the choices for which it seeks to prepare disclosees are unfamiliar, complex, and ordinarily managed by specialists, novices cannot master them with the disclosures that lawmakers usually mandate."[84]

The fundamental problem with mandatory disclosures—too complex for consumers to understand even if they were inclined to read—applies equally to standardized terms generally, even in the most common consumer contracts. Ben-Shahar identified the challenges in detail:

> [D]isclosures are unreadable and unread because you can't describe complexity simply. The problem is not just illiteracy and innumeracy.[85] It is also the "quantity question," which comprises the "overload" problem and the accumulation problem. The overload problem arises when a disclosure is too copious and complex to handle. The accumulation problem arises because disclosees daily confront so many disclosures and yearly confront so many consequential disclosures that they cannot attend to (much less master) more than a few. Decisions are complex because so much must be learned well and used capably. But it is hard to organize and present masses of information cogently.[86]

The argument *against* mandatory disclosures undermines too the inference of consent from the consumer's having notice of standard contract terms. If the no-reading problem makes mandated disclosure futile,[87] it is not clear why the Consumer Contracts Restatement has built much of its contract formation doctrine on the premise that consumers will read boilerplate terms that empirical evidence has shown they will not read.

The Easier It Is to Contract . . .

In a world where contracts "roll"[88] and consumers "click" and "browse" their way into agreement, it is easier to contract than it has ever been. Yet there is a difference between the twenty-five-page paper contract that you hold in your hand and the twenty-five-page electronic document you agree to without ever possessing. And contextual differences intimate bases for legal distinctions, at least if we take the situationist perspective seriously. At the same time, technical developments in the way we conduct business and enter into contracts reduce some risks inherent in the contract formation setting (e.g., increasing access to information may increase regulatory accountability).[89] Even then, problems arise when consumers become too comfortable with easy, online contracting and ignore the cautionary role of familiar formal requirements.

Just as it is easier to enlighten by way of the Internet, it also is easier to obfuscate. As Gabaix and Laibson discovered, the Internet's tendency to increase the amount of information available to consumers does not prevent shrouding. For instance, it is quite difficult to find on the Internet the per-page printing cost of various printers,[90] a figure crucial to determining the true relative costs of competing printers.[91] Gabaix and Laibson's conclusions suggest that technology will lead to *more* rather than *less* obfuscation by streamlining the contract formation process and encouraging the proliferation of more settings in which constructive consent will suffice. Firms may suppress information in order to manipulate situations. It is the intersection of shrouding and situationism that challenges contract in the twenty-first century, a juncture where the decisional dynamic, at most, delivers only the illusion of consent.

Neuroeconomics and the Decisional Dynamic

Neuroeconomics is the study of how decisions are made and, in particular, how the brain operates to go from perception of phenomena to action. It may not be a striking observation to point out that other actors affect (even deliberately manipulate) the decisions we make. Indeed, were that not the case, advertising would be rendered pointless and so might Shakespearean sonnets. Yet neuroeconomics enables us to decipher and particularize the extent to which other actors and various situations affect our ability to consent. Understanding the neural constituents of decision making will enable us to better comprehend contract formation (manipulability and all), empowering us to select more justly those contracts to be enforced.

We know that education, formal or informal, and experience help us make better decisions.[92] Here, "better" refers to decisions that we are less likely to regret sometime later. And better decisions are not just those that result in more wealth, or even greater welfare, unless we define welfare in terms that take into account hedonistic considerations. Instead, better decisions are those that result in our being happier, perhaps in the long run. Blinding us to the better decisions, perceptions that result in our making decisions we would regret in the future (or even presently, if we understood them in the moment) may be the perceptions that identify those decisions that should not support a finding of real consent.

If you agree to buy my defective car for $10,000 when it is worth only $3,000 (taking into account the defect), we might conclude that you have been defrauded if I misrepresented the fitness of the car and in so doing convinced you to pay me $10,000 rather than $3,000 for the car. Yes, in a way you consented to pay $7,000 too much, but your consent was the product of my malfeasance. If I sold the car to you as is, thereby imposing on you the risk of determining the car's fitness, then we may conclude that you have sufficiently consented by effectively assuming the risk that the car was worth significantly less than you paid for it. The two circumstances involve the same actors and the same car, but different perceptions of risk.

Contract law has ways to deal with the actions of sellers who innocently or otherwise mislead their unsuspecting buyers, generally through the sales law's provision of warranties.[93] In both the sales warranty and products liability law, we impose the risk of failing to adequately communicate relevant product information on sellers. A disclaimer of warranties must be clear,[94] perhaps even conspicuous,[95] and even when a seller warns about the dangerous propensity of the goods it sells, the buyer of those goods may still have a cause of action if he is able to establish that the warning was insufficient to communicate the dangers the product actually presented.

Warranties and product warnings may be exceptional. Generally, contractual consent is inferred from the mere fact of the transaction—that is, if Judge Easterbrook and the courts that have followed his lead are right. (And they are right as long as they construe consent to mean something less than an understanding of the assumption of legal duties or surrender of legal rights—that is, consent in a wholly inauthentic sense as regards typical human agency.) There may be good reason to impose on buyers of consumer goods and other subordinate parties the obligation to make themselves aware of the consequences of their choices.

Have you consented, really, when you click "agree" but never read what it was you agreed to? You probably have not, at least not in any meaningful way. But you probably have chosen to go forward with the transaction, relying on something other than actual consent to provide you with sufficient comfort to go forward. Maybe you trust the market (assume that the terms agreed to must not be too egregious or "someone would have done something about it by now"), or maybe you just assume nothing will go wrong. (Be honest, how often have you really been victimized, beyond the level of mere annoyance, by terms to which you formally agreed but to which you did not substantially consent?)

The calculus changes, though, if we have reason to conclude that dominant parties are exploiting means and methods to undermine consent in ways that will (perhaps over a broad cross section of the consuming public) lead to inefficient or unfair results. The conclusions of Gabaix and Laibson confirm that dominant party contracting practices are *designed* to exploit variations in sophistication, created by educational and social disparities, in ways that result in inefficiencies at equilibrium. So it is not just a matter of the frustration of distributional aspirations; consent corrupted may result in real inefficiencies. That is, the rent-seeking of dominant parties may provide them monopoly profits at the greater expense of us all.

If a store were to introduce a substance[96] into its ventilation system that made customers more likely to buy its products, or less discriminating about the price charged for those products, should the law enforce contracts entered into under the influence of that substance? Would it matter if the store told customers they would be exposed to such a substance, perhaps even told about the likely effects of the substance? These are normative questions, a matter of what the law *should* do. Answers to them would be a matter of doctrine, a matter of how the contract doctrine delimits consent. Neuroscience might well provide insight into such normative consent doctrine considerations.

The Constituents of Consent

There are many ways to get to "yes," and those in the business of getting there will thrive if they discover and take advantage of alternate routes. Whether *A* will agree, will promise, is determined by a combination of factors, some contextual (the setting in which the promise is sought) and some quite personal (*A*'s mood, stress level, "state of mind"). It has long been recognized that at least some of the decisions we reach are the product of factors beyond our control. That is not to merely acknowledge that we are sometimes constrained to buy something that we do not want to buy (unpleasant medications, for example). The more subtle point is that there are (at least) two of us making each decision we make: There is the part of our brain that thinks fast and the part that thinks slow.[97] Fast thinking

(thinking that relies on quick emotional response) may be cast as less reliable than slow thinking (depicted as coolly rational), but in fact that is not necessarily true. Indeed, if it were, it is not likely that evolution would have endowed us with such a dual-stage decisional process.

When we think fast, trust our hunches, "go with our gut," we are relying on heuristics to frame our reactions and to make decisions. And heuristics are valuable things: They enable us to *usually* (or at least much more often than not) reach the most efficacious result with the least (or lesser) expenditure of cognitive energy. Preservation of cognitive energy is not to be understood as laziness; it is an imperative in a world where "he who hesitates is lost." Neuroscience tells us that rational thought resides in the dorsolateral prefrontal cortex (dlPFC)[98] and that the limbic system, particularly the amygdala, is the site of emotional reaction. The amygdala "lights up" (in the fMRI sense) when the subject is confronted with phenomena that we understand to evoke an emotional reaction,[99] such as pushing a large person off a bridge and in the path of an oncoming trolley to save the lives of five innocents.[100] Meanwhile, a brain image will demonstrate heightened activation of the dlPFC when the same subject is asked instead to reflect on the decision to throw a switch to divert the same trolley onto a side track, killing one innocent rather than five.[101] Indeed, there must be something different about the two choices—throwing the person versus throwing the switch—that very literally resonates in the brain and is manifest in distinguishable patterns of neural activity.

Much may be made of the fact that the brain reacts differently at the neuronal level to distinguishable phenomena that precede a similar result: pull a lever or push a man to save (net) four lives. It appears that we enlist the emotional system by presenting the brain with hot facts, those more prone to trigger a fast-thinking reaction.[102] If, however, a particular emotional reaction leads to a decision that is not vindicated upon further reflection (that is, would not result from slow thinking), then we may have caused the agent to reach the wrong conclusion, or, at least, to reach a conclusion at time T_1 with which he would not be comfortable at time T_2.[103]

It would be simplistic, and ultimately mistaken, though, to strictly bifurcate the decision process; simplistically distinguishing the emotional from the rational presents a false dichotomy.[104] There are cognitive processes—chemical, electrical, and structural—that operate and cooperate when agents (human or otherwise)[105] make a decision, but there is nothing fundamentally different about the two that justifies elevating one over the other. And there are several consequences of coming to that realization.

First, sometimes decisions that are the product of those neural systems we associate with affect, emotions so-called, yield the better if not the only acceptable results. Confronted with threatening circumstances, it is quite often much better to run or fight than to deliberate over the comparative advantages and disadvantages of doing either. Indeed, doing nothing long enough to deliberate may prove fatal. Therefore, we should rely on the cues provided us by our limbic system[106] when fast thinking is the only viable option, and there will be time for thinking slow when there is time for slow thinking.

Second, our emotions may actually facilitate rational thought, and, indeed, rational thought may be less effective without emotional input.[107] When we attribute some decision to our intuitions, we are, in a very real sense, acknowledging our debt to emotional valence in reaching ostensibly rational conclusions.[108] We also may recognize the role that experience plays in forming intuitions, and experience is codified in memory, which is more vivid when it includes an emotional component. Those events that are most emotionally, even viscerally, salient are more surely burned into our memories.[109] We are more likely to remember our first significant car accident than our first oil change, and not just because accidents are less frequent than oil changes. Emotion decides what wisdom we most clearly retain from experience and, thus, determines what experiential knowledge is most available for our use in rational decision making.

Third, the emotional–rational dynamic actually describes poles on a continuum rather than the confluence of distinct systems, and that truth is borne out in the range of emotional–rational reactions of different agents. Some of us are more emotional than others, some less. Women, it turns out, are less vengeful than men,[110] and that is confirmed by fMRI, suggesting, contrary to popular lore, that men are more emotional than women in at least some settings.

Fourth, notwithstanding the complexity of the relationship between emotional and rational responsiveness, it is clear that "by changing emotions, we can also change choices."[111] Because emotion responds to phenomena at a more primal level (thinking fast), to accommodate the type of fight-or-flight response that might be crucial to survival (or at least was on the savannah about twenty-five thousand years ago), we can manipulate choice, and sometimes do so in ways that will not inure to the benefit of the agent making the choice or to the broader welfare.

We make errors—defined as T2 regret (arising at time T2) on account of a T1 decision (made at time T1)—in part because of the way we are wired. We also may err because our perceptual system has failed us (that was not your spouse crossing the street) or because, as facts develop subsequent to T1, the calculus at T2 was different from what your T1 self expected. If the weather report tells you it is not going to rain, you have not erred by deciding not to carry an umbrella even though it does in fact rain. The decision not to carry the umbrella turned out to be less convenient, but you did not err in doing the calculus at T1. The type of error that would be pertinent to the consent calculus would be dissonance between a T1 decision and the decision you would have made at T1 if your cognitive processes had not relied on an inappropriate heuristic.

Further, not all of us are endowed with the same heuristic sense. That is, A's T1 decision, the product of fast thinking and appropriate reliance on heuristics, may be vindicated at T2, while B would have reason to regret his T1 decision at T2. Some just are able to think faster on their feet than others, through no fault and to no credit of their own. That better use of heuristics may be the product of practice. Or A may just have greater cognitive capacity or skill than B. Now it would seem that the Bs of the world could compensate, at least to a degree, for some cognitive limitations. On the whole, though, it is probably better to be smarter. Indeed, one of the surest indicia of greater intelligence seems to be what we describe as maturity, the ability to slow down our thinking, to trust heuristics enough, but not too much.

Surely the contract law can do nothing about the range of cognitive capacities that consent doctrine must accommodate. Absent the most extreme circumstances,[112] the capacity to contract is assumed among the adult population. But there always will be those who are myopic in certain circumstances as well as those who generally make unfortunate choices. The contract law cannot protect all improvident actors from the consequences of their own improvidence. But the consent doctrine could rely upon neuroscientific insights to distinguish those actions of dominant parties we will brook from those we will not.

The Manipulation of Consent

Once we understand the neural structures and functions that result in the assumption of obligation, we get a sense of consent that challenges the extant doctrine. Insofar as consent may be the product of manipulation, we are hard pressed not to take that possibility into account when determining the sufficiency and normative power of the consent criterion as a determinant of obligations that the law should enforce. Because sophisticated economic analysis undermines the simplistic conclusion that the market sufficiently polices unconscientious behavior, we have reason not to defer to the formal indicia of consent and refuse to see consent manifested in the subordinate party's clicking "agree" or not objecting to terms that the dominant party had good reason to assume the subordinate party would not read (or would not understand if she had read).

Neuroscience can describe the cooperation of chemical, electrical, and structural properties that could merge to result in such agreement or acceptance. And giving legal effect to such apparent agreement or acceptance is often not problematic. It may not matter that a plaintiff taking a cruise agrees to try his slip and fall claim against the cruise line in a forum more convenient to the cruise line than the plaintiff. But if all subordinate parties, all consumers who do business with Ajax International agree to resolve any claims against Ajax in binding, single-party arbitration, then we might want to take into account the mechanics of that consent and how the dominant party produced it, before we would enforce the subordinate party's alleged choice. In deciding whether there has been consent, whether the promise is enforceable, it makes sense to consider the substance and not just the fact of the promise.

If the agent simply regrets the deal he made because it did not, in the event, provide the payout the agent anticipated, that is not problematic from the perspective of the consent doctrine. We must be free to make bad deals if we want to be free to make good deals. Instead, consider the case in which a dominant party has manipulated the context at T_1, knowing full well that at T_2, when the subordinate party realizes that he has been manipulated, he will recognize that the transaction was not as advantageous as it seemed to be at T_1. If the agent simply miscalculated, and the dominant party was not responsible for that miscalculation, there may be no good normative reason to protect the agent at T_2 from her perceptions at T_1. But if the dominant party framed the transaction in order to take advantage of an informational disequilibrium that would result in certain or probable regret at T_2 as a result of that disequilibrium, then there may well be good reason to question whether the normative object of the doctrinal consent criterion is satisfied. Consent is an impotent doctrine if it cannot capture that crucial distinction.

That type of analysis is arguably at the heart of much consumer protection doctrine and may as well animate the unconscionability calculus[113] in the contract law. Disclosure requirements in legislation go only so far;[114]

regulators also specifically proscribe certain practices that are designed to take advantage of the informational and sophistication disequilibria.[115] We must understand contemporary consent doctrine in the contract law in terms of such proscriptions. Neuroscience can reinforce limitations on consent and demonstrate that those limitations are not the product of political commitments but are instead the product of good science.

Conclusion

Consent, the crux of contract, is a dubious concept once we take seriously the nature of human agency revealed by neuroscience, in terms refined by neuroeconomics and behavioral economics and psychology. This chapter has described the consent doctrine in the context that challenges most profoundly its dissonance with the realities of human agency. And it has demonstrated too how that dissonance may be exploited to frustrate rather than serve the normative objects of the contract law. So just as we saw with regard to the criminal law and tort law, the doctrine's failure to attend to and take account of the realities of human agency as revealed by neuroscientific insights does not just impair the law's operation; it actually undermines, and even subverts, it.

The next chapter turns to noninstrumental theory that would make sense of the doctrine, or at least (or most) provide bases for critique around the edges. But that theory fails for the same reason extant noninstrumental theory fails in the criminal and tort contexts: It misunderstands the human agent, its ultimate subject.

CHAPTER SEVEN

Neuroscience and Contract Law Theory

Introduction

This chapter applies the neuroscientific template to theories of consent in contract, the crucial challenge to the contract law in its most common context: contracts between consumers and businesses, the boilerplate problem.[1] It will reveal that consent, in the sense that the concept is afforded normative force in the consumer contract law, is a chimera.[2] The dominant normative theories do not engage consent in a way that reveals the concept's folk psychological infirmity; indeed, they generally afford consent, its reality, its substance, little attention; they assume its substantiality. But as long as contract is based on the voluntary assumption of legal obligation (which is what distinguishes contract from the criminal and tort law), any worthwhile noninstrumental normative apology for the doctrine must come to terms with consent and do so in a manner that is considerate of the authentic human agency that neuroscience reveals. Philosophical inquiry that proceeds without taking account of the normative limits of human agency could be nothing more than an insubstantial intellectual enterprise, unlikely to describe or prescribe very well at all.

The chapter will consider exemplary noninstrumental normative theories of contract. There are some commonalities among the theories, but the currency of the realm would seem to be the discovery that one unique moral story can explain all, or virtually all, of the contract law. Each posited thesis recognizes a safety valve. The efforts are interpretive, designed to both describe the law as it is and to prescribe the law as it should be. So to the extent that any of the theories cannot explain the law as it is, their authors can say "Ah, ha: Here's where I have proved my normative thesis; I have predicted exactly where the doctrine fails, and see, it does!" That suggests a certain nonfalsifiability, which should be intellectually suspect. The instant inquiry will maintain focus on and will evaluate each of the theories by reference to its ability to make sense of consumer contracting. The theory's proponent either must explain why we do enforce boiler-plate or why we do not (when we rarely do not). If the theories do not at least acknowledge the issue, they can ultimately have very little to say about contract, insofar as consumer contracts are, by far, the most common form of contract. A normative theory of contract that fails to engage, quite directly, the challenges such contracts present for the doctrine is not a normative theory of contract in any meaningful or worthwhile sense.

A common preoccupation of the theories considered here is the focus on promise, a conventional *moral* undertaking that seems to be a foundation for noninstrumental normative discourse about the nature and incidents of the *legal* undertaking that is the basis of contractual obligation: the obligation to which a promisor consents. So the two ideas, consent and promise, are inextricably intertwined in these exemplary theoretical treatments. The commentators do not agree on the fit between consent and promise, but they do appreciate the integration of the two ideas. There is no question that a party who consents does so by promising; the question that remains is the extent to which extralegal conceptions of promise should determine the substance of the contractual undertaking consensually assumed (and provide a measure of the integrity of the doctrine). It is not clear that promise has the sharp indisputable sense that some of the theorists imagine that it does. We might conclude that on the one hand there are moral promises and on the other hand there are legal promises, as well as a range in between. And along that range there may be as well various shades of moral commitment, some of which could or should have legal significance and some of which would not. The important point to recognize here is that just as consent or the lack thereof seems to describe either an obligation or no obligation, promise theories of contract seem to afford the same type of binary sense to promise. In fact, though, it is almost certainly error to understand either consent or promise in such binary terms, and doctrine that does so is ultimately unlikely to fit well with human agency as we now know it: human agency without moral responsibility. Interpretive theories that begin from that misunderstanding too are not likely to describe or prescribe well what the law of voluntary undertakings is or should be. The problem here, though, is not the same problem as we encountered in criminal and tort law: The breaching promisor is generally not directly punished in the way the criminal or tortfeasor is sanctioned. The promisor, though, is bound to an adjustment of legal rights on account of actions that are afforded significance beyond the appreciation of human agents. And just as cigarette manufacturers know to a certainty that some of its customers will not be able to stop using their product,[3] purveyors of consumer form contracts know to a certainty that some (indeed, virtually all) of its customers will make promises on the basis of only the most ephemeral consent.

Further, most of the theories focus on or at least assume some conception of the will, a folk psychology concept for which we may struggle to find a reality referent in the neuroscience. That does not mean that the theory that relies on will is necessarily infirm—it could still say something once we distill the will out of the calculus—but to the extent that a normative theory of contract attributes significance to the will that is not vindicated by the conception of human agency that neuroscientific insights reveal, this chapter will argue that the theory necessarily fails.

The neuroscientific perspective need not underwrite a theory that would replace any of the normative theories considered here, or, for that matter, any extant.[4] There is no reason to believe that contract realizes more normative consistency over time than any other human endeavor has realized. The doctrine is often, simply, a mess: the product of a series of historical and political and intellectual accidents.[5] The first step in trying to improve (or even just coherently critique) the law would be to better understand it: where, how, and when it fails and succeeds. But that effort could be worthwhile only if it proceeds from an authentic conception of human agency. It is the failure to do just that that undermines the theoretical inquiry so far. The contribution of this effort would be to establish the premises from which a worthwhile inquiry could proceed. It is fundamental to suggest the common ways that the noninstrumental theories fail in terms that reveal the power of the neuroscientific perspective.

The Folk Psychology of Assent and Misunderstanding Human Agency

As explored in depth in chapter 3, folk psychology denotes the conclusion, or at least assumption, that familiar terms describing psychological states have the type of reality referent that justifies deontological responses to the actions those mental states are understood to entail. *Belief*, for example, would describe a particular mental state not reducible to certain cognitive or empirical measurement, a mental state that may, perhaps, even be a product of the same type of will that is substantially different from (and yet acts upon) the network of neural material that constitutes the brain. When we, and the law, deal with individuals by reference to and on account of their beliefs, we are dealing with them in an authentic way (or so the story goes), thoroughly considerate of what makes them human. For those who champion folk psychology, to question the reality of beliefs, desires, and intentions would be to undermine the essential humanness of human agents. So ignoring that essential reality, in turn, would result in the promulgation and application of laws and social policies generally that would not serve, and would perhaps even undermine, the normative objects of social institutions, such as law.

Recall that it is certainly clear that some folk psychological concepts work at some level of inquiry: We can infer beliefs, desires, and intentions in many contexts when it is helpful to do so; they can, at the appropriate level of acuity, provide a helpful heuristic.[6] We are able to draw such inferences when there is sufficient coincidence between the folk psychological state and the cognitive neuroscientific state. But to the extent that there is no such coincidence, the heuristic—folk psychology—fails and leads us to make normative mistakes because folk psychology obscures the authentic normative calculus. The error is to confuse that mere coincidence with some fundamental identity; the error is to attribute social and legal significance to the folk psychological state and to ignore the cognitive constituents and elements of that state.

The problem with folk psychological terminology, then, is the same as the problem with all heuristics: Just as heuristics provide a shorthand, a means to capture more with less, they also are as problematic as all rules of thumb. They are, or may be, wrong because they are insufficiently accurate when more accuracy matters. We lose sight of that fact when we assume normative significance from the attribution of mental state, in folk psychological terms, from cognitive processes, brain states, that may coincide with the imposition of a folk psychological label. That is, we attribute to folk psychology terms a normative reality that they cannot support. Worse: From the imposition of the folk psychology label we justify noninstrumental legal reactions (sometimes criminal sanctions, sometimes imposition of obligation) as though the folk psychological label had captured a reality that cognitive neuroscience cannot: "Brains don't kill people; people kill people."[7] Neuroscience, though, reveals that there is no ineffable reality that we need folk psychological terminology to disclose; in fact, we make normative errors when we assume such an ineffable reality exists.

The criminal law chapters described how folk psychology misdirects the normative inquiry by finding room for retribution. Once folk psychology is set loose in that way, defenders of the status quo will develop elaborate justifications for retributive punishment, and in the case of some (notably Michael Moore)[8] will rely on an emotion, guilt, to impose blame, which in turn will both support retribution and (we can imagine) distinguish revenge. But retribution will ultimately undermine rather than serve the normative objects of the criminal law because it can justify punishment for punishment's sake and so actually increase the level of crime in a community. It also treads upon, rather than serves, human dignity in the process. It is curious, then, but not surprising, that law built upon folk psychology may actually achieve ends that are inimical to its object. Surely the object of the criminal law must be to reduce the level of crime and to encourage correction of antisocial activities and lifestyles. So if, retribution, as a theory of punishment, actually *increases* crime and *perpetuates* antisocial activities and lifestyles and is based on folk psychology, we can at least wonder if not immediately certainly conclude that folk psychology fails as a normative matter.

Just as retribution is, in the estimation of at least a substantial portion of noninstrumentalists, a fundamental aspect of criminal law doctrine and theory vindicated by folk psychology, the basis of contract, assent, is a fundamental aspect of contractual liability; assent determines when a voluntarily assumed obligation is enforceable by the obligee. Assent is here designed to be normatively neutral and could be synonymous with bargain, agreement, or consent, as well as other imaginable terms, all of which would capture the voluntariness of the obligor's (or promisor's) undertaking. By using *assent* and *obligor*, I am trying to avoid, for the time being, the use of any term that has assumed, in the noninstrumentalist accounts, normative significance.

A thesis of this chapter is that assent (and its cognates in the contract law) is a folk psychological term. Just as the heuristic significance of a term is a matter of degree and context, assent is a folk psychological conception in the contract law just insofar as it obscures, even undermines, the normative inquiry that supports contract liability. That is, once we use the assent heuristic to describe a normative setting that it misdescribes, once we confuse something that is not essentially assent for real assent, the indication of knowing and voluntary agreement, the folk psychology risk arises: the risk that we will impose legal liability in ways that undermine rather than serve the normative object. Although we certainly could decide that the normative object is not related to real, voluntary undertaking, but instead is an instrumental object such as efficiency, assent might have nothing to do with the realization of that object, and so use of the term would only mislead, much the way attribution of belief, desire, or intent may mislead.

So in reviewing the noninstrumental normative theories that have been posited to interpret contract, it will be important to focus on how the theories engage the assent requirement, whether they use it in a way that reflects cognitive reality or whether they use it in a folk psychological, misleadingly heuristic sense. The second principal way in which the dominant noninstrumental theories ignore neuroscientific reality is by invoking the will, as if there were such a thing independent of (and not reducible to) neural function. Imagining such a normatively significant will provides the means of indulging the supernatural in ways that would support results vindicating an inauthentic sense of human agency. The analysis that follows, then, focuses on the will theories of contract, as that term has come to be broadly construed.

The Cacophony

To suggest that the contract doctrine, even limited to the assent criterion, is merely the product of historical accident is to oversimplify. History, and temporal setting, may play a part in the development of doctrine, but the doctrine is at least as much a matter of moral accident, or the moral inconsistency that we see in all human endeavors. We would be hard pressed to expect our institutions to be unaffected by the neural and cognitive infirmities that condition our individual moral predispositions and conclusions. A naturalistic explanation of morality in contract is an elaboration of the moral genealogy generally: That is, morals are just a label imposed on a particular stage of the normative progression from instinctual reaction, to emotional reaction, to rationalization of the emotional reaction in moral terms, and, in the legal setting, to laws that validate that morality.[9] That is not to say that law institutionalizes the worst missteps of our instincts, but the case could be made that legal reform is much taken up with adjustments that conform our normative reactions to contemporary realities rather than to realities of life on the savannah twenty-five thousand or so years ago.[10]

That genealogy of law's normativity, such as it is, supports the conclusion that legal doctrine may describe sometimes overlapping, sometimes
inconsistent, sometimes even outright contradictory normative reactions
to problems that recur in the course of coordinating human activity. That
is most obvious, of course, in the doctrinal variations among bodies of law
that govern the same or very similar transactions (broadly construed) in
different jurisdictions. In the contract setting, the most striking example
of that divergence might be the different treatment afforded the terms of
consumer form contracts in the common law and civil law systems.[11] There
is no obvious reason why geography (or perhaps even political systems)
should afford diametrically opposed treatment to expressions of assent to
boilerplate: The reality of the voluntary undertaking is no more real in the
case of US consumers than it would be in the case of European consumers. But that is noted here just to indicate that the law, the legal doctrine's
elaboration of moral conclusions (derived, at some level, from instinctual
and emotional reactions), can take divergent paths even from the same
starting points.[12] The ultimate significance of small variations in patterns
at the outset has been borne out in neural properties of human agents.[13]
(Indeed, that truth underlies much of the significance of Adrian Raine's
work discussed at length in chapter 2.)[14]

The Necessary Limitations of Doctrine

Insofar as the development and application of legal doctrine is a human af-
fair, subject to the vicissitudes of human agency (including all-too-common
human error), it should not surprise that over the course of doctrine's development and application, inconsistencies would arise. Further, the nature of doctrine itself almost guaranties that inconsistencies, or, at least,
ostensible differences in result would eventuate. Doctrine does not, could
not, perfectly reflect the underlying moral contours and dimensions of a
legal controversy. That is, in part, why "hard cases make bad law." Some
seem to know instinctively that, for example, *Peevyhouse*[15] was wrong and
Jacob & Youngs[16] was right. The doctrine just did not inquire sufficiently
into the underlying normative balance to reflect those differences in result.
The doctrine in fact obscures the normative calculus, fails in the way heuristics are wont to do.

Further, and similarly, even were doctrine perfectly calibrated to cut
precisely at the normative joints, there is an insurmountable overlap between and among normative objects. That is evidenced in the consumer

contracting setting: We want consumers to have the capacity, the power, to bind themselves contractually. At the same time, in some settings we want to protect consumers from the type of informational asymmetries that could lead them to make improvident choices. It is one thing to police (or not police) regret: the conclusion at time T2 that a choice taken at time T1 was not prudent (even when that T1 choice was fully reasoned); it is wholly another to provide the consumer at T2 the power to avoid a decision at T1 that turned out to be mistaken in light of developments subsequent to T1. Without resolving that tension for all circumstances, it is enough for now to recognize that the normative lines distinguishing those two situations are uncertain and are subject to being obscured by more sophisticated parties. The response of the doctrine to that tension, and others like it, has been to rely on deal-policing mechanisms, such as the requirement of good faith in the performance and enforcement of contracts,[17] and the unconscionability doctrine.[18] And the amalgam of such and similar mechanisms[19] further contributes to normative drift.

Even the best (read "most carefully drawn and applied") doctrine could not surely bear the normative weight imposed on it. For present purposes, though, the significance of that truth is the effect it has on interpretive theories of contract, normative theories of what contract is and should be. So once the challenge is understood, asserting the moral impossibility of contract doctrine is not pejorative, is not an indictment of the law's inability to arrange coherently the normative values at stake. It is, instead, an objective observation that fixes the lens through which we evaluate noninstrumental interpretive contract theory. The object of such theory would be to explain the law as it is and, at the same time, provide the bases to critique that status quo and suggest doctrinal adjustment. It is not surprising, then, that these (and instrumental normative theories too) alternate between asserting their positive (that is, descriptive) power and urging reconsideration of the current doctrine as inconsistent with the underlying, usually latent, normative object of the law that the interpretive theory posits. But because of the nature of human agents and human agency, that doctrine can do only so much to formulate on the head of a pin the full normative calculus contemplated by (or resulting from) deference to promise and application of the consent criterion.

The focus here is on the fit, or, rather, lack of fit between noninstrumental normative theories and the authentic conception of human agency that neuroscience provides. If human agents are not the type of moral actors certain normative contract theories assume human agents to be, the normative theories fail, necessarily. The fundamental thesis of this chapter is that all noninstrumental (and largely deontic) normative theories of contract do necessarily fail because they miscomprehend human agency; they fail for the same reason that those noninstrumental theories that assert efficiency-based normative theories fail: The foundation of each is fictional, an abstraction (a heuristic) that more often misleads than reliably guides. Human agents are not *homo economicus*, and human agents are not morally responsible.

The neuroscientific critique of noninstrumental theories of contract in this chapter does not just reproduce the analyses of the criminal or tort law provided in the preceding chapters. Contract is different from the criminal and tort law, just as criminal and tort law are significantly different from one another with regard to their understanding of human agency. Contract is based on consent and the consent of human agents. But contract doctrine and the extant noninstrumental theories in fact assume a uniquely inauthentic human agency. We will see that the theories all are wrong about some of the same things, but each is wrong about different things too. Initially, though, to best describe the (inaccurate sense of) human agency the theories assume, it is worthwhile to focus on their common errors before turning to their idiosyncratic errors by considering how contract is distinct from criminal and tort law.

Fit among Criminal, Tort, and Contract Noninstrumental Normative Theories

This part of the chapter considers two questions: First, what distinction or distinctions between and among contract and criminal and tort law reveal normatively significant aspects of the doctrine and the doctrine's (mis-)fit with an authentic conception of human agency? Second, how does focus on the bargain concept reveal or at least illustrate the noninstrumental theories' normative incoherence in light of the perspective neuroscientific insights support? The questions are treated seriatim.

What Distinguishes Contract: Strict Liability

Criminal law and tort liability, broadly speaking, are things that *happen* to human agents, not something they necessarily *choose* to have happen to them. That is not to say that there are not circumstances in which parties recognize that their actions will entail unpleasant consequences—speeding home while willing to pay the fine, if caught; or creating a nuisance because the cost of doing so, the damages payable, are less than the cost of the opportunity otherwise foregone—but it is to say that the parties do not desire those consequences the same way they desire the salary for which they have contracted to do even unpleasant work. Even that summary statement suggests the continuum: The line between paying for emergency dental surgery and suffering the physical injury that necessitated it is a fine one on the desire continuum. For that matter, many common contracts rely on a thin conception of consent; utility and rent payments, for example, suggest the tension.

Putting aside that (pesky) continuum issue, we may agree, at least, that contract is a typically (even generally) significant normative distance from the criminal and tort law on the voluntariness scale. Even though the power may be limited, the parties to a contract undeniably have *some* power to determine the extent and contours of their legal liability. Indeed, insofar as the terms of the contract are provided by the parties, or even just by the dominant party, that liability may be fixed in a manner generally independent of doctrinal intervention. It is not far-fetched to understand contract as a system of default rules that the parties are free to adjust according to their whim.[20] Contract doctrine does police the parties' bargain, though that intervention is generally only at the margins: By "freedom of contract" we do mean the parties' autonomy to do what they will with the power to assume obligations backed by the enforcement powers the state makes available to private litigants. In fact, the premise of contract is that posited power, even if the parties do not share it equally. There is also "freedom from contract," which may have varying and various meanings, but which itself empowers the parties or either of them to simply walk away, to choose not to accept the deal (though, again, in some settings that ability to walk may be severely circumscribed by circumstance). For purposes of drawing the distinction pertinent here, it suffices to say that parties can avoid contract liability in ways that they could not avoid criminal or tort liability.

Not unrelated to the ways contract obligations may be assumed and circumscribed is the nature of the liability that assumption of obligation contemplates. To distinguish contract from criminal and most forms of tort liability, it often is observed that liability in contract is "strict." That is to say that a party may be found in breach of contract even through no fault of her own. Although that basis of distinction from the criminal and tort law may not stand up to an analysis of human agency informed by neuroscientific insights, the point is clear enough: We may expect that contract would operationalize a different set of liability rules in that someone may breach a contract faultlessly,[21] notwithstanding his best efforts to perform. Here, again, we are on a continuum because under current tort doctrine the naturally less attentive person, that is, someone who is not as attentive as the norm, may be liable for actions that the reasonable person would not have performed, and for not performing actions that the reasonable person would have performed: The usually attentive driver who becomes flustered when others generally would not may be liable for the damages he does to others.

The premise of the criminal and tort doctrine, on the one hand, is that the actors subject to the law have failed to act in a manner consistent with a general norm, the specifics of which are fixed by reference to typical characteristics of typical actors. Contract, on the other hand, does not, and need not, presume determination of liability by reference to the capacities of the contracting parties, though again that is subject to exceptions at the margin.[22] If you say you will do X, and do not do X, you will be liable for breach of contract and the damages flowing therefrom. (The "expectation" or "benefit of the bargain" contract damage measure too is pertinent to the distinctiveness of contract doctrine and liability.)[23] And the fact that contract liability is strict may make sense of the doctrine in ways that distinguish contract from criminal and tort law.

Most pertinently for the present account, though, is the fact that the strictness of contract liability tells us something about the contract doctrine's conception of human agency: If your agreement is the product of bargain, you will be liable for the loss flowing from your breach, irrespective (again, generally) of the reasons for your failure to perform. Though the idiosyncratic capacities of the parties at the time of contract formation at the margin may be pertinent to the contract liability calculus,[24] in the general course we assume that you are strictly liable for deficiencies in your performance, even if your counterparty exploited your obvious (or predictable) and typical cognitive limitations. That was the point of the doctrinal analysis in the foregoing chapter,[25] and that suggests the limitations of the contract law's understanding of human agency.

The contract law does make some allowance, at least in statutory iterations of the doctrine, for transactor sophistication,[26] but those are statutory interventions in common law jurisdictions and constitute exceptions to the background doctrine. The interventions take two general types: special provisions for consumer protection, generally requiring measures to enhance (or at least make marginally more likely) consumers' understanding of the liability they are assuming[27] and laws specifically limiting what the substance of the terms of the contract may be.[28]

There would seem to be a necessary connection with the damages available for the breach of a social obligation—criminal, tortious, or contractual—and the normative valence of the doctrine describing the contours of that obligation. So we would know (or find grounds to infer) something essential about the normative commitment(s) of contract when we appreciate how the remedial scheme, the measure and form of damages, responds to breach of contract. We might conclude that damages should be limited in contract because contract liability is, after a fashion, strict. In fact, because tort is based on fault, intuition might suggest that tort damages should at least generally be greater than contract damages. But that is not the case: The standard contract measure of damages is expectation, or benefit of the bargain, focused on putting the disappointed, nonbreaching party in the position he would have been in had there been no breach. Tort law measures damages by the out-of-pocket loss, broadly construed,[29] suffered by the plaintiff-victim. Criminal law is in tension: Sentences are limited by instrumental theories to remediation of the circumstances that gave rise to the crime (rehabilitation or correction) and extended by noninstrumental theories to retribution, or revenge. What is true across all three areas of law, though, is that we may draw inferences about the best interpretive theories by reference to the consequences of violating the duties fixed by the doctrine.

There is, of course, an extensive theoretical literature that posits normative connections between contract damages and the objects of contract. Some of those interpretations will be considered in this chapter, but for now it suffices to recognize that even those interpretive theories that do not focus on the contract damage measure(s) must ultimately account for them in any conclusions that would formulate contract precisely. And with the strict liability basis of contract, we might expect that what distinguishes tort from contract damages is focus on the victim of the breach, or the relationship between the breaching party and the victim of the breach: There must be some significant normative difference between the tort and contract relationships. It should be clear, though, that insofar as tort and contract describe a continuum—degrees of voluntariness or assent—any convincing distinction between the normative conclusions of the two doctrinal areas must account for the differences of degree that matter and how we can rationalize the resulting distinctions. That is, contract and tort theories could formulate convincingly distinguishable conceptions of human

agency at work in the two legal settings in order to justify the different damage measures. Or we might rationalize differences in the damage measures
by focusing on the agency of the breaching party: Tort liability requires
fault, some kind of wrongdoing, broadly construed; contract imposes strict
liability. All the bona fides and care in the world will not generally insulate
a promisor from the consequences of his breach. The focus seems to shift
back to the victim of the breach, then, when we impose greater liability
than would be imposed in the tort setting: Expectation damages are generally greater than tort out-of-pocket damages.

Whether or not we can make sense of the expectation measure of contract damages, it is clear (*clear* being a relative term, of course) that one
aspect of contract damages is distinguishable from the consequence of
breaching criminal law or tort duties. It is axiomatic that the criminal law
punishes: That is, indeed, the identifying characteristic of the criminal law.
Now the particular punishment may be designed to serve an instrumental
(e.g., deterrence) or noninstrumental (e.g., retribution/revenge) object, but
there is no question that the penalty (tautologically) punishes: It does not
compensate the victim of the crime in anything like the way tort or contract
damages compensate. Tort law's imposition of liability on tortfeasors also
discourages behavior, and so, from the perspective of the tortfeasor, imposes punishment for tortious action. Indeed, in some cases, punitive damages also may be awarded against certain egregious tortfeasors. Contract
damages are not punitive. Although that axiom is either shaky or rife with
exceptions,[30] it is nonetheless a foundation of the doctrine and a relatively
certain distinction between contract on the one hand and tort and criminal law on the other for which interpretive theory must account. And, once
again, normative theory must account for that essential distinction in terms
that resonate with an authentic conception of human agency; there must be
something about the contract relationship (a relationship between human
agents) that makes punitive damages inconsistent with the normative commitments of the contract doctrine. We just do not understand the contract
law as doing the same normative work as the criminal and tort law does. The
basis of the distinction seems to be in the way contract liability is assumed.

Promise, Bargain, and Agreement

This part of the chapter considers aspects of contract that must be essential to the noninstrumental normative theories that would interpret the
law. Unlike the preceding chapter, it does not describe the doctrine and its
operation. The contract formation constituents assume that human agents

have capacities and pervasive characteristics that human agents just do not, simply cannot have. Further, the law draws distinctions. Insights drawn from neuroscience, broadly construed, can reveal that dominant noninstrumental contract theory relies on assumptions that can find no purchase in the authentic human agency that careful empirical inquiry would reveal.

Contract emerges from the coincidence of promise, bargain, and agreement, though the contours of that coincidence are, at best, murky, and even more than occasionally inconsistent. Promise, we are to assume, has the meaning generally afforded it in the lay understanding: Promise is a moral commitment to do or not do something upon which the promisor intends the promisee to rely; indeed, the fact of that reliance may be—and, through the complementary doctrine of promissory estoppel,[31] often is—a sufficient basis of promise enforcement on its own. Contract emerges from the exchange of promises that indicate a bargain, the bargain that results from agreement. Bargain is in fact a theory of promise enforcement that accounts for the consideration (quid pro quo) requirement of the doctrine and may take the place of benefit or detriment theories of consideration that would premise promise enforcement on the promisor's receipt of a benefit (perhaps the return promise) or the promisee's detriment (also a return promise or prejudicial change of position, in the estoppel sense).[32]

Bargain fixes time in a way that may not be consistent with the nature of human agency. That is not to say that we should not have a concept such as bargain, or that contract *should* not bind agent A at T_2 for what he undertook at T_1; it is, though, to suggest that doctrinal mechanisms that provide means to revisit T_1 decisions, bargains, at T_2 may well be consistent with the normative realities of human agency revealed by neuroscientific insights. The extent to which we want to defer to T_1 decisions at T_2 may be a function of some normative judgment (perhaps efficiency, by some measure) that deemphasizes the variability and manipulability of human agency in favor of some other value, instrumental or noninstrumental. Bargain, then, as a measure of normative obligation, can provide a lever: Find bargain and you enforce it, find reason not to enforce and you may find there never was a bargain, or at least not the bargain one of the parties is now trying to enforce.

It is axiomatic, and even expressly part of the doctrine, that contract is about the enforcement of promises.[33] And promise is a concept familiar to those with no legal training at all: It is an understanding between two or more parties that contemplates some moral opprobrium for breach.[34] That is pretty much all there is to it, though. The promisee may think less of you (would be justified by some moral metric in thinking less of you), in a sense, if you fail to fulfill a promise. The extent of the appropriate moral cost, the measure of opprobrium, if any, will be socially constructed and will vary in the estimation from one to another judge of that conduct. But for the most extreme cases, and perhaps even then,[35] the reactions will vary. Promise, though, certainly captures something short of contract, and the significance of that may matter to noninstrumental theorists.

The consequences of breaching a "mere" promise are generally extra-legal. The party who breaks a promise will (perhaps) be subject to criticism but will not be subject to formal legal sanction. So from that fact we could infer that social systems intend the two conventions, promise and contract, to provide distinguishable instruments for the coordination of activity: Those who would be willing to make a promise and suffer moral criticism for its breach might not be willing to assume the legal liability that would attend a breach of contract. We also could recognize a wholly practical basis for the distinction: There may be promises that it is best the legal machinery just stay away from. For example, do we really want to provide a cause of action for the promisee when the promisor breaks a dinner date? Alternatively, you could find that contract is just another iteration of promise. However you resolve that tension, you may or may not find reason to applaud or criticize the law's at least ostensible distinction between the two. Individual human agents, though, not appreciating the nice legal distinctions, may confuse contract and promise in ways that tap into the emotional valence of each. If economy of judicial resources and the ill will that attends their use explains distinguishing casual promises from formal contracts, then we would need some readily accessible indicator of the parties' appreciation and deployment of the distinction, some way for the parties to certainly indicate when the courts are to become involved and when they are not. That distinction is generally captured in the consideration requirement: the provision of a quid pro quo as the basis to make a promise enforceable.

That quid pro quo is not difficult to come by, though much has been made of the possible legal enforceability of promises not attended by consideration: cases of so-called promissory estoppel. The fact that promissory estoppel, whether it is a creature of contract or not,[36] has some vitality and that consideration itself is a fragile doctrine,[37] suggests that the legal distinction does not do important normative work, at least not consistently and reliably. That is, there are promises that the parties, and society generally, want to treat as contracts even though we can distinguish mere promises from contracts in the normal course. And just as there are mere promises that may be given legal effect, there are also nonlegal sanctions for breach of contract that may have greater power to affect promisor behavior than the damages that breach of contract normally entails. Karl Llewellyn recognized the significance of reputational considerations when he drafted article 2 of the Uniform Commercial Code in a fashion that took advantage of commercial parties' concern for the real market value of their reputations.[38]

The bottom line, then, is that promise and contract can describe distinct social devices that share importantly common normative incidents. We may make best sense of the doctrine, and its operation, when we are sensitive to human agents' appreciation of those normative commonalities even when doctrinal distinctions would seem to draw fixed and certain lines. The doctrine may fail to reflect human agents' understanding, and that may compromise the doctrine's coherence in important recurring contexts. Certainly a contract you know would not be legally enforced, albeit enforceable, may have even less binding effect than a mere promise. Indeed, an argument could be made that such contracts are dominant: It would not be rational to engage the legal system to enforce a contract for less than, say, $100, or anytime the cost of enforcement would exceed the damages recoverable. So what do we make of the boilerplate in a $25 or 99 cent (iCloud services)[39] purchase? Do such terms really constitute a contract, in the sense of a legally enforceable promise? Would the consumer who agrees to such terms ever expect to enforce them, or have them enforced (other than by the vendor's self-help termination of service)? It is not clear that such contracts would, in the common understanding, have the binding moral force of a bare promise not attended by legal enforceability. Indeed, as discussed in chapter 6 surveying the consent doctrine particularly in consumer form contracts,[40] it is highly unlikely that the subordinate party to such common contracts has ever even read the boilerplate terms; it would be irrational to do so. What then to make of the agreement criterion, which operates with the bargain or promise elements?

The promise and bargain requirements of the contract doctrine complement the agreement requisite: The contours of the parties' mutual promises are fixed by the coincidence of their agreement. If there is no agreement in fact, there is no contract: promises or bargain wholly apart. And agreement contemplates, at least in the jargon of contract, a meeting of the minds.[41] Where bargain responds to the consideration requirement, agreement really represents an alternative element of enforceable promise formation: You are not bound to a bargain unless you have agreed to it. *Agreement* certainly contemplates a meeting of the minds, even if that phrase captures the sense more than the reality of the formation requirement. There is a fundamental problem with agreement that doctrine obscures but does not really eviscerate: How can third parties, finders of fact, reach reliable conclusions about the contracting parties' subjective states of mind? For that matter, how can the two contracting parties be certain that there is enough coincidence of their mental states to conclude that there is the type, or extent, of agreement that the contract law will enforce?

There is a necessary inscrutability, or inaccessibility of mental state. And, neuroscience would suggest, that dilemma is even more problematic when we have reason to lack confidence in parties' understanding of their own mental states. Recall, state of mind can be manipulated; even the most ostensibly rational actors may be subject to situational influences that overcome general dispositional characteristics.[42] That manipulation may be the object of marketing in many settings. Indeed, it is not too cynical to imagine that contract boilerplate assumes that situational influences may overcome prudent resolve. Further, it is important to keep in mind that because of the nature of neural development, agent A at T_1 is in a literal sense *not* the same agent at T_2; the differences, of course, will be a function of the time elapsed between T_1 and T_2, but it will depend too on situational adjustments that may well be within the control of the dominant contracting party. This is a dimension of the pervasive regret problem in contract.[43]

Moreover, we could not find true assent to the terms of a contract if there were not agreement to them, and there could be no agreement if there were not understanding. That is, if I offer to sell you my Mustang (the car), and you agree to buy my mustang (the horse), we do not have a contract, and all the promising, bargaining, and ostensible assent in the world will not provide the basis to find the enforceable promise that could be a contract. In that way the promise, bargain, and agreement criteria are distinguishable. The deal would fail on account of misunderstanding, a species of mistake.[44]

"Meeting of the minds," though, seems to contemplate a strict test. But in fact how frequently do minds actually and completely meet? More likely, and certainly in practice, the criterion describes a range of overlapping understanding, the dimensions of which are probably measured by reference to the nature of the subject matter and even normative considerations such as the parties' relative sophistication. Neuroscience (again, broadly construed) can reveal the contours of such a coincidence of understanding. The sources surveyed in chapter 6 demonstrate, though, that substantial agreement fails in many (even perhaps most) common consumer contract settings. Indeed, it is clear that the constituents of agreement may be manipulated (in the pejorative sense) to accomplish the pretense of actual agreement that the law might construe as substantial agreement.

What we are left with, then, when the dominant contracting party manipulates the formation process to create evidence of agreement where there actually is none, is a manifestation of assent that, it seems, may do just as well to bind the subordinate party. Keep in mind that within the standard boilerplate are terms to which the subordinate party could never truly agree because the average consumer (or even the consumer of more than average sophistication) has absolutely no idea what the terms mean. How many consumers (or law students or lawyers, for that matter) appreciate the consequences of a choice of law, consent to venue or jurisdiction, or binding arbitration clause? The number who do understand is likely a miniscule percentage of all those who sign such contracts. That, of course, is one of the reasons that consumer advocates are most troubled by the standard boilerplate.[45] Indeed, it is now quite clear that in many consumer contracting settings, the dominant party, the party drafting the boilerplate, is in a better position to appreciate the consumer's circumstances than is the consumer.[46] So if mistake is a basis to find that there was no agreement, and mistake is a basis to avoid contract liability when the nonmistaken party was aware of its counterparty's mistake, an understanding of the doctrine could put in question the enforceability of many common consumer contracts. But that would not likely be the result of the mistake doctrine's application. For present purposes, the point remains that empirical evidence on subordinate parties' state of mind in recurring consumer contexts calls into question the essential reality of the agreement criterion. And if actual agreement is not requisite, what will do just as well? Apparently manifestation of assent (or, cynically, the pretense of it) will.

There is an at least ostensibly attractive normative argument to be made to the effect that a consumer, or any less sophisticated party, who signs a writing indicating agreement to its terms assumes the risk that any lack of understanding of those terms will ultimately prove prejudicial to her interests. The naive transactor is simply assuming a risk, and by assuming that risk, effectively obtains the goods or services contracted for at a lower price. If you were not comfortable assuming greater risk, you would impose more of the risk on your counterparty and pay the higher price therefor. Because contract is to a large measure the effort to fix risk, if a risk does not materialize, or does, someone may well have suffered a loss. Certainly the common fixed-price sales contract is exemplary: It imposes the risk of price increases on the seller and price decreases on the buyer. In the event the price fluctuates in the way one of the parties had not anticipated and the other had, the disappointed party will have reason to regret having assumed the risk that materialized, but that regret is not reason to relieve the disappointed party from his deal. That situation is distinct from the case in which one of the parties is aware of a risk and the other is not, even if the risk pertains to a matter that would seem to be within the contemplation of the ultimately disappointed party. The consumer who really agrees to a contractual choice of venue should not be heard to complain when she suffers an injury and must bring her action in a distant forum. The normative calculus depends on whether there was in fact agreement, and the doctrine has struggled with that calculus.

Less sophisticated, though philosophically pretentious, approaches would make much of the ostensible expression of consent,[47] and that perspective certainly resonates with libertarian conceptions of free will, at least akin to the libertarian free will that is refuted by the materialism that neuroscientific insights vindicate. The simple, and simplistic, idea is that we respect the agency of the party who consents to be bound, and we question that consent at our peril: To do so would be to undermine the integrity of the promisor. That argument falters, though, when we indulge a more authentic sense of human agency, recognizing that dominant parties—who would champion the right of the subordinate party to consent—are in the position to exploit informational asymmetries to the prejudice of the vulnerable promisor who knowingly consents. And the suggestion that more careful review of the incidents of such agreement will lead us down a road to serfdom ultimately fail to convince.

Nonetheless, there is doctrinal support for the conclusion that the manifestation of assent is sufficient, and so reason to conclude that actual, knowing consent is not necessary.[48] That reading of the doctrine may be strained and, in any event, would lead to results that undermine both instrumental[49] and noninstrumental objects, but it is currently ascendant. The curious reasoning of the United States Supreme Court in *Carnival Cruise*[50] and Judge Easterbrook in *ProCD*[51] and *Hill*[52] represents, by some counts,[53] the majority rule (even if perhaps not the best reading of the doctrine that preceded the decisions).[54] The proposed Consumer Contracts Restatement, following and elaborating on the lead provided by the *Principles of the Law of Software Contracts*,[55] would weave sufficient normative consent from ostensible assent, disengaging the doctrine from making the difficult but crucial inquiry into agreement. It is easier to find or not find agreement if we can rely on easily observable indicia, but there is no reason to believe, in this context at least, that the easiest device is the most reliable: once we conclude that *reliable* contemplates either instrumental or noninstrumental values.

This part of the chapter has demonstrated that contract theory that would assume the essential reality of promise, consent, bargain, and agreement would misconceive the nature of human agency. Contractual consent is a folk psychological fiction, an ill-fitting heuristic, which ultimately does more harm than good. Human agents just do not consent in the way the doctrine and interpretive noninstrumental theory assume (imagine) that they do. Similarly, the distinctions between promise and contract—a focus of many noninstrumental interpretive theories—are obscured by the doctrine and rely on nuance beyond the appreciation of typical transactors. Although we know that there is a difference between promise and contract, and can find good reason to maintain the distinction, theory that would ignore the distinction's normative significance must cash out a conception of promise in and out of contract that will make sense of the consent criterion when it is most strained, in the context of boilerplate in consumer transactions. Comprehensive normative theories of contract that rely on nuance generally inaccessible to human agents will frustrate rather than serve human thriving, in pretty much any terms you would construe human thriving.

The object at this juncture is to consider the integrity of a few illustrative noninstrumental interpretive theories as they would relate to the consent criterion and promise in consumer contracts, the most common contract form. If the noninstrumental normative theories cannot account for such contracts, they are infirm. And, in fact, they cannot account for such contracts precisely because they lack a coherent conception of human agency, the conception that neuroscientific insights would provide. The conclusion here, too, is that if the noninstrumental normative theories fail to account for consumer form contracts, they likely will fail to explain much else about contract more generally.

"Will" Theories

It should be clear by now that normative theory that depends upon the will to explain anything fundamental about human agency would be profoundly suspect (at least), insofar as the will is a folk psychological construct that lacks independent substance. But the leading noninstrumental normative theories of contract rely on free will, which requires a conception of the will as an autonomous uncaused cause (or at least sufficiently uncaused to satisfy apologists for compatibilism). That is true of the promise theories that have elaborated upon Kantian deontology as well as libertarian consent theory (and probably even contractualist formulations).[56] The limited scope of this book's inquiry precludes exhaustive survey of all the extant theoretical approaches, even if that survey were limited to noninstrumental approaches.[57] But the argument is that what undermines the theories considered here undermines all noninstrumental theories that rely on an inauthentic sense of human agency. So any interpretation of the doctrine such approaches provide must fail.

Kant and Autonomy: Fried's Promise Theory

Charles Fried's *Contract as Promise* is a watershed: The short book described a Kantian theory of contract that proceeds from the premise that breaking a promise is normatively akin to lying and, from the Kantian perspective, in the same way violates the categorical imperative to treat others as an end in themselves and not as a means to an end.[58] From there, Fried did his best to make sense of contract in terms of promise, demonstrating how an understanding of what promise entails would provide as well an understanding of the incidents of contract, as revealed in the doctrine. The will is central: "the conception of the will binding itself [is] the conception at the heart of the promise principle[.][59] . . . [S]o long as we see contractual obligation as based on promise, on obligations that the parties have themselves assumed [that is, *consented to*], the focus of the inquiry is on the will of the parties."[60] Promise empowers, and Fried focused in the first instance on the way promise and the enforceability of a promise in contract empowers the promisee: "By promising we put in another man's hands a new power to accomplish his will, though only a moral power: What he sought to do alone he may now expect to do with our promised help, and to give him this new facility was our very purpose in promising. . . . Morality, which must be permanent and beyond our particular will if the grounds for our willing are to be secure, is itself invoked, molded to allow us better to work that particular will."[61] Now we need not yet get carried away with Fried's use of the concept of will; there is no necessary reason to attribute to his sense anything more than the expression of intention, at least so far. And there also is no reason to make too much of his suggestion that morality is "permanent and beyond our particular will." But he was surely wrong about morality: Morality certainly changes as the social settings in which human agents find themselves shift[62] and as we come to appreciate that, what was once perceived as (and may at some level have actually been) a threat is not in fact threatening. That of course is the story of our overcoming racial and sexual preference bias. It is sufficient to understand Fried's point here as nothing more than identifying that there is value to a promisor in recognizing a way for the promisor to transfer a right to the promisee. Fried then made clear just how and why the promisor benefits from, so would agree to, a system that accommodates such transfers: "In order that I be as free as possible, that my will have the greatest possible range consistent with the similar will of others, it is necessary that there be a way in which I may commit myself. . . . If it is my purpose, my will that others be able to count on me in the pursuit of their endeavor, it is essential that I be able to deliver myself into their hands more firmly than where they simply predict my future course."[63] There is nothing here that the instrumentalist could not recognize: The promisor increased his "autonomy by providing means for restricting it."[64] At the close of this portion of his interpretation, though, Fried posited a notion of human agency that is compelling and suggestive of the limits of the morality he will develop further: "unless one assumes the continuity of the self and the possibility of maintaining complex projects over time, not only the morality of promising but also *any coherent picture of the person becomes impossible.*"[65] We can appreciate how Fried's conception of human agency, then, is central to his theory and how if he is wrong about that, he is wrong about a great deal. Fried takes no account of the manipulability of consent and the chimerical nature of consent in recurring contractual contexts.

For now, focus on what is most prescient about Fried's conclusion: The morality of promising, in the Kantian deontological terms that Fried will embrace, depends on the immutability of the self—assumes, that is, the dispositionist conception of the transactor that situational realities undermine.[66] Fried recognized the demands that his understanding of the morality of promising imposes on the notion of human agency. It would follow, then, that if Fried's assumption that a coherent picture of the human agent depends on the stasis that neuroscience rejects in favor of an authentic conception of human agency that instead recognizes the ubiquity and power of manipulable situational dynamics, then Fried's morality of promising is assailable. Morality of promising and dispositional continuity of agency operate in tandem; when we expose the insubstantiality of such dispositional continuity (and recognize the power of manipulation), the morality of promising Fried claims would fail: And it is in the context of consumer form contracts, where consent is most attenuated, that that failure is most often evident.

Fried's discovering the morality of promising in Kantian autonomy actually further undermines his argument, again most noticeably in the context of consumer contracting: "The obligation to keep a promise is grounded not in arguments of utility but in respect for individual autonomy and in trust."[67] Fried focused on the liar and promise-breaker who use another person, immoral actors certainly under Kantian autonomy theory. But there would seem to be no normative difference between someone who uses another by breaking a promise to them and one who manipulates another to enforce a promise obtained without actual, rather than just formal, assent. (Barnett's consent-based response to that suggestion is treated in a later discussion.)[68] At this point, once again, Fried relied on a depiction of human agency that is simplistic and ultimately undermines his normative argument: "If we decline to take seriously the assumption of an obligation because we do not take seriously the promisor's prior conception of the good that led him to assume it, to that extent we do not take him seriously as a person. We infantilize him, as we do quite properly when we release the very young from the consequences of their choices."[69] That analysis and conclusion is redolent of Moore's suggestion that we owe it to the guilty to punish them; to fail to do so would be to denigrate their humanity (apparently in the "this is going to hurt me more than it is going to hurt you" sense). But the perversity of such arguments is revealed when we confront the moral responsibility fiction. That perversity is particularly stark in the consumer contracting setting, in which sophisticated dominant parties take advantage of the naive, take advantage of the very elements of the human agency that Fried's Kantian theory obscured. In the criminal law setting we could see how Adrian Raine's appreciation of the environmental and genetic forces that mold individuals constrain their choices,[70] and so undermine their moral responsibility. In contract the problem is even more acute: It is the dominant contracting party that presents and then exploits the frailties that undermine veridical consent, the consent that contract assumes contrary to evidence developed in the preceding part of this chapter, an assumption that is misplaced and that would be the basis of the moral responsibility Fried requires.

Elaboration of Promise-Based Contract Theory

Charles Fried found fertile ground in promise theory to rationalize contract, but his appreciation of the parameters of the identity he described did not exhaust the normative incidents of comparing promise and contract, that is, his idea that the more seriously we take the promise-contract relationship the better we understand promise, contract, and the relationship between them. There are or may be important normative differences between promise and contract, and when we understand those differences we may better understand the contract doctrine and its incidents.

Kimel began his comparison of promise and contract by positing that promises engender trust and create a relationship that does not otherwise exist between strangers:

> The keeping of the promise . . . closes a circle through which we establish a bond of trust and respect, thus somewhat removing ourselves, so to speak, from the domain of strangerhood. Stated plainly, the point is that the practicality of promises between strangers hinges on the actors' very disposition to treat each other, for the matter of the promise and as far as the promise is concerned, the way people do not normally, or at least cannot generally be expected to, treat strangers. If you like, effective promises between complete strangers are possible, but the more effective, the more likely a promise is to render its parties no longer complete strangers.[71]

That psychological insight is the key to Kimel's understanding of the normative role of promising. The promise shifts the state of mind of the promisor and promisee: the creation of a relationship, presumably of a unique kind.

All communication between strangers, verbal or not, of course creates a relationship of sorts. That relationship assumes normative significance based on what is said, what is not said, the appearance of the strangers to one another, and the context in which the communication takes place. But Kimel concluded that promise is uniquely suited to create trust. That is an empirical conclusion that would seem necessarily to entail some psychological acuity. Indeed, it is not clear that a philosopher would be in the best position to confidently make that type of assertion. Would it not be more convincing if the assertion, that promise creates trust that somehow converts strangers into something closer than strangers, were supported by experimental evidence? Is there really reason to believe that the consumer signing a package of mortgage documents to evidence a loan that is then sold several times before finally reaching the ultimate servicer actually feels anything like the trust that could not exist between strangers? It is, at least, not clear that on any authentic psychological level Kimel's conclusion is plausible, much less convincing. But the reply to that skepticism may be that it is unfair (and unhelpful) to take Kimel too literally here. What he may have been after was no more than a description of what promise tends to do in familiar interpersonal contexts so that we might better appreciate the relationship between promise and contract. So we may continue, if not with nodding agreement, at least with indulgent curiosity.

Kimel based his conception of the power of promise on the familiar Kantian imperative of autonomy: Promising, and contract, in turn, increases the autonomy of promisor and promise. He drew on Raz and then connected Raz and Fried:

> Raz went on to explain that the moral principles governing the binding force of promises "can only be justified if the creation of such special relationships between people is held to be valuable." And personal autonomy can indeed play a central role in explaining why it is valuable. . . . When someone promises willynilly, makes promises that it is not in her interests to make, or makes promises to the wrong people, or at the wrong time (etc.), she may be authoring more of her own obligations, but authoring them badly—and, most likely, "badly" precisely in the relevant sense: to the detriment of her personal autonomy.[72]

This is still the stuff of psychology, for how we would measure much less conceive of autonomy and its benefits unless we appreciate its effect on the psychological and ultimately cognitive welfare of the "autonomous" human agent? But even if Kimel's terms are vague, he did eventually afford them some substance in a portion of his argument that engaged Fried's speculation about "the continuity of the self and the possibility of maintaining complex projects over time":[73] "can the freedom and the capacity to change one's mind, at least in circumstances where the change in question is relative to a past commitment of the agent's own making, ever have a distinct value, appreciable by those who see personal autonomy as key to the good life? I think so." It is here that Kimel threw off the inauthentic conception of the human agent as static, fixed in time and circumstance, and replaced it with the more robust understanding of human agency such as that which emerges from neuroscientific insights: "the continuity of past commitments may exceed that of the relevant self. The continuing self, that is, may also continue to evolve. Particularly over extended periods of time, tastes develop, priorities change, temperaments mellow or intensify, new passions are ignited, old ones extinguished."[74] Kimel recognized that what is at stake is, indeed, the authenticity of our account of human agency: "Perhaps the quality at stake is authenticity. . . . Particularly when it is remembered that personal autonomy consists primarily in long-term endeavors and relationships, it becomes natural to think of it as involving not only *original* authorship of fundamental aspects of one's life, but also a continuing sense of authorship—a striving, if you like, towards authenticity in one's pursuits."[75] Kimel accurately recalibrated the normative calculus by correctly coming to terms with the nature of human agency and its relationship to conceptions of autonomy that, alas, remain obscure or at least unfocused conceptions. He recognized that the morality of contracting is not cashed out solely in terms of fictional conceptions of disposition but instead must take account of the agent's circumstances, his situation at the time he assumes the contractual obligation, and his situation sometime later, when the consequences of that assumption eventuate.

What would seem to fall squarely within Kimel's sensitivity to the normative dynamics of human agency in contract is appreciation of the manipulability of ostensible assent to contract liability. It is cynical in the extreme to build an interpretation of contract on such manipulation, which, though, is just what the libertarianism of Barnett's "consent theory of contract" accomplished. That is considered more fully later, but for now it suffices to recognize that as Kimel's more authentic understanding of human agency refines what promise can mean to contract, less authentic conceptions would undermine any true morality of contracting. Shiffrin explored what the relationship between promise and contract can mean for both promise and contract and found Fried's treatment of the two as symbiotic troubling. Her thesis lacks legal sophistication, though, and relies on logical leaps that undermine any normative point she might make.

Contract Is Not Promise, or Promise Is Not Contract

Shiffrin concluded that promise is diminished if we say that contract is promise, and that we, human agents, are in turn diminished when we equate the two: If we conclude that promise is no more than contract, then promise is not a very attractive social convention because contract would seem to fix dubious moral parameters. And she based that conclusion on what are essentially the infirm retributionary principles that inform noninstrumental theories of criminal punishment:

> By claiming that contract diverges from promise, I mean that although the legal doctrines of contract associate legal obligations with morally binding promises, the contents of the legal obligations and the legal significance of their breach

do not correspond to the moral obligations and the moral significance of their breach. For instance, the moral rules of promise typically require that one keep a unilateral promise, even if nothing is received in exchange. By contrast, contract law only regards as enforceable promises that are exchanged for something or on which the promisee has reasonably relied to her detriment. When breach occurs, the legal doctrine of mitigation, unlike morality, places the burden on the promisee to make positive efforts to find alternative providers instead of presumptively locating that burden fully on the breaching promisor. Morality classifies intentional promissory breach as a wrong that, in addition to requiring compensation, may merit punitive reactions, albeit sometimes minor ones; these may include proportionate expressions of reprobation, distrust, and self-inflicted reproofs, such as guilt. Contract law's stance on the wrongfulness of promissory breach is equivocal at best, manifested most clearly by its general prohibition of punitive damages.[76]

Setting aside the suggestion that we could not accommodate a meaningful normative relationship between contract and promise,[77] the important point to be gleaned here from Shiffrin's thesis is her conclusion that promise is a device that contemplates punishment and so contract fails, as a normative matter, because contract does not incorporate punishment; indeed, as a matter of doctrine if not strict practice, contract abjures punishment. But we might conclude that it is in fact promise that fails as a normative matter because promise does, if Shiffrin is right, incorporate noninstrumental punishment. It might be that contract rather than promise has the better of the normative contest. Whether that is true, whether retribution is normatively coherent in the case of human agents, depends on our conception of human agency. For the same reasons that retribution fails in the criminal law setting, you could conclude that punitive reactions to breach of promise fail: All punishments, for noninstrumental purposes, are immoral given the nature of human agency. Indeed, insofar as contract liability, *distinct from criminal liability*, is strict liability (at least after a fashion), contract makes more normative sense than legal doctrine that depends on the moral responsibility of human agents, or at least it could by instrumentalist lights. Shiffrin completely ignored important distinctions between the normative commitments of the primary areas of legal doctrine premised on the bases of liability.

Shiffrin's error is more pronounced when proceeding toward her avowed object in demonstrating the difference between contract and promise. Focus now on her argument that when we equate contracts with promises we diminish promising:

The basic concern begins with a background supposition about good behavior and forms of habituation in thought, emotion, and behavior. Namely, a great deal of morally virtuous behavior depends upon cultivating sound instincts and habits and allowing these to guide one's behavior. Morally good agents do not and cannot consciously redeliberate about all the relevant considerations bearing on a decision on every occasion. For everyday matters, agents must often depend on past deliberations that have become encoded into their general cognitive, emotional, and behavioral reactions to moral choices. Much of this deliberation and encoding is supported directly by social institutions and influenced more indirectly by the behaviors they encourage and render salient or standard. This may be especially true when the law plays (or is meant to play) a leadership role in shaping social practice. If this abbreviated account is plausible, then we should be concerned about law's assigning significantly different normative valences and expectations to practices that bear strong similarity to moral practices, especially if we expect both practices to occur frequently and often alongside each other. That is, we should be concerned that the one will influence the other, making it more difficult to maintain those habits and reactions that are essential to the moral behavior. To expect otherwise, one would need to rely heavily on a clear delineation of the different behaviors and their proper contexts, as well as on our abilities to compartmentalize tightly.[78]

Balanced against this concern (but not to admit it is a valid concern *ab initio*)[79] would be the moral failure of undermining the normative coherence that contract doctrine might vindicate in *not* incorporating retributionary, noninstrumental ideas that are inconsistent with an authentic sense of human agency. To the extent that the punishment Shiffrin would promote assumes the moral responsibility of human agency, her argument fails.

Kimel, relying on Raz[80] (and in contrast with Shiffrin), concluded that promising is a unique institution and has a value distinct from legal enforceability:

> [T]he link which Raz highlights between promising and relationships hints at another proposition. Special relationships between people, relationships the parties to which are united by bonds that do not exist between people in general, can be said to be valuable in themselves, regardless of the possibility of cooperation or the co-ordinated pursuit of various projects which are essentially external to the relationship. If so, then promising, not only as a practice by which people undertake obligations to others, but particularly as a practice grounded, as it is, in trust and respect, may be valuable—intrinsically valuable—for its capacity to promote and reinforce personal relationships.[81]

For present purposes it does not matter that Kimel's conclusion there relied on psychological insights for which he did not cite authority; what does matter is that there may be a plausible psychological and normative reason to distinguish the undertakings represented by promise and contract: We need not find identity between the two. Now that does not mean that the two conventions could not be related in some way. Indeed, promise may be a constituent of contract and may be so without our needing to confuse the two as Shiffrin suggested we necessarily could, and would at our moral peril. Common usage would suggest that we are able to appreciate a wide range of normative undertakings within the term *promise*, and one constellation of dependent promises might well be afforded legal significance that another constellation would not. But the primary point remains that both Kimel (more coherently) and Shiffrin (less so) infer philosophical truth from unsubstantiated psychological conjecture. That is, of course, just easier from the armchair.

Waller offered a scenario that grounds an understanding of promise, and its psychological incidents, on a more empirical base (though also from the armchair).[82] He told the story of Rachel and Sarah, both of whom make a promise to a friend and both "under at least some circumstances" will keep their promise. But after making their promises, each receives an attractive alternative opportunity, which would preclude their keeping their promise. Rachel keeps her promise but Sarah does not. Why?

> Rachel is a "chronic cognizer" who often and easily switches into careful rigorous "slow" deliberation,[83] while Sarah is a "cognitive miser" who finds careful deliberate thought more aversive. Rachel has a strong sense of cognitive self-efficacy, and is confident that she can effectively work out the right path through her own critical powers, while Sarah has much less confidence in her ability to rationally reach the right conclusion. Rachel has a strong internal locus-of-control, and believes that what happens in her life is largely under her own effective control; Sarah has an external locus-of-control, and regards herself as largely under the control of external powers. Rachel has not been faced recently with any challenging decisions, while Sarah has struggled with a number of very difficult problems over the past couple of hours and is currently in a state of severe ego-depletion . . . that makes her particularly likely to avoid hard deliberation in the present case. Rachel deliberates carefully and thoroughly about her present choice, reflecting fully on the importance of friendship, honoring her commitments, exercising self-control and thus becoming even stronger.[84]

It does not matter whether Waller captured all the dimensions of the decisional dynamic, whether his presentation of the two women was an exhaustive psychology profile. What does matter is whether you can recognize
either Rachel or Sarah (perhaps, to an extent, in yourself, from time to
time?). The point is that Rachel and Sarah are not normatively identical
because each is the product of forces that have acted upon them, nature
and nurture: There is no independent "I" (or "they") to push back against
those forces. And there is no ideal "I" either against whom it makes sense
to measure the psychological dimensions of promising in any meaningful
way. That is not to say we cannot conceive of ideal conditions, decisions,
and promise makers; it is to say that in formulating what *promise* means,
as a psychological matter, to individual promisors, to authentic human
agents, we are not able to gain traction if we ground conclusions in the
inauthentic ideal.

To different degrees, promise theorists such as Fried, Kimel, and Shiffrin could not really understand the relationship between promising and
contract because their premises deny the relationship between promising
and authentic human agency. When each builds a conception of contract,
and its normative substance and aims, on such arid conceptions of promise, it is unlikely that they would be able to interpret the doctrine coherently, and certainly could not account for contracting contexts in which the
"most human" incidents of human agency are fundamental, such as boilerplate in consumer form contracts. *Promise* describes a range of normative
relationships, and that range is obscured by simplistic normative accounts
of what the fictional unitary formulation of contract might be. It is only a
conception of human agency informed by neuroscientific insights that can
make clear the limitations (failure, actually) of such sterile promise theory,
and only a more nuanced and rich conception that could found an interpretation of the doctrine that would matter to the ordering of the affairs of
authentic human agents.

In fact, it is when authenticity is most sacrificed for the fiction that
would support a political agenda that interpretations of contract doctrine
most egregiously fail.

The Consent Pretense

Barnett's consent theory of contract has, perhaps curiously, been afforded
significant prominence in the theoretical dialogue. His thesis is quite
straightforward: You are bound by what you ostensibly agree to. "Consent"

corrects will theories, such as Fried's "promise" theory of contract, because
it must do so to facilitate contract's role in accommodating commerce:
"because the subjective approach [vindicated by will theories of contract]
relies on evidence inaccessible to the promisee, much less to third parties,
an inquiry into subjective intent would undermine the security of transactions by greatly reducing the reliability of contractual commitments."[85]
He relied on Hume to provide some normative legitimacy for that conclusion.[86] An objective theory *must* operate in lieu of "A rigorous commitment
to a will [including a promise] theory."[87] Barnett's thesis, though, did not
rest on such dubious practical necessity.[88] He grounded it in morality, and
a sense of morality that departs from Fried's Kantian conclusion. Rather
than vindicating freedom of contract, Barnett concluded, Fried's "contract
as promise" would undermine that freedom: "To consent to contract is to
commit to be legally responsible for nonperformance of a promise. So consent is a commitment in addition to whatever moral commitment inheres in
a promise. . . . Requiring a manifested intention to be legally bound facilitates the will or private autonomy of the parties, because one's manifested
intention is highly likely to reflect an underlying subjective assent."[89] That
is facially absurd, at least as it would pertain in the context of some forms
of contracting. It most certainly strains credulity in the most common contracting context: consumer form contracts. And the moral argument, based
on autonomy, is cynical in that boilerplate actually undermines rather than
serves autonomy in just the way dominant contracting parties intend it to
(or, at least, know it will). Barnett conceded that to an extent in his "highly
likely" qualification, and his statement that "*To the extent that parties themselves are normally the best judge of their own interests*, the substance of the
agreements that result from the parties' consent are also likely to be fair."[90]
In the end it is not the existence of actual consent but "the manifestation of
consent to be legally bound [that] facilitates this relational quality of contract law."[91] Barnett took much more seriously, and far too seriously, the
ostensible reality of the inauthentic conception of the human agent that
emerges from the contract formation fictions. As a result, whatever he was
writing about, it had little to do with human agents as they are actually constituted (and exploited by those who appreciate their vulnerability and the
doctrine's susceptibility to unconscientious manipulation).

Barnett offered an interpretation of contract, that is, a description
both of what the doctrine *is* and what the doctrine *should be*. He argued
that contract as consent was a better interpretation of contract than was
Fried's contract as promise, and therefore a superior theory. But from a

perspective informed by the materialism of neuroscientific insights that undermines both Fried's conception of promise and even more certainly Barnett's arguments for the morality of contract by manifestation of consent, neither theory takes human agency seriously and so, ultimately fails as a worthwhile normative contribution.

What is most striking, indeed, discouraging, is that some combination of the promise and consent theories *does* seem to offer the best positive interpretation of the current doctrine. The weight of current authority[92] and the direction of the law rely on manifestation of consent, with some provision for formal confirmation of that manifestation.[93] There also continues to be reliance on deal-policing mechanisms to provide courts the means to avoid overreaching,[94] in those cases where subordinate contracting parties have the sophistication to resist contract enforcement on such grounds (and the circumstances in which such resistance would make much sense are likely quite few, particularly insofar as the United States Supreme Court has insulated even the most egregious terms from court challenge).[95] That most recent iteration of the restatement of the contract doctrine was drafted by three scholars whose own very fine prior work had identified the insubstantiality of manifestations of consent.[96] And their Consumer Contracts Restatement may soon be officially promulgated by the American Law Institute.[97] So the doctrine actually continues to diverge, at an accelerated rate, from the more authentic conception of human agency that neuroscience depicts. Contract doctrine is headed in precisely the wrong direction, and the noninstrumental interpretative theory on offer is impotent to stop it.

Conclusion

This chapter completes the neuroscientific critique of the noninstrumental normative theory supporting the three primary legal areas: criminal, tort, and contract law. This and the preceding chapter have demonstrated that neuroscientific insights, broadly construed, can affect contract law and theory as profoundly as they can upset the familiar and comfortable depictions of the human agent at the foundation of the criminal and tort law. Indeed, the neuroscientific critique may even be more devastating in contract than it is in the other two primary areas, though certainly concepts of moral responsibility are pervasively suspect. But at least in the criminal law there is no reason to deny the reality of the essential delict: the crime. We can acknowledge that crime is substantial, not a fiction, not a creation of the law "from whole cloth." And torts too are real: People suffer casualty to their persons and property because of the actions of others. But contract creates liability out of something(s) that we do not understand well at all: consent and promises.

We can engage moral responsibility, and correctly deny it, in the case of crimes and torts because we can at least make sense of causal responsibility and then distinguish the two. But it is more difficult to engage that same idea in contract, where the very basis of responsibility *simpliciter* is suspect: Can we be sure that consent has the kind of fixed and determinant essence that the law can capture and operationalize? If we can only make sense of consent in terms of promise or will, then, at least in the context of consumer form contracts, are we standing, precariously, on the parapets of sand castles in the air? A vandalized home or totaled car are certainly real. But consent and promise are not fixed and determinant; the nature of human agency being what it is, they could not be. At the end of the day, then, the greater understanding of human agency provided by neuroscientific insights could well be more threatening to contract, at least in the context of consumer form contracts, than it could be to either the criminal or tort law.

The following, and final, chapter of the book takes aim at perspectives that would deny, perhaps because they misunderstand, the assertions and implications of the neuroscientific critique of legal doctrine and noninstrumental normative theories of law.

CHAPTER EIGHT

An Age of Realization

The idea that our own understanding of ourselves, our conception of what it means to be human, may be wrong (or an illusion) is, to say the least, disquieting. How could we have been so wrong for so long about something that is literally right under, or above and behind, our very noses? It would be like learning that the earth is not flat, not the center of the universe, that stars are not fixed, that we are descended from "lower" life forms, that there is no God: very dangerous ideas. For most of human history and even down to the present time many people have been killed for even entertaining such thoughts. The fact that such a reconceptualization of human agency would change something so pervasive as our law and legal systems makes the matter even more salient.

Fodor famously concluded that "if commonsense intentional psychology really were to collapse, that would be, beyond comparison, the greatest intellectual catastrophe in the history of our species; if we're wrong about the mind, then that's the wrongest we've ever been about anything."[1] The sky surely would be falling. Of course, to compare how wrong we may be about human agency with how wrong we have been about other things ("innate" racial superiorities, "innate" gender superiorities, heliocentrism, intelligent design, bloodletting, occultism, magic, alchemy, just for examples) might be a tad alarmist. To discover the extent, or even existence, of the catastrophe, it would be necessary to focus on a context and then to measure the effect of reconceptualizing more accurately human agency in that context. That is what this book has endeavored to do. It has taken seriously the materialism and monism, the hard determinism established by emerging neuroscientific insights that reveal the deficiencies of the folk psychological heuristics that have animated and actually determined the contours of our social system of law and its three primary fields: criminal, tort, and contract law.

This final chapter will take up the challenge Fodor presented by knocking down the strawmen that founded his fears and the fears of like-minded legal theorists. It also will draw a telling parallel between the challenge the current age presents to our thought and the challenge that Bacon, Descartes, Copernicus, Galileo, and Newton confronted in their time. If the seventeenth century was the Age of Genius[2] and the eighteenth century was the Age of Enlightenment, then the twenty-first century may be ushering in a new Age of Realization, the realization that we are not what we thought we were. If we are able to marshal that better understanding, the realization will not be the catastrophe Fodor feared, but a time when our social institutions, most prominently our laws, will necessarily become more humane because they will better reflect what it means to be human.

The primary foil of this final chapter will be ideas recited by many of the meek compatibilists who, by definition, are unable to come to terms with the consequences of hard determinism for conceptions of moral responsibility. A recent book chapter and a recent article written by Stephen Morse, the leading defender of the legal status quo in the face of the hard deterministic critique,[3] provide at several junctures an accessible rendition of those arguments in favor of keeping on the blinders that neuroscientific insights vindicating the materialism of hard determinism would remove. Assault on each of six strawmen that perspective has set up demonstrates the failure of current legal conceptions and foreshadows the Age of Realization.

Legal Doctrine Is Folk Psychological

It is, though, only fair, and completely accurate, to begin by noting the important idea that Morse got completely right: Our extant law—criminal, tort, and contract—depends on commonsense folk psychological conceptions. Morse was correct to recognize, as this book also has emphasized, that it is not just the criminal law that depends on folk psychology and its misconception of human agency; *all* of law, to some extent, depends on folk psychology and folk psychology–supported fictions. Further, normative apologies for noninstrumental interpretations (normative or positive) of the law depend on the same mistake. (Instrumental theories too may be mistaken, as developments in behavioral psychology and economics would reveal,[4] but those errors are empirical rather than conceptual and more remediable therefor.) Yet it is one thing to observe that the law *is* based on folk psychology (a correct observation) and wholly another to say that it must (dubious) or should (wrong) be based on folk psychology.

The language of legal doctrine across the three primary areas considered in this book certainly supports the conclusion that the law assumes the rectitude of folk psychology. Even if instrumental interpretations of the law seem to some more congenial, those instrumental interpretations may be accomplished only by translating moral terms—such as *responsibility*, *fault*, *culpability*—into instrumental terms—such as *efficient*—to achieve instrumental, that is, utilitarian, ends. *Retribution*, *blame*, *fault*, *desert*, *consent*, and *promise* really have no meaning unless afforded one in noninstrumental terms or translated into instrumental objects. As a historical matter, though, it would seem clear that the fundamental concepts were designed to resonate with deontological, probably primarily Aristotelian and Kantian, ideas. It would follow, then, that such folk psychological concepts are only as correct as Aristotle and Kant and noninstrumentalism generally could be about human agency.[5] It is a thesis of this book that deontological normative theory fails to understand the nature of human agency, and that failure is confirmed by the insights neuroscientific findings provide into the way the behavior of human agents is determined. Although it is not necessary (and beyond the scope of this book) to engage the breadth of the deontological tradition, it is the case that if deontology fails, then legal doctrine premised on deontological principles fails along with it. And, of course, deontological normative interpretations of the doctrine would be incoherent.

Strawman 1: "Common Sense"

Fodor referred to "commonsense intentional psychology."[6] Morse built on Fodor's conclusion that that commonsense notion is accurate (as well as indispensable to the current law, not the same thing): "our commonsense understanding of agency and responsibility and the legitimacy of law generally and criminal law in particular are not imperiled by contemporary discoveries in the various sciences, including neuroscience and genetics."[7] The refutation of that summary conclusion was the object of the preceding six chapters. Each of the chapters demonstrated that crucial conceptions supporting pivotal portions of the doctrine and noninstrumental normative theory fail when neuroscientific insights, including the increasing number of studies that demonstrate the effect of nature and nurture on individual human agents, are deployed to reveal the failure of moral responsibility. So it is just not correct that "our commonsense understanding . . . [is] not imperiled." In fact, our commonsense notions are conclusively undermined, at least insofar as criminal law embraces retribution, tort law's negligence takes some sense of wrong or moral responsibility seriously, and contract law depends on consent and promise. Those legs of the doctrine and theory supporting the three primary legal areas are indispensable to our commonsense notions and are imperiled.

Further, though Fodor and Morse may have understood this, commonsense is not an argument. Keep in mind that common sense has explained many clearly wrong and even evil theories and social practices. Wootton recounts the common sense of a typical well-educated European in the seventeenth and eighteenth centuries:

> [W]e will take someone from England, but it would make no significant difference if it were someone from any other European country as, in 1600, they all share the same intellectual culture. He believes in witchcraft and has perhaps read the *Daemonologie* (1957) by James VI of Scotland, the future James I of England, which paints an alarming and credulous picture of the threat posed by the devil's agents. He believes witches can summon up storms that sink ships at sea—James had almost lost his life in such a storm. He believes in werewolves, although there happen not to be any in England—he knows they are to be found in Belgium (Jean Bodin, the great sixteenth-century French philosopher, was the accepted authority on such matters). He believes Circe really did turn Odysseus's crew into pigs. He believes mice are spontaneously generated in piles of straw. He believes in contemporary magicians: he has heard of John Dee, and perhaps of Agrippa of Nettesheim (1486–1535), whose black dog, Monsieur, was thought to have been a demon in disguise. If he lives in London he may know people who have consulted the medical practitioner and astrologer Simon Forman, who uses magic to help them recover stolen goods. He has seen a unicorn's horn, but not a unicorn.
>
> He believes that a murdered body will bleed in the presence of the murderer. He believes that there is an ointment which, if rubbed on a dagger which has caused a wound, will cure the wound. He believes that the shape, color and texture of a plant can be a clue to how it will work as a medicine because God designed nature to be interpreted by mankind. He believes that it is possible to turn base metal into gold, although he doubts that anyone knows how to do it. He believes that nature abhors a vacuum. He believes the rainbow is a sign from God and that comets portend evil. He believes that dreams predict the future, if we know how to interpret them. He believes, of course, that the earth

stands still and the sun and stars turn around the earth once every twenty-four hours—he has heard mention of Copernicus, but he does not imagine that he intended his sun-centered model of the cosmos to be taken literally. He believes in astrology, but as he does not know the exact time of his own birth he thinks that even the most expert astrologer would be able to tell him little that he could not find in books. He believes that Aristotle (fourth century BCE) is the greatest philosopher who has ever lived, and that Pliny (first century CE), Galen and Ptolemy (both second century CE) are the best authorities on natural history, medicine and astronomy. He knows that there are Jesuit missionaries in the country who are said to be performing miracles, but he suspects that they are frauds. He owns a couple of dozen books.[8]

Fortunately, though, "common sense" over time becomes more sensible:

But now let us jump far ahead. Let us take an educated Englishman a century and a quarter later, in 1733[.] . . . Our Englishman has looked through a telescope and a microscope; he owns a pendulum clock and a stick barometer—and he knows there is a vacuum at the end of the tube. He does not know anyone (or at least not anyone educated and reasonably sophisticated) who believes in witches, werewolves, magic, alchemy or astrology; he thinks the *Odyssey* is fiction, not fact. He is confident that the unicorn is a mythical beast. He does not believe that the shape or colour of a plant has any significance for an understanding of its medical use. He believes that no creature large enough to be seen by the naked eye is generated spontaneously—not even a fly. He does not believe in the weapon salve or that murdered bodies bleed in the presence of the murderer.

Like all educated people in Protestant countries, he believes that the Earth goes round the sun. He knows that the rainbow is produced by refracted light and that comets have no significance for our lives on earth. He believes the future cannot be predicted. He knows that the heart is a pump. He has seen a steam engine at work. He believes that science is going to transform the world and that the moderns have outstripped the ancients in every possible respect. He has trouble believing in any miracles, even the ones in the Bible. He thinks that Locke is the greatest philosopher who has ever lived and Newton the greatest scientist. . . . He owns a couple of hundred—perhaps even a couple of thousand—books. . . . The only name we have for this great transformation is "the Scientific Revolution."[9]

More recently (but no less frighteningly) common sense has provided argument in favor of limiting the rights of racial minorities and women[10] (a common sense built into leading religious beliefs and systems as well)[11] and criminalizing sodomy.[12] Common sense, then, does not have an excellent track record and quite often has significantly retarded the progress of human understanding. In only the last few years, what was once common sense is now condemned as wrong, and perniciously so.

It is common sense, Fodor and Morse have confirmed, that supports the notion of human free will. Through scientific advances we have unveiled many of the governing laws of the physical universe, yet the human brain *seemingly* cannot be reduced to any set of rules. Wolfram equated this to the behavior exhibited by a cellular automation model.[13] Although the underlying rules are entirely definite, when carried through enough iterations, the overall system appears to follow no obvious laws. Wolfram identified this as the source of our tendency to believe in free will. Although individual neural cells follow discrete laws, they are too far removed from the complex behavior of the human brain, creating an illusion of freedom. This "irreducible computation" phenomenon indeed continues to fool the typical well-educated citizen of the twenty-first century, an error that perhaps the Age of Realization will allow us to overcome. All this is not to say that common sense is *always* wrong; it is just to make the point that common sense is not an argument: It is a description of accepted thinking that may be right, but may be *very* wrong as well.

Strawman 2: Overconfidence in Science

Morse expressed concern that neuroscience overclaims, and so neuroscience is likely to be enlisted to support arguments it is actually impotent to make. Certainly there are examples of some who would stretch the emerging neuroscience beyond its breaking point, even for apparent pecuniary gain.[14] But the important theoretical critique proceeds on a different level, a more profound level that challenges naive notions of human agency. It is in the elaboration of his overclaim argument that Morse failed to appreciate the nature of the fundamental materialistic challenge to the erroneous conception of human agency vouched safe by the extant legal doctrine and noninstrumental theory:

> Many people intensely dislike the concept and practice of retributive justice,
> thinking that they are prescientific and harsh. Their hope is that the new neuroscience will convince the law at last that determinism is true, no offender is
> genuinely responsible, and the only logical conclusion is that the law should

adopt a consequentially based prediction/prevention system of social control guided by the knowledge of the neuroscientist-kings who will finally have supplanted the platonic philosopher-kings.[15] Then, they believe, criminal justice will be kinder, fairer and more rational. They do not recognize, however, that most of the draconian innovations in criminal law that have led to so much incarceration—such as recidivist enhancements, mandatory minimum sentences, and the crack/powder cocaine sentencing disparities—were all driven by consequential concerns for deterrence and incapacitation. Moreover, as C. S. Lewis recognized long ago, such a scheme is disrespectful and dehumanizing (Lewis 1953). Finally, there is nothing inherently harsh about retributivism. It is a theory of justice that may be applied toughly or tenderly.[16]

It is worthwhile to consider each of those points seriatim.

First, it would, I suppose (with some confidence), be difficult to accurately describe the reasons that those who dislike noninstrumentalist retribution have for disliking it. The problem is not certainly with retribution's being "prescientific and harsh." Compassion and mercy are prescientific in just the same way, and withholding either may seem harsh. But it is not obvious that those who are considerate of compassion and mercy see them as scientific rather than prescientific. They are strategies or dispositions that may or may not be efficacious. It does not strain credulity to believe that those who dispense or withhold compassion, mercy, or retribution, for that matter, do so in a manner considerate of the consequences of their action or inaction. Their object may be behavior modification, and they show compassion or mercy or might even seek retribution, in the instrumentalist sense, because they believe that doing so will yield the best consequences by some measure. Such instrumentalism is not what Morse had in mind. He would defend noninstrumental retribution, the variety that depends on deontological premises, based on something like an ephemeral moral realism.[17] The problem with that perspective is not that it is prescientific; the problem with it is that it is wrong, even morally wrong.

A second point is related. Morse's conclusion is essentially cynical and attributes intellectual dishonesty (or self-delusion) to those who question his conception of common sense. When Morse based his response to determinism on an unease with retribution, he had the reasoning process perfectly reversed: Determinists do not deny compatibilism and embrace determinism because it provides the means to avoid retributionary punishment; they are compelled to the deterministic conclusion and the rejection of noninstrumental retribution that entails because that is the only way they can make sense of human agency in light of neuroscientific realities. Morse opined that those uncomfortable with retribution are hopeful that neuroscience will vindicate determinism. That puts the cart before the horse; it would seem more generous and, likely, more accurate to postulate that science will confirm determinism is true and so retribution based on free will, or at least compatibilism, is ultimately inefficacious. It is not the case that those who have a negative emotional reaction to retribution are *looking* for an instrumental argument against it. Hard determinists take human agents as they find them and *then* conclude that noninstrumental retribution is inefficacious. Proof of determinists' intellectual honesty and Morse's cynicism would be found in the consequences that determinism would discover from a conception of human agency that makes no room for retribution.

And this entails the important third point: While the rejection of retribution would likely lead to the abrogation of some forms of punishment, including capital punishment, that is because determinism reveals the ultimate immorality and inefficacy of some punishment regimes (e.g., the death penalty and isolation). It would be immoral to put to death someone who could not have ultimate moral responsibility or to isolate an incorrigible youth, who may prove to be corrigible as a young adult, who would be profoundly impaired by the isolation experience.[18] At least that conclusion is open to the determinist. But there is nothing in the deterministic understanding of human agency that would actually preclude any form of punishment that did not, as an instrumental matter, result in more desirable consequences. As a matter of fact, though, it would seem that few could argue, once retribution was removed from the calculus, that the death penalty results in more desirable consequences. There are voluminous studies that confirm capital punishment's ultimate inefficacy as a means of reducing crime and criminality.[19] And can we be sure that retribution does not justify the harsh treatment of at least some juvenile offenders?[20]

Further, there is nothing in the deterministic conclusion that necessarily results in shorter sentences for those convicted of engaging in antisocial activities. Indeed, the deterministic conclusion might result in essentially indeterminate sentencing, the kind of thing we are already seeing in the case of sexual predators.[21] The deterministic perspective, which entails the conclusion that human agents are not morally responsible, would be neither more nor less harsh then. Sentencing would be a function of the continuing risk the convicted criminal presents to social welfare. That is why the neuroscientifically sophisticated response to the curious case of Mr. Oft[22] concludes that once the offending tumor is confirmed to have been the efficient cause of Mr. Oft's aberrational behavior and is certainly completely removed, there is no reason to incarcerate him. But retributive principles dependent on an inauthentic conception of human agency might well reach the opposite conclusion, as would Morse on conceptual rather than empirical bases.[23]

Indeed, if Mr. Oft's tumor could not have been thoroughly excised, so as to render him no longer a threat,[24] the determinist who understands human agency as not entailing any notion of moral responsibility would find good reason to detain Mr. Oft, for the same reason public health officials would limit the movement of those with communicable diseases.[25] But what would a punishment system based on retribution do? It would seem that you are to be set free when your sentence ends, when you have "paid your debt to society." It would, of course, be serendipitous in the extreme if the risk were ameliorated precisely when (and not years before or after) you have paid that debt. The problem is not just that retribution is based on a fiction, the fiction of morally responsible human agents; it is that societal control by reference to retributionary principles will almost certainly lead to punishment practices in the criminal law that will frustrate rather than serve the interests of societal security. Retribution is a guess, based (for those who would follow, for example, Moore)[26] on an emotional reaction but ultimately measured by something other than that emotional reaction, or so it would seem.[27] One could certainly make the argument that all noninstrumental normative theory, at least since the time of Kant, has been an effort to rationalize emotional reactions, to give some ostensibly sophisticated and more than occasionally opaque reason for feeling the way you do.[28]

It would seem that Morse recognized, to his credit, the ultimate indeterminacy of retribution: "there is nothing inherently harsh about retributivism. It is a theory of justice that may be applied toughly or tenderly."[29] But there is ambiguity there: Is that an argument in favor of retribution (it seems to be) or is it a criticism of retribution's arbitrariness? Actually, it may be an argument in favor that fails to appreciate its own built-in refutation. Curiously, Morse offered no citation in support of that conclusion. Perhaps he considered it self-evident. But recognize that what Morse seemed to find lacking was any connection between retribution and instrumental purpose. So Morse's retribution would often not just fail to track instrumental objects; it might ultimately undermine them. Would he really endorse a retributive punishment that results in *more* crime, perhaps because the perpetrator had no reason to *feel* enough, or a certain quantum of, guilt? And, if Morse would reject that measure of retribution, what measure would he endorse? (Moore seems to have undermined the other contestants.)[30] Morse would have to posit and defend a theory of retribution before he could conclude that retribution could make any claim to even cooperate with instrumental bases of punishment. He offered no such defense: He just assumed that some sense of retribution makes sense within his incomplete conception of human agency.

Strawman 3: Ought from Is

Morse made the point that criminal law especially, but tort and contract law as well, is based on folk psychology.[31] That is certainly true. But from that accurate positive observation he seemed to proceed to the normative conclusion that legal doctrine consistent with folk psychology is necessarily correct. That conclusion is built on the idea that folk psychology is correct because folk psychology explains how humans reason. Morse correctly observed that human agents respond to reason. (Of course it is true that all living organisms respond to reasons, broadly construed: Your dog wags its tail when it sees you or sits on your command for its own reasons.) From that observation, Morse concluded that it is folk psychology that, in fact, explains human agency: "Unless people are capable of understanding and then using legal rules to guide their conduct, the law is powerless to affect human behavior. The law must treat persons generally as intentional, reason-responsive creatures and not simply as mechanistic forces of nature."[32] It is one thing to say that people respond to reasons, the kind of reasons for which folk psychology seems to account. But it is wholly another to premise sufficient free will, via compatibilism, to premise moral responsibility on the ability to respond to reasons. And that is the crucial point that compatibilists miss; it is the point that demonstrates the insufficiency, and ultimate malignancy, of folk psychological justifications for the imposition of moral and, for that matter, legal responsibility.

It is at this point that Morse revealed the vacuity of the normative case he tried to make in favor of folk psychology: "Virtually everything for which agents *deserve* to be praised, blamed, rewarded or punished is the product of mental causation and, in principle, is responsive to reasons, including incentives."[33] Certainly, desert only makes sense in a system of moral responsibility. That conclusion is pertinent if all human agents were uncaused causes, *equally*, or even roughly equally, competent to respond to reasons, including incentives. But we are not, as the work of Adrian Raine, considered at length in chapter 2, made clear. We do not all start out at the same place, and we do not all receive the same material and immaterial gifts that determine our ability to respond to reasons. The compatibilist argument that it all equals out in the end is just wrong.[34] Indeed, it is compatibilism's insistence that there is enough free will that makes the compatibilist conclusion incoherent. Small, seemingly insignificant discrepancies in competence at the outset (perhaps as a function of genetics or epigenetics) may determine significant differences in life choices and opportunities.[35]

While Morse was correct that criminal, tort, and contract doctrine is certainly based on folk psychological conceptions of human agency, he was wrong to assume that the law *should be* based on folk psychology. As a matter of fact, it would not even be correct to assume that the law *must* be based on folk psychology. Instrumental theories of criminal, tort, and contract liability could easily be built on a normative system that wholly rejects folk psychology, and, contrary to Morse's conclusion, doing so would not undermine human dignity; rather, it would recognize the accurate conception of human agency and allocate loss in the manner most likely to encourage human thriving.

In the meantime, though, we can and likely will limp along making determinations of legal responsibility by reference to folk psychology. At many junctures, such as with regard to the doctrine considered in this book, we could apply the law even if we reject the premise of folk psychology. Morse has recognized that "Folk psychology does not presuppose the truth of free will, it is consistent with the truth of determinism, it does not hold that we have minds that are independent of our bodies (although it, and ordinary speech, sound that way), and it presupposes no particular moral political view."[36] But would folk psychology make any sense if there is no such thing as free will? What would be the point of law based on folk psychology if there is no free will? It is surely the case that law can proceed from an inaccurate conception of human agency, but should it? To do so is ultimately to frustrate human thriving and the morality that even noninstrumentalists themselves seem to consider sacrosanct: treating like cases alike but recognizing differences that matter.

Morse thought that folk psychology works well enough as he conceived it as long as "human action is in part causally explained by mental states."[37] So "close enough is good enough"? But even for that conception of folk psychology to make normative sense, to connect human action to mental states in a way that would support moral responsibility, folk psychology would need to do the very work that cognitive psychology, informed by neuroscientific insights, reveals more and more often that folk psychology cannot do. Curiously, Morse did not seem to appreciate the troublesome nature of his own observation that "demonstrating that an addict has a genetic vulnerability or a neurotransmitter defect tells the law nothing *per se* about whether the addict is responsible."[38] That seems to be correct: The reason why someone is an addict does not matter to legal doctrine that relies on folk psychology. We can take no issue with that as a positive matter. But Morse again at this juncture seemed to infer the *ought* from the *is*. If we are not uncaused causes, but, instead, the sum total of forces that have or have not acted upon us, the fault or credit, such as it would be, is not ours; the responsibility lies entirely outside us for the simple reason that there is no place within us independent of those forces, over which "we," as separate reflective entities, have any control. Ultimately, then, law that is inconsiderate of that fact, law that blithely goes on assuming we are something that we are not will not fail to operate; it will just not operate very well.

It is here that we must find most curious Morse's argument, considered above,[39] that because of some lack of capacity *not necessarily manifest in behavior* psychopathy should be an excuse to criminal liability. What could possibly be the normative difference between the psychopath who lacks the ability to empathize because of trauma, genetic vulnerability, or structural brain defect and the addict who has the genetic vulnerability or neurotransmitter defect that is the efficient cause of her addiction? On what basis could Morse draw a normative distinction if all he has to work with are the rough tools, the too often misleading heuristics of folk psychology?[40]

Strawman 4: Compatibilism Is True

Now we come to what Morse correctly understood to be the central foundation of his apology for the folk psychological basis and interpretation of legal doctrine: Compatibilism, the means to make sense of moral responsibility in a deterministic world, is the majority view among philosophers and is consistent with common sense, and so it must be true. There are several responses to that conclusion, but it is worthwhile to note that Morse seems to be a self-avowed determinist;[41] after all, only those who accept the essential truth of determinism need compatibilism to try to make some sense of the moral responsibility of human agents in a deterministic world.

Assuming that nose counting is of value, Morse discovered support for his conclusion in a poll that found that 59 percent of philosophers are compatibilists.[42] "[P]lausible 'compatibilist' theories suggest that responsibility is possible in a deterministic universe.[43] Indeed, this is the dominant view among philosophers of responsibility and it most accords with common sense."[44] Morse then relied on that conclusion to maintain that any contrary metaphysical argument would face a very high threshold to deny effectively the possibility of responsibility (which, he said, none does).

Initially, it is not clear what it means to say that 59 percent of philosophers share any conclusion about compatibilism and responsibility if no more than a small number of them agree on what compatibilism must be in order to overwhelm arguments against moral responsibility. It is not enough for Morse to just count the number of philosophers who agree with his conclusion: He would have to find sufficient noncontradictory reasons among them to support that belief. And it is difficult to find any two much less as many as 50 percent of philosophers who subscribe to the same conception of compatibilism and who reach the same conclusion about the effect of their conclusion on moral responsibility. Further, many of the arguments for compatibilism were developed before the dawn of the Age of Realization, when neuroscience could reveal aspects of human agency that philosophers could not even imagine, even from the armchair. It also is likely that the percentage of philosophers who endorse compatibilism may not be static; it is certainly true that naturalistic perspectives have grown in influence in recent years, largely as a consequence of neuroscientific insights.[45] It also would not be surprising that philosophers as a group need a worldview that makes room for moral responsibility, else they are out of business. Certainly much of the practice of philosophy depends on a view of human agency that accommodates the deontology of Kant and noninstrumentalism generally. Where would champions of those perspectives be if the human agent did not have some moral responsibility, if free will did not exist at all? It is not clear, then, why we would take the word of a divided, even fractured, corner of an endangered intellectual perspective and afford the view held by that group as entitled to extraordinary deference, especially on account of something as ephemeral as common sense.

The fact remains, though, that Morse's description of legal doctrine as wholly unconcerned with the fact—and, I would argue, ultimate undermining truth—of determinism is certainly accurate. But he, again, muddied the waters by misconstruing the materialistic critique. He assumed that determinists are looking for a way to excuse criminal acts: "the claim is that causation *per se* is an excusing condition. This is sometimes called the 'causal theory of excuse.'[46] Thus, if one identifies genetic, neurophysiological, or other causes for behavior, then allegedly the person is not responsible."[47] At that juncture, Morse certainly confused, though certainly not intentionally, important and divergent senses of responsibility and the normative significance of them.

Surely determinism does not entail excusing antisocial behavior; the determinist would be most interested in reducing the menace of those who upset the social equilibrium by instrumentally changing their behavior or, failing that, reducing the risk that those engaging in such behavior present to human thriving. So if there had been no surgery that could have removed the tumor that was the efficient cause of Mr. Oft's sexually predatory behavior, he would have to have been incarcerated and perhaps removed from situations in which he posed a threat for as long as he posed that threat. Understanding what causes an effect is not the same as excusing it, at least if excuse means something like forgive.[48] Morse's apparent failure to appreciate that crucial point is quite clear:

> Non-causation of behavior is not and could not be a criterion for responsibility because all behaviors, like all other phenomena, are caused. Causation, even by abnormal physical variables, is not per se an excusing condition. Abnormal physical variables, such as neurotransmitter deficiencies, may cause a genuine excusing condition, such as the lack of rational capacity, but then the lack of rational capacity, not causation, is doing the excusing work. If causation were an excuse, no one would be responsible for any action. Unless proponents of the causal theory of excuse can furnish a convincing reason why causation per se excuses, we have no reason to jettison the criminal law's responsibility doctrines and practices just because a causal account can be provided.[49]

Of course no one is responsible, in the moral sense that would support such as retribution, for anything. That is the point. But (virtually) everyone is responsible in a causal sense,[50] and that is all instrumental theory needs in order to deal with them and to protect society.

Part of the problem, illustrated by Morse's understanding of responsibility, may be that our vocabulary depends on the same misunderstanding that empowers misconceptions of human agency. Our ability to correctly conceptualize human agency is undermined by "[t]he powerful and ubiquitous presence of our moral responsibility system [that] makes the truth of moral responsibility seem obvious, and objections to moral responsibility seem silly or incoherent. Our vast moral responsibility system has been developed and refined over many centuries, and its elaborate network of rules and principles makes it difficult to step outside the system and level criticisms against it."[51] It is as though we have been wearing virtual reality headsets that depict moral responsibility by denying the virtual nature of the image projected and cannot come to terms with authentic—materialistic, determined—human agency revealed by neuroscientific insights (again, broadly construed to include epigenetics, cognitive psychology, and behavioral economics). But once the normative vocabulary better comports with the realities of human agency, once moral responsibility is appreciated as the error it is in the case of human agency, we may approach normative questions generally and questions of legal doctrine and theory specifically in a way that will accommodate a more coherent system of social regulation. That assertion leads ineluctably to confrontation with another argument in favor of the status quo.

Strawman 5: Neuroscience Does Not Explain Why Folk Psychology Is Wrong

Keep in mind that folk psychology is not so much persistently wrong as it is persistently awkward, and so too often compels results and conclusions that are inconsistent and even incoherent. Punishing Mr. Oft because, in folk psychological lights, he is culpable, to blame for the actions a tumor in his brain provoked, is a mistake, from any coherent normative perspective that would not subscribe to insubstantial syllogisms such as "Brains don't convince each other; people do."[52] Surely that offers no real argument. We hope that Morse would agree that people use their brains, and only their brains, to convince other people (insofar as everything everyone has ever done or ever could do is the product of brain activity). Certainly even those who believe that the mind is an uncaused cause of thought and action understand that the brain must at least instantiate what the mind somehow causes. We would not imprison someone for having bad eyesight; we would give them corrective lenses. So you could understand the brain as just as much of a physical system as are our senses generally, and just as or even more prone to error, which is sometimes correctible.

When neuroscience discovers the source of aberrant behavior in a neural anomaly, neuroscience has indeed explained why folk psychology is wrong in the particular instance. When neuroscience isolates (if and when it can) the efficient source of the neural aberration, demonstrates the specific chemical, electrical, or structural anomaly that triggered the aberrant behavior, neuroscience has indeed supplanted folk psychology. Think of it this way, and the analogy, though worn, is apt: Your car fails to start one morning, and that causes you great distress and inconvenience. Applying folk psychological principles of blame and culpability to your car, you would sentence it to the garage for one week after replacing the starter. You will be without a car for the week, but you will have taught your car a lesson. Alternatively, after replacing the starter, you could just continue to drive the car. That would be better for you and for the rest of society that depends on you and your car.

It would certainly make sense to sentence someone to a prison term for correction, if, contrary to fact, correction were actually the result of serving that sentence,[53] just as it would make sense to sentence your car to a few hours' "rest" if its malfunction were related to overuse. It would make no more sense to sentence your car to rest for a week, though, if that would not improve your car's performance, than it would to sentence a juvenile to isolation for his blameworthiness and culpability, if so doing actually resulted in his becoming a greater threat to society.

So every time neuroscientific insights help us isolate the cause of aberrant behavior and correct them in a way that would be obscured (or missed entirely) by the operation of folk psychological principles, neuroscience explains why folk psychology is wrong. Now that does not mean that the folk psychological response will not feel good, on some primitive level. We are, in fact, wired (or at least predisposed) to feel in just that way; that predisposition is fed by the very basic emotional reaction that proved adaptive. But it would be grave error to continue to rely on folk psychological conceptions that depend on those emotions and that will mislead in a society only vaguely like the one encountered by our forebears on the savannah. The problem is not so much that belief, desire, motivation, and similar folk psychological conceptions are wrong; the problem is that they are so imprecise that they too often mislead us into making decisions that are now actually maladaptive. Folk psychology worked well enough long enough, just as Newtonian physics worked well enough to explain what needed to be explained up to the twentieth century. But when Einstein explained time and relativity in terms that demonstrated Newton's mistakes, we used Einstein's conceptions to make better sense of our universe.

Strawman 6: Neuroscience Cannot Explain Justice

Belief, desire, and intent[54] are the stuff of folk psychology: We infer something about other people's motivations, and so their actions and predispositions, from inferences we draw from surrounding circumstances about their beliefs, desires, and intents. We like or dislike others on the basis of folk psychology states; we punish based on such states; and we impose civil liability too on the basis of inferences about beliefs, desires, and intents. That makes a measure of sense to even instrumentalists because a certainly accurate judgment about another's beliefs, desires, and intents would be an excellent indicator of their dangerousness, or at least the extent to which they present a threat to the generally favored social order. So the greatest problem with folk psychology is not just that it focuses on the wrong thing or that it supports the imposition of legal liability on insubstantial bases. Folk psychological judgments may well be coincident with accurate judgments of dangerousness, of sociopathy. The problem with folk psychology is that it is a system of heuristics that actually obscures the important normative calculus by relying on imprecise inferences and then sanctioning (in the sense of punishing) the indicator—belief, desire, and intent—rather than the source of the indicator. We *knew* that Mr. Oft *intended* to sexually assault his stepdaughter. If we stop at the folk psychology conclusion and do not go further, to the level of cognitive neuroscience, we miss the opportunity both to more effectively respond to the risk his behavior presents and to conserve the resources that would otherwise be wasted on his incarceration (including the cost to him as well). In that way folk psychology leads to suboptimal, inefficacious—even bad—results, results that are not normatively defensible either.

But what about conceptions such as justice, or fairness, or even reasonableness, conceptions that seem to entail necessarily a normative calculus? Could neuroscience unpack in any meaningful way what justice is? Yes, once we understand that justice, fairness, reasonableness, and their cognates generally describe not morally real things but emotional reactions. When used in a noninstrumental sense, those terms are best understood as exclamations: captured by moral pronouncements such as "that's just wrong." Now that assertion, or observation, does not entail emotivism or even noncognitivism. Indeed, nothing suggested here is inconsistent with some conceptions of cognitivism. The observation merely reflects an empirical conclusion about the typical fit between the use of terms such as *justice* and the underlying constituents of what might in fact be just once we identify a basis of morality, perhaps in terms of human thriving. Whether that observation moves the needle or even pertains to meta-ethical questions about the nature, existence, or identity of moral properties is quite beside the point made here.

Neuroscience certainly can explain emotional reactions and the exclamations that proceed therefrom. To the extent that such exclamations are used to describe an instrumental result, we can either do the mathematics to determine whether the challenged result in fact serves or disserves the instrumental object, or decide that the mathematics cannot be done or would be too expensive to do.[55] If we use *justice* and its cognates in their most familiar noninstrumental sense, and understand that such terms describe emotional reactions, neuroscience can help us make sense of them.

Kandel won the Nobel Prize in Physiology or Medicine for his research into memory storage.[56] Memory is learning. Kandel studied very simple organisms[57] and discovered how they *and we* learn.

> [Nerve cells] have been conserved . . . through millions of years of evolution. Some of them were present in the cells of our most ancient ancestors and can be found today in our most distant and primitive evolutionary relatives: single cell organisms such as bacteria and yeast and simple multicellular organisms such as worms, flies, and snails. These creatures use the same molecules to organize their maneuvering through their environment that we use to govern our daily lives and adjust to our environment. . . . [T]he human mind evolved from molecules used by our lowly ancestors and . . . the extraordinary conservation of molecular mechanisms that regulate life's various processes also applies to our mental life.[58]

So such neural phenomena—chemical, electrical, and structural—are *all* that we are, all that any living thing is. That is why Kandel's work was able to generalize from the simplest organisms to the human agent. All we are, all we can be, is a collection of neural material and an array of neural reactions.[59]

There is, then, nothing mysterious, nothing holy, about justice. Winners generally find that the result was just; losers reach the opposite conclusion. That is probably adaptive, even if the perceptual mechanics seem suspect. Deontological ratiocination aside, it is no more complex than the disagreement between the opposing fans over a close call at home plate, or even between many of the judgments made by supporters of one rather than another political party. Neuroscience explains justice as it explains everything else about human agency, in mechanical terms, and that is true even if we have not yet figured out all the mechanics. That may be, for some, even for most, the cause of an awakening. And the coming age, an Age of Realization, may be the rudest awakening yet.

Notes

Chapter One

1. Stephen J. Morse, "Lost in Translation? An Essay on Law and Neuroscience," in *Law and Neuroscience*, Current Legal Issues 13, ed. Michael Freeman (Oxford, UK: Oxford University Press, 2011), 559–62.

2. Allen v. Bloomfield Hills School Dist., 760 N.W.2d 811, 815–16 (Mich. Ct. App. 2008).

3. Van Middlesworth v. Century Bank & Trust Co., 2000 WL 33421451, 1, 6–7 (Mich. Ct. App. 2000).

4. Roper v. Simmons, 543 U.S. 551 (2005): execution of individuals under age 18 at time of crime is violation of Eighth and Fourteenth Amendments; Graham v. Florida, 560 U.S. 48 (2010): Eighth Amendment prohibits sentence of life without parole for juveniles guilty of nonhomicide crimes; Miller v. Alabama, 132 S. Ct. 2455 (2012): Eighth Amendment proscribes mandatory life sentences without possibility of parole for juveniles.

5. Tor D. Wager et al., "An fMRI-Based Neurologic Signature of Physical Pain," *New England Journal of Medicine* 368 (2013): 1388, 1396, showing a functional MRI (fMRI) can discriminate between physical and emotional pain.

6. Portions of this section are developed from Peter A. Alces, "Naturalistic Contract," in *Commercial Contract Law: Transatlantic Perspectives*, ed. Larry A. DiMatteo, Qi Zhou, Séverine Saintier, and Keith Rowley (Cambridge: Cambridge University Press, 2013), 85, 110–12.

7. Richard Joyce, *The Evolution of Morality* (Cambridge, MA: MIT Press, 2006).

8. G. E. Moore, *Principia Ethica* (Cambridge: Cambridge University Press, 1903).

9. For thoughtful elaborations, see, e.g., William K. Frankena, *Ethics* (Englewood Cliffs, NJ: Prentice Hall, 1973), 85–86; Allen Gibbard, "Normative and Recognitional Concepts," *Philosophy and Phenomenological Research* 64 (2002): 151–53; Ridge, "Moral Non-Naturalism," *Stanford Encyclopedia of Philosophy*, ed. Edward N. Zalta, http://plato.stanford.edu/archives/spr2010/entries/moral-non-naturalism/.

10. See, e.g., Ridge, "Moral Non-Naturalism," 8; Bernard Williams, *Ethics and the Limits of Philosophy* (Cambridge, MA: Harvard University Press, 1985), 121–22.

11. See generally Sam Harris, *The End of Faith* (New York: W. W. Norton, 2004); Sam Harris, *Letter to a Christian Nation* (New York: Knopf, 2006). Harris is perhaps best known for his strident atheism.

12. Sam Harris, *The Moral Landscape: How Science Can Determine Human Values* (New York: Free Press, 2010), 158.

13. Ibid., 43.

14. Patricia S. Churchland, *Braintrust: What Neuroscience Tells Us about Morality* (Princeton, NJ: Princeton University Press, 2011), 200.

15. See Vilanyan S. Ramachandran, *A Brief Tour of Human Consciousness* (New York: Pi Press, 2005), 59: "The solution to the problem of aesthetics, I believe, lies in a more thorough understanding of the connections between the thirty visual centers in the brain and the emotional limbic structures (and of the internal logic and evolutionary rationale that drives them)."

16. See, e.g., Alvin I. Goldman, "Theory of Mind," in *Oxford Handbook of Philosophy of Cognitive Science*, ed. Eric Margolis, Richard Samuels, and Stephen Stitch (Oxford, UK: Oxford University Press, 2012), 402–3, referring to folk psychology as another name for theory of mind; Ian Ravenscroft, "Folk Psychology as a Theory," *Stanford Encyclopedia of Philosophy*, ed. Edward N. Zalta, http://plato.stanford.edu/archives/fall2010/entries/folkpsych-theory/: "Psychologists rarely use 'folk psychology,' preferring the phrase 'theory of mind.'"

17. Folk psychology relies on manifestations of beliefs, desires, motivations, and the like to draw inferences about the causes and consequences of those dispositions. So it is like induction from the specific to the general, and it is as fallible (or, at least, incomplete).

18. Much as behaviorism works well, much of the time; but each proceeds from a level of abstraction that alternative perspectives reveal to be insufficiently articulated, some of the time.

19. And before we knew this to be true, we just assumed *A* was evil.

20. Portions of this section are developed from Peter A. Alces, "Naturalistic Contract," 85, 110–12.

21. See, e.g., M. R. Bennett and P. M. S. Hacker, *History of Cognitive Neuroscience* (Malden, MA: Wiley-Blackwell, 2008), 241: "The central theme of our book was to demonstrate the incoherence of brain-body dualism"; Morse, "Lost in Translation?," n5: "I do not mean to imply dualism here. I am simply accepting the folk-psychological view that mental states . . . play a genuinely causal role in explaining human behaviour"; Michael S. Pardo and Dennis Patterson, *Minds, Brains, and Law: The Conceptual Foundations of Law and Neuroscience* (New York: Oxford University Press, 2013), xiii: "We are not dualists."

22. See, e.g., M. R. Bennett and P. M. S. Hacker, *Philosophical Foundations of Neuroscience* (Malden, MA: Blackwell, 2003), 3: "An alternative conception of mind—the one we believe is more plausible—is as an array of powers, capacities,

and abilities possessed by a human being. These abilities implicate a wide range of psychological categories including sensations, perceptions, cognition . . . cogitation . . . emotions and other affective states . . . and volition." A reader may note that the evidence suggests that every example of a mental state corresponds with a neural state, even if that exact relationship is beyond our current understanding.

23. Stephen J. Morse, quoted in Stuart Fox, "Laws Might Change as the Science of Violence Is Explained," *Live Science*, June 7, 2010, http://www.livescience.com /6535-laws-change-science-violence-explained.html.

24. See Bennett and Hacker, *Foundations of Neuroscience*.

25. Ibid., 72.

26. Bennett and Hacker, *Foundations of Neuroscience*, 29.

27. For example: "But [neuroscience's] discoveries in no way affect the conceptual truth that these powers and their exercise in perception, thought and feeling are *attributes of human beings*, not of their parts—in particular, *not of their brain*" (Bennett and Hacker, *Foundations of Neuroscience*, 3); "[C]urrent neuroscientists (and others) ascribe a multitude of psychological (especially cognitive and volitional) functions to the brain. . . . To do so is in effect to ascribe to a part of an animal attributes which it makes sense to ascribe only to the animal as a whole" (ibid., 15); "[O]nly human beings—not brains (or non-language-using creatures)—can *construct models* of anything, and use models to *predict events*" (ibid., 313).

28. Pardo and Patterson, *Minds, Brains, and Law*.

29. Ibid., 2–4. Pardo and Patterson also protested, unconvincingly, that they are not dualists. They made their argument only by recourse to nonphysical substance, the defining characteristic of dualism. Hence Glannon pointed out that Pardo and Patterson came "dangerously close to . . . substance dualism" (Walter Glannon, "Brain, Behavior, and Knowledge," *Neuroethics* 4 [2011]: 194). Their method was linguistically distinct but functionally identical to Cartesian dualism; they substituted *person* and *brain* for *mind* and *body*, but dualism by any other name is still dualism.

30. For Bennett and Hacker, see Maxwell Bennett et al., *Neuroscience and Philosophy: Brain, Mind, and Language* (New York: Columbia University Press, 2007), 4: "Empirical questions about the nervous system are the province of neuroscience. . . . By contrast, conceptual questions . . . are the proper province of philosophy." For Pardo and Patterson, see Michael S. Pardo and Dennis Patterson, "Philosophical Foundations of Law and Neuroscience," *University of Illinois Law Review*, no. 4 (2010): 1211, 1222: "We explain why the question of what the mind is and what the various psychological categories under discussion are (e.g., knowledge, intention, rationality), are conceptual rather than empirical questions."

31. See Pardo and Patterson, "Philosophical Foundations," 1211, 1222n62.

32. Thomas Nadelhoffer, "Neural Lie Detection, Criterial Change, and Ordinary Language," *Neuroethics* 4 (2011): 210 (emphasis added).

33. David L. Wallace, "Addiction Postulates and Legal Causations, or Who's in Charge, Person or Brain?," *Journal of the American Academy of Psychiatry and the Law* 41 (2013): 93.

34. See, e.g., Oliver Sacks, *The Mind's Eye* (New York: Knopf, 2010); Sacks, *Seeing Voices* (Berkeley: University of California Press, 1989); Sacks, *The Man Who Mistook His Wife for a Hat* (New York: Summit Books, 1985); Sacks, *A Leg to Stand On* (New York: Summit Books, 1984); Sacks, *Awakenings* (New York: Harper Perennial, 1973).

35. Sacks, *Man Who Mistook* (visual agnosia).

36. Sacks, *Leg to Stand On* (body integrity identity disorder).

37. Sacks, *Mind's Eye* (prosopagnosia, or "face blindness").

38. John Colapinto, "Brain Games: The Marco Polo of Neuroscience," *New Yorker*, May 11, 2009, 76; Vilanyan S. Ramachandran, D. Rogers-Ramachandran, and S. Cobb, "Touching the Phantom Limb," *Nature* 377 (1995): 489.

39. See R. C. Coghill, "Pain: Neuroimaging," in *Encyclopedia of Neuroscience*, ed. Larry R. Squire (New York: Academic Press, 2009); Rolf-Detlef Treede et al., "The Cortical Representation of Pain," *Pain* 79 (1999): 105–11.

40. Vilanyan S. Ramachandran, D. Rogers-Ramachandran, and M. Stewart, "Perceptual Correlates of Massive Cortical Reorganization," *Science* 258 (1992): 1159–60.

41. See Antonio R. Damasio, *Descartes' Error: Emotion, Reason, and the Human Brain* (New York: Harper Perennial, 1995), 35–50.

42. *Location* here must be understood in the sense of "network" too; while a network may imply process rather than location, the constituents of the network have a physical location or locations.

43. See generally Stephen J. Morse, "Avoiding Irrational NeuroLaw Exuberance: A Plea for Neuromodesty," *Mercer Law Review* 62 (2011): 837–60.

44. Morse, "Lost in Translation?," 529–62.

45. Ibid., 559.

46. Ibid.

47. Ibid., 560.

48. For Morse's source of the account of Mr. Oft, see Jeffrey M. Burns and Russell H. Swerdlow, "Right Orbitofrontal Tumor with Pedophilia Symptom and Constructional Apraxia Sign," *JAMA Neurology* 60 (March 2003): 437–40, http://archneur.jamanetwork.com/article.aspx?articleid=783830. (Note added.)

49. Morse, "Lost in Translation?," 560.

50. Ibid.

51. Robert M. Sapolsky, "The Frontal Cortex and the Criminal Justice System," *Philosophical Transactions of the Royal Society of London B: Biological Sciences* 359 (2004): 1787–88.

52. Morse, "Lost in Translation?," 560–61.

53. See Daniel Kahneman, *Thinking, Fast and Slow* (New York: Farrar, Straus and Giroux, 2011); Daniel Kahneman, Paul Slovic, and Amos Tversky, eds., *Judgment under Uncertainty: Heuristics and Biases* (Cambridge: Cambridge University Press, 1982).

54. See A. David Redish, *The Mind within the Brain: How We Make Decisions and How Those Decisions Go Wrong* (Oxford, UK: Oxford University Press, 2013).

55. See, e.g., Roper, 543 U.S. 551 (2005); Graham, 560 U.S. 48 (2010); Miller, 132 S. Ct. 2455 (2012).

56. See Roper, 543 U.S. 551, 617 (2005) (Scalia, J., dissenting), pointing out that although American Psychological Association (APA) argued that "persons under 18 lack the ability to take moral responsibility for their decisions" in a murder case, APA had previously argued that "juveniles are mature enough to decide whether to obtain an abortion without parental involvement."

57. See, e.g., M. E. Thomas, "Confessions of a Sociopath," *Psychology Today*, May 7, 2013, in which a law professor, using a pseudonym, describes her life as sociopath: "Despite having imagined it many times, I've never slit anyone's throat."

58. See, e.g., Scott O. Lilienfeld and Ashley Watts, "Not All Psychopaths Are Criminals—Some Psychopathic Traits Are Actually Linked to Success," *The Conversation*, January 26, 2016, https://theconversation.com/not-all-psychopaths-are-criminals-some-psychopathic-traits-are-actually-linked-to-success-51282.

59. See Robert Hare, "Focus on Psychopathy," *FBI Law Enforcement Bulletin*, July 2012, https://leb.fbi.gov/2012/july/focus-on-psychopathy, which estimates that 10 to 15 percent of criminal offenders are psychopaths.

60. See generally Sapolsky, "Frontal Cortex."

61. See American Law Institute, *Restatement (Third) of Torts* (St. Paul, MN: American Law Institute, 2012), § 2(b) and 2(c), showing that even with regard to "strict products liability," the language of the restatement still appeals to fault-based theories of liability.

62. For more on this conception of tort law as described by Ernest Weinrib's theory of corrective justice in tort, see the section "Weinrib's Corrective Justice" in chapter 5.

63. See Jules Coleman, *The Practice of Principle: In Defence of a Pragmatist Approach to Legal Theory* (Oxford, UK: Oxford University Press, 2001), 13–24, discussing bilateralism in tort law.

64. See Christopher H. Schroeder, "Corrective Justice and Liability for Increasing Risks," *UCLA Law Review* 37 (1990): 439–78, challenging the requirement that "an individual must have caused harm before he or she can be held liable in tort" and suggesting instead, under corrective justice theory of tort, "that we be liable when we have increased the risk of harm occurring, whether or not it eventually does."

65. See John C. P. Goldberg and Benjamin C. Zipursky, "Civil Recourse Defended: A Reply to Posner, Calabresi, Rustad, Chamallas, and Robinette," *Indiana Law Journal* 88 (2013): 569, 571–74; see also Jason M. Solomon, "Civil Recourse as Social Equality," *Florida State University Law Review* 39 (2011): 243, 244, justifying the right to civil recourse in terms of distributive justice and social equality.

66. See A. Mitchell Polinsky, *An Introduction to Law and Economics* (New York: Wolters Kluwer Law and Business, 1989). Strict liability's goal may be to achieve "right" amount of activity: "In essence, the rule of strict liability results in efficient

behavior because it forces the injurer . . . to take into account all of the adverse effects of his behavior on the victim" (ibid., 41).

67. See American Law Institute, *Restatement (Second) of Contracts* (St. Paul, MN: American Law Institute, 1981), § 19 cmt. b; American Law Institute, *Restatement (Second) of Torts* (St. Paul, MN: American Law Institute, 1965), § 283.

68. For example, criminal assault has civil equivalents in battery and assault; trespass can carry both civil and criminal liability; illegal terms in a contract will not be enforced and may result in criminal charges; and fraudulent misrepresentation can result in a contract being nullified or, in some cases, in criminal charges.

69. See, e.g., David Grey, "Punishment as Suffering," *Vanderbilt Law Review* 63 (2010): 1619, 1682–93; Adam Kolber, "The Subjective Experience of Punishment," *Columbia Law Review* 109 (2009): 182–236. And it may be too cruel or immoral, in some noninstrumental sense, to exact more punishment than is necessary to specifically deter.

70. Jeremy A. Blumenthal, "Law and the Emotions: The Problems of Affective Forecasting," *Indiana Law Journal* 80 (2005): 155. Affective forecasting refers to one's predictions as to the nature, intensity, and duration of future emotional states. See also George Loewenstein, "Affect Regulation and Affective Forecasting," in *Handbook of Emotion Regulation*, ed. James J. Gross (New York: Guilford Press, 2007), 180. Much of the law assumes that people are able to predict accurately their future emotions, but recent research indicates that our affective forecasts often fail to match our actual reactions (ibid.).

71. Richard A. Posner, *Economic Analysis of Law*, 8th ed. (New York: Aspen, 2010), 279, noting that, according to empirical literature, criminals generally do, in fact, behave as if they are "rational calculators" capable of appreciating such factors.

72. See Polinsky, *Introduction to Law and Economics*, 80: "in deciding whether to [take a risk], [a person] will compare his benefit [from taking the risk] to the expected fine—the fine multiplied by the probability of detection."

73. See generally Aristotle, *Nicomachean Ethics*, ed. Roger Crisp (Cambridge: Cambridge University Press, 2000).

74. See John Braithwaite, "Restorative Justice: Assessing Optimistic and Pessimistic Accounts," *Crime and Justice* 25 (1999): 1; Howard J. Vogel, "The Restorative Justice Wager: The Promise and Hope of a Value-Based, Dialogue-Driven Approach to Conflict Resolution for Social Healing," *Cardozo Journal of Conflict Resolution* 8 (2007): 565, 572. Restorative justice focuses on restoring victims, offenders, and communities through a collaborative process that promotes social healing.

75. See Kolber, "Subjective Experience of Punishment," arguing the need to take into account subjective factors, such as offenders' dispositions toward certain punishments, and not just objective factors, when sentencing in order to justify punishment regimes.

76. Although sometimes, the phrenologists got lucky. One memorable example is with Nobel prize-winning economist Ronald Coase. When taken to a phrenologist at age eleven, he was told he possessed "considerable mental vigour." Soon after,

his parents worked so that Coase could take the secondary school scholarship examination, which enabled him to transfer out of a "school for physical defectives" (Coase, "Biographical," in *Nobel Lectures: Prize Lectures in Economic Sciences*, vol. 3, *1991–1995*, ed. Torsten Persson [Singapore: World Scientific, 1997]), 7.

77. See John Fleischman, *Phineas Gage: A Gruesome but True Story about Brain Science* (Boston: Houghton Mifflin, 2002), 1–10, recalling the improbable survival of Phineas Gage, a railroad construction foreman impaled through brain by large metal rod.

78. See "The Many Dangers of Brain-Based Lie Detection," *Center for Science and Law*, November 19, 2012, http://www.neulaw.org/blog/1034-class-blog/4131-the -many-dangers-of-brain-based-lie-detection. For more examples of the current limitations of the neuroscience and some companies' attempts to make claims potentially exceeding those limits, see Chapter 8, note 14.

79. Although deontological perspectives are not helpful in conceptualizing morality, utilitarian moral philosophies may be. See "The Immorality of Moral Responsibility" in chapter 3.

80. Benjamin Libet, "Unconscious Cerebral Initiative and the Role of Conscious Will in Voluntary Action," *Behavioral and Brain Science* 8 (1985): 529–39.

81. Daniel M. Wegner, *The Illusion of Conscious Will* (Cambridge, MA: MIT Press, 2002), 2.

82. See, e.g., Tim Bayne, "Libet and the Case for Free Will Scepticism," in *Free Will and Modern Science*, ed. Richard Swinburne (Oxford, UK: Oxford University Press, 2011); Eddy Nahmias, "Why We Have Free Will," *Scientific American*, January 2015, 77–79; Alfred R. Mele, *Effective Intentions: The Power of Conscious Will* (Oxford, UK: Oxford University Press, 2009), 21–91.

83. See Eddy Nahmias, "Is Free Will an Illusion?," in *Moral Psychology*, vol. 4, *Free Will and Responsibility*, ed. Walter Sinnott-Armstrong (Cambridge, MA: MIT Press, 2014), 22–23n6 (citing http://philpapers.org/surveys/results.pl), finding that 59 percent of philosophers identified themselves as compatibilists; David Bourget and David J. Chalmers, "What Do Philosophers Believe?," *Philosophical Studies* 170 (2013): 476, finding that of the 59 percent of philosophers who identify as compatibilists, 35 percent fully accept compatibilism and 24 percent "lean toward" compatibilism.

Chapter Two

1. Roper v. Simmons, 543 U.S. 551 (2005).
2. Ibid., 569–70.
3. Ibid.
4. Retributivists recommend equal punishments for offenders who commit equally blameworthy crimes. But retributivists fail to take into account other pertinent aspects of punishment that make punishment unequal. Adam Kolber ("Punishment

and Portfolio of Beliefs" [unpublished manuscript, 2016], 35) offered the example of two equally culpable rapists, one of whom develops testicular cancer and is required to undergo major surgery and drug treatments, which diminish his ability to commit another sexual offense. Retributivists would recommend the same punishment because the two committed the same crime; instrumentalists would recommend a lesser sentence for the prisoner who is less likely to reoffend.

5. On the relationship between retribution and revenge, see, e.g., Ken Levy, "Why Retributivism Needs Consequentialism: The Rightful Place of Revenge in the Criminal Justice System," *Rutgers Law Review* 66 (2014): 660: "When the state avenges a crime, we call it one thing—retribution; when a private citizen avenges a crime (or perceived transgression), we call it something else—simply revenge"; Richard Lowell Nygaard, "Crime, Pain, and Punishment," *Dickinson Law Review* 102 (1998): 363: "Retribution is revenge plain and simple. We punish offenders who violate the law because we are angry and want to get even. Retribution is about power. It is about force. It is about repression."

6. Certain biblical passages, particularly those in the Old Testament, state plainly that justice demands an eye for an eye, while other passages in the New Testament instruct believers to turn the other cheek. Compare Exod. 21:24 (King James): "Eye for eye, tooth for tooth, hand for hand, foot for foot . . ."; Deut. 19:21: "[L]ife shall go for life, eye for eye, tooth for tooth, hand for hand, foot for foot"; Lev. 24:20: "Breach for breach, eye for eye, tooth for tooth . . ."; with Matt. 5:38–39: "Ye have heard that it hath been said, An eye for an eye, and a tooth for a tooth; But I say unto you, That ye resist not evil: but whosoever shall smite thee on thy right cheek, turn to him the other also"; Luke 6:28–29: "Bless them that curse you, and pray for them which despitefully use you. And unto him that smiteth thee on the *one* cheek offer also the other; and him that taketh away thy cloke forbid not to take thy coat also."

7. See, e.g., Matthew R. Ginther, "The Language of *Mens Rea*," *Vanderbilt Law Review* 67 (2014): 1327–72, finding that individuals do not straightforwardly perceive a moral distinction between knowingly and recklessly as presupposed by the Model Penal Code; Ian P. Farrell and Justin F. Marceau, "Taking Voluntariness Seriously," *Boston College Law Review* 54 (2013): 1545–1612, advocating a voluntariness requirement over a voluntary act requirement because the former concept emphasizes that the offense (not just one constituent act) must be voluntary; Elizabeth Nevins-Saunders, "Not Guilty as Charged: The Myth of *Mens Rea* for Defendants with Mental Retardation," *UC Davis Law Review* 45 (2012): 1419–86, analyzing assumptions implicit in *mens rea* through discussion of awareness of social norms, awareness of likely consequences, capacity to perceive and evaluate choices on offer via social norms and likely consequences, and deliberateness in pursuit of selected choices; Douglas Husak, "Rethinking the Act Requirement," *Cardozo Law Review* 28 (2007): 2437–60, proposing that the concept of controlled act might better accomplish the normative work of *actus reus* requirement than does concept of voluntary act.

8. See, e.g., Sarah Blaffer Hrdy, *Mothers and Others: The Evolutionary Origins of Mutual Understanding* (Cambridge, MA: Belknap, 2009), 273. Mammals, unlike other classes of animals, are born with underdeveloped brains and require protection and guidance past gestation. Humans, in particular, are born with arguably the least matured brains. See, e.g., Sapolsky, "Frontal Cortex": "No mammal in the world has produced young that take longer to mature or depend on so many others for so long as did humans in the Pleistocene" (1787, 1792). For this reason, human children act characteristically uninhibited and often exhibit poor judgment and social skills.

9. See Sapolsky, "Frontal Cortex": "If someone with epilepsy, in the course of a seizure, flails and strikes another person, that epileptic would never be considered to have criminally assaulted the person who [*sic*] they struck" (1788).

10. See, e.g., Sam Harris, *Free Will* (New York: Free Press, 2012): "If a man's choice to shoot the president is determined by a certain pattern of neural activity, which is in turn the product of prior causes—perhaps an unfortunate coincidence of bad genes, an unhappy childhood, lost sleep, and cosmic-ray bombardment— what can it possibly mean to say that his will is 'free'?," 5–6; Derk Pereboom, *Living without Free Will* (Cambridge: Cambridge University Press, 2001): "if hard incompatibilism is true, it would appear unacceptable to blame criminals for what they have done, and we would therefore seem to have inadequate justification for punishing them," 158.

11. See, e.g., Michael Moore, *Placing Blame: A Theory of the Criminal Law* (Oxford, UK: Oxford University Press, 2010): "Retributivism . . . is the view that punishment is justified by the desert of the offender" (87).

12. State v. Kirkbride, 185 P.3d 340 (Mont. 2008).

13. Ibid., 342. Kirkbride attacked Paul Rafferty in order to obtain money for drugs. As Rafferty screamed for help, Kirkbride stabbed Rafferty in the back with a large hunting knife, puncturing his left lung, pericardial sack, and pulmonary artery. Kirkbride took Rafferty's wallet and left him to die in the street.

14. Ibid., 343, holding that although "retribution is not specifically among the sentencing policies articulated in [Montana's correctional and sentencing policy statute,] . . . [t]he Court has repeatedly said retribution is a component of punishment."

15. For other cases reaching similar conclusions about the role of retribution, see, e.g., Norgaard v. State, 339 P.3d 267, 276 (Wyo. 2014): "The American criminal justice system has two general objectives—to punish wrongdoers and prevent future harm"; Doe v. State, 111 A.3d 1077, 1098 (N.H. 2015): "The fourth factor we must examine is whether the statute promotes the traditional aims of punishment— retribution and deterrence. 'Retribution is vengeance for its own sake. It does not seek to affect future conduct or solve any problem except realizing "justice."'"; U.S. v. Irey, 612 F.3d 1160, 1206 (11th Cir. 2010): "Because the punishment should fit the crime, the more serious the criminal conduct is the greater the need for retribution and the longer the sentence should be."

16. See generally Immanuel Kant, *Groundwork of the Metaphysics of Morals*, ed. Mary Gregor and Jens Timmermann (Cambridge: Cambridge University Press, 2011), 1785, introducing the categorical imperative. In a similar vein, see Moore, *Placing Blame*: "the tension that exists between crime-prevention and any other punishment goal . . . is due to retributivism's inability to share the stage with any other punishment goal" (28).

17. Peter H. Rossi and Richard A. Berk, *Just Punishments: Federal Guidelines and Public Views Compared* (New York: Aldine de Gruyter, 1997), 19, 21. The *United States Sentencing Guidelines* are drafted by the United States Sentencing Commission, "a bipartisan, independent agency within the federal judicial branch" (Rossi and Berk, *Just Punishments*, 19) tasked with addressing "issues of disparity, dishonesty, and undue leniency in sentencing" (ibid., 21).

18. 18 U.S.C.A. § 3553(a).

19. See Ernest J. Weinrib, "Corrective Justice in a Nutshell," *University of Toronto Law Journal* 52 (2002), "Corrective justice is the idea that liability rectifies the injustice inflicted by one person on another. This idea received its classic formulation in Aristotle's treatment of justice in *Nicomachean Ethics*, Book V. More recently, it has become central to contemporary theories of private law" (349; internal note omitted). Compare Simon Connell, "What Is the Place of Corrective Justice in Criminal Justice?," *Waikato Law Review* 19 (2011): "Although corrective justice may have a role in the criminal law, it is a limited one" (134).

20. See, e.g., U.S. v. Marshall, 736 F.3d 492, 500 (6th Cir. 2013): "Congress instructs district courts to impose sentences 'sufficient, but not greater than necessary, to comply with' several enumerated purposes in the statute" (citing 18 U.S.C. § 3553(a)); U.S. v. Smith, 325 Fed. Appx. 77, 79 (3d Cir. 2009); U.S. v. Williams, 314 Fed. Appx. 888, 890–91 (7th Cir. 2009).

21. 350 F. Supp. 2d 910 (D. Utah 2005).

22. In November 2007, Judge Cassell resigned from the U.S. District Court for the District of Utah to return full time to the University of Utah, College of Law, to teach, write, and litigate on issues relating to crime victims' rights and criminal justice reform. Paul G. Cassel Faculty Profile, University of Utah (last visited September 12, 2015), https://faculty.utah.edu/u0031056-PAUL_G._CASSELL/biography/index.hml.

23. Judge Cassell supported this conclusion by reviewing the legislative history accompanying the Sentencing Reform Act:

> [Just punishment]—essentially the "just deserts" concept—should be reflected clearly in all sentences; it is another way of saying that the sentence should reflect the gravity of the defendant's conduct. From the public's standpoint, the sentence should be of a type and length that will adequately reflect, among other things, the harm done or threatened by the offense, and the public interest in preventing recurrence of the offense. From the defendant's standpoint

> the sentence should not be unreasonably harsh under all the cir-
> cumstances of the case and should not differ substantially from the
> sentence given to another similarly situated defendant convicted of
> a similar offense under similar circumstances.

Wilson, 350 F. Supp. 2d at 916–17 (quoting S. Rep. No. 98-225).

24. Ibid., 917.

25. Judge Cassell used data reported in a book comparing the Sentencing Guide-lines recommendations with public opinion to support this proposition. See Rossi and Berk, *Just Punishments*. For example, Rossi and Berk found both the guide-lines and the public recommended 39.2 years for a kidnapping in which the victim is killed (92). There were also disparities, however. For example, the guidelines recom-mended 22 years for trafficking crack cocaine, while the public recommended only 10 years (92).

26. See, e.g., Francis T. Cullen and Karen E. Gilbert, *Reaffirming Rehabilitation* (Cincinnati, OH: Anderson, 1982), 155.

27. Model Penal Code § 6.11A (Tentative Draft No. 2, 2011; approved by in-stitute membership in 2011). In 2007, the American Law Institute revisited the sentencing portion of the Model Penal Code for the first time since the code was developed in 1962. Since that time there had been dramatic changes in sentencing philosophy and practices, as well as an unprecedented rise in the US prison popu-lation, making the old code obsolete. See American Law Institute, "Model Penal Code: Sentencing Foreword," Tentative Draft No. 1 (St. Paul, MN: American Law Institute, 2007).

28. Model Penal Code (Tentative Draft No. 1), § 1.02(2) (emphasis added).

29. Model Penal Code(Tentative Draft No. 1), § 1.02 cmt. b.

30. See, e.g., Moore, *Placing Blame*: "The tension between retribution and any other goal of punishment such as crime-prevention prevents us from any comfort-able 'mix' of goods for punishment. We cannot happily seek both goals and kill two birds with the proverbial one stone, for by aiming at one of the birds we will neces-sarily miss the other" (28).

31. Paul H. Robinson and Markus D. Dubber, "The American Model Penal Code: A Brief Overview," *New Criminal Law Review* 10 (2007): 319–20: Though some provisions have not been widely accepted, "the Model Penal Code is the closest thing to being an American criminal code" (320).

32. See, e.g., U.S. v. Kosma, 749 F. Supp. 1392, 1402 (E.D. Pa. 1990), finding de-fendant's paranoid schizophrenia precluded the government from demonstrating beyond reasonable doubt that the defendant intended his statements to President Reagan to be threats; State v. Shank, 367 S.E.2d 639, 643 (N.C. 1988), holding trial court erred by excluding expert testimony tending to show that defendant was inca-pable of premeditation and carrying out plans as a result of mental and emotional disturbances.

33. See 28 U.S.C.A. § 994(a)(2). Policy statements aid federal courts in

determining which circumstances warrant greater or lesser punishment and further the goals stated in 18 U.S.C.A. § 3553(a)(2) as a matter of public policy.

34. U.S.S.G. § 5K2.13 (emphasis added).

35. See generally Sarah Hyser, "Two Steps Forward, One Step Back: How Federal Courts Took the 'Fair' Out of the Fair Sentencing Act of 2010," *Pennsylvania State Law Review* 117 (2012): 503, 504–13. See also John Schwartz, "Thousands of Prison Terms in Crack Cases Could Be Eased," *New York Times*, June 30, 2011. The Anti-Drug Abuse Act of 1986 established a 100:1 cocaine-to-crack ratio. Under this regime, a person convicted of possessing five grams of crack cocaine was subject to the same mandatory minimum sentence—five years in prison—as someone convicted of possessing 500 grams of powder cocaine. In 2010, the US Congress passed the Fair Sentencing Act, which lowered the ratio to 18:1. In June 2011, the United States Sentencing Commission unanimously voted to retroactively apply the law to those imprisoned for crack offenses (ibid.). See also Deborah E. Milgate, "The Flame Flickers, but Burns On: Modern Judicial Application of the Ancient Heat of Passion Defense," *Rutgers Law Review* 51 (1998): "a man who killed the lover of his estranged wife after finding her in bed with him was convicted of manslaughter, while a woman who killed her husband, who had repeatedly abused her, is spending the rest of her life in the penitentiary" (193–194); compare Lucy Fowler, "Gender and Jury Deliberations: The Contributions of Social Science," *William and Mary Journal of Women and the Law* 12 (2005): 1–4. Women were not permitted to serve on juries for most of US history.

36. Sarah-Jayne Blakemore and Suparna Choudhury, "Development of the Adolescent Brain: Implications for Executive Function and Social Cognition," *Journal of Child Psychology and Psychiatry* 47, nos. 3–4 (2006): 296. The brain continues to develop well beyond birth. In early childhood, the brain proliferates synapses ("synaptogenesis") between neurons (ibid., 297). This proliferation is followed by a period of synaptic elimination (or "pruning") in which infrequently used synapses are pruned and frequently used synapses are strengthened (ibid.). This experience-dependent process takes several years and results in a net decrease in synapses, but an increase in specialization and efficiency (ibid.). After puberty the adolescent brain undergoes synaptogenesis and synaptic pruning primarily in the prefrontal cortex (ibid.). Those synapses are pruned into specialized, efficient networks, which in turn improve executive function (e.g., inhibition and decision making) and social cognition (e.g., self-awareness and empathy; ibid., 301–2). Much like the fine-tuning of senses during infancy, adolescence may be the critical period to develop executive functions and social cognition skills after which the brain may be much less malleable (ibid., 307).

37. In fact, research shows that critical neural development related to decision making continues into a person's twenties, and the adult brain may not be fully formed until age 22 or later. See Nico U. F. Dosenbach, "Prediction of Individual Brain Maturity Using fMRI," *Science* 329, 1358–61; Kathryn L. Mills et al.,

"Developmental Changes in the Structure of the Social Brain in late Childhood and Adolescence," *Social Cognitive and Affective Neuroscience* 123 (2014): 123–31; V. J. Schmithorst and W. Yuan. "White Matter Development during Adolescence as Shown by Diffusion MRI," *Brain and Cognition* 72 (2010): 16–25.

38. Sarah-Jane Blakemore, "The Mysterious Workings of the Adolescent Brain," *TED*, June 2012, http://www.ted.com/talks/sarah_jayne_blakemore_the_mysterious _workings_of_the_adolescent_brain?language=en (citing Jason Chein et al., "Peers Increase Adolescent Risk Taking by Enhancing Activity in the Brain's Reward Circuitry," *Developmental Science* 14 [2011], F1–F10; Iroise Dumontheil et al., "Online Usage of Theory of Mind Continues to Develop in Late Adolescence," *Developmental Science* 13, no. 2 [2010]: 331–38; Jay N. Giedd et al., "Brain Development during Childhood and Adolescence: A Longitudinal MRI Study," *Nature Neuroscience* 2 [1999]: 861–63).

39. Formerly referred to as mental retardation, the definition in American Psychiatric Association, *Diagnostic and Statistical Manual of Mental Disorders*, 5th ed. (Washington, DC: American Psychiatric Publishing, 2013), for intellectual disability specifies an IQ two standard deviations below average (approximately 70). See American Psychiatric Association, *Diagnostic and Statistical Manual*, 33–37 (hereafter cited as *DSM-V*). Recently, Justice Kennedy took note of this change in *Hall v. Florida*: "Previous opinions of this Court have employed the term 'mental retardation.' This opinion uses the term 'intellectual disability' to describe the identical phenomenon" (citations omitted). Hall v. Florida, 134 S. Ct. 1986, 1990 (2014). See also Rosa's Law, Public Law No. 111-256, 124 Stat. 2643 (2010), changing all references of "mental retardation" in federal law to "intellectual disability."

40. Compare Ward Farnsworth and Mark F. Grady, *Torts: Cases and Questions*, 2nd ed. (Austin, TX: Wolters Kluwer Law and Business, 2009), 129, explaining that as matter of policy, the reasonable person standard in tort liability is applied to mentally deficient to incentivize prudent caretakers as well as to ensure that law does not force an injured party to unjustly internalize externalities.

41. See, e.g., Md. Code Ann., Crim. Law § 10-108(c), prohibiting minors from using or possessing tobacco products; N.D. Cent. Code § 5-01-08(1), prohibiting minors from purchasing, consuming, or possessing alcohol.

42. See, e.g., Wash. Rev. Code Ann. § 9A.04.050: "Children under the age of eight years are incapable of committing crime"; N.Y. Penal Law § 30.00(3): "In any prosecution for an offense, lack of criminal responsibility by reason of infancy, as defined in this section, is a defense."

43. See, e.g., Ind. Code § 35-38-1-7.1: "In determining what sentence to impose for a crime, the court may consider the following aggravating circumstances: (3) The victim of the offense was less than twelve (12) years of age or at least sixty-five (65) years of age at the time the person committed the offense"; N.J. Stat. Ann. § 2C:44-1(a)(2): "In determining the appropriate sentence . . . the court shall consider the following aggravating circumstances: (2) The gravity and seriousness of

harm inflicted on the victim, including whether or not the defendant knew or reasonably should have known that the victim of the offense was particularly vulnerable or incapable of resistance due to advanced age . . . or extreme youth."

44. See, e.g., Tenn. Code § 40-35-113 ("[M]itigating factors may include . . . (6) The defendant, because of youth or old age, lacked substantial judgment in committing the offense"). See also State v. Pearson, 836 N.W.2d 88, 97 (Iowa 2013): "under *Miller* and *Null*, a juvenile's culpability is lessened because the juvenile is cognitively underdeveloped relative to a fully-developed adult. This lessened culpability is a mitigating factor that the district court must recognize and consider" (citations omitted).

45. "*Tout comprendre c'est tout pardonner*" ("To understand all is to forgive all"). Michael Moore rejected that commonsense view because, he opined, it leads to biased expert testimony, causes legal theorists and judges to "slant their interpretations of the existing legal excuses," and "may lead to an unfortunate cynicism about the moral basis of the criminal law" (*Placing Blame*, 488–90).

46. Hooks v. Thomas, 2011 WL 4542901 (M.D. Ala. July 1, 2011).

47. Roper v. Simmons, 543 U.S. 551 (2005).

48. *Hooks*, WL 4915840, at *1 (M.D. Ala. Oct. 17, 2011). In 1985 Hooks was convicted of murdering Donald and Hannelore Bergquist during the course of a robbery and was sentenced to death.

49. *Hooks*, WL 4542901, at *3. Hooks claimed that brain-imaging evidence was necessary to prove that executing individuals with frontal lobe dysfunction is unconstitutional under the "cruel and unusual punishments clause" of the Eighth Amendment.

50. Ibid.

51. See also Graham v. Florida, 560 U.S. 48, 82 (2010): The cruel and unusual punishments clause "does not permit a juvenile offender to be sentenced to life in prison without parole for a nonhomicide crime"; *Hooks*, WL 4542901, at *4, construing *Roper*, *Atkins*, and *Graham* to limit imposition of sentences based on discrete categories: defendant's age or defendant's possibility of intellectual disability.

52. See Solemn v. Helm, 463 U.S. 277, 286 (1983), finding that proportionality principle applies to the cruel and unusual punishment clause of Eighth Amendment.

53. Terrie E. Moffitt, "Adolescence-Limited and Life-Course Persistent Antisocial Behavior: A Developmental Taxonomy," *Psychological Review* 100 (1993): 674–701: "When official rates of crime are plotted against age, the rates for both prevalence and incidence of offending appear highest during adolescence; they peak sharply at about age 17 and drop precipitously in young adulthood. The majority of criminal offenders are teenagers; by the early 20s, the number of active offenders decreases by over 50%, and by age 28, almost 85% of former delinquents desist from offending" (674–75).

54. Robert Plomin, Michael J. Owen, and Peter McGuffin, "The Genetic Basis of Complex Human Behaviors," *Science* 264 (1994): 1733–39: "Heritabilities range

from about 40 to 50% for personality, vocational interests, scholastic achievement, and general intelligence" (1733–34).

55. See Y. Gao et al., "Association of Poor Childhood Fear Conditioning and Adult Crime," *American Journal of Psychiatry* 167 (2010): 58.

56. Ibid. (citing D. Tranel and B. T. Hyman, "Neuropsychological Correlates of Bilateral Amygdala Damage," *Archives of Neurology* 47 [1990]: 349–55: "Unlike some other brain regions (e.g., the polar prefrontal and temporal cortices), the amygdala is rarely susceptible to illness and injury, and hence amygdala dysfunction may be more likely to have an early neurodevelopmental basis," 58).

57. Jeffery Rosen, "The Brain on the Stand," *New York Times Magazine*, March 11, 2007: "If adolescent brains caused all adolescent behavior, [Stephen Morse stated,] 'we would expect the rates of homicide to be the same for 16- and 17-year-olds everywhere in the world—their brains are alike—but in fact, the homicide rates of Danish and Finnish youths are very different than American youths.'" Of course, the fact that adolescent homicide rates differ by country may have something to say about nurture's effect on adolescent brains.

58. These questions are considered further in the next chapter.

59. See, e.g., Stephen J. Morse, "Failed Explanations and Criminal Responsibility," *Virginia Law Review* 68 (1982): 971–1083; Morse, "The Ethics of Forensic Practice: Reclaiming the Wasteland," *Journal of the American Academy of Psychiatry and the Law* 36 (2008): 206–17; Morse, "Brain Overclaim Syndrome and Criminal Responsibility: A Diagnostic Note," *Ohio State Journal of Criminal Law* 3 (2006): 397–412.

60. See Stephen J. Morse, "Psychopathy and Criminal Responsibility," *Neuroethics* 1 (2008): 205–12, arguing that psychopaths should not be held criminally responsible for their actions because "the psychopath is not a member of the moral community" (209); "Moral concern plays no role [for the psychopath] in understanding and obeying [societal] prohibitions" (209), and the law should broaden the insanity defense to include psychopathy.

61. *DSM-V*, 661. See also Craig S. Neumann et al., "Psychopathic Traits in Females and Males across the Globe," *Behavioral Science and the Law* 30 (2012): 557–74, showing greater proportion of individuals with high self-reported psychopathy responses in males than in females.

62. See R. D. Hare, *Hare Psychopathy: Checklist—Revised (PCL-R)* (North Tonawanda, NY: Multi-Health Systems, 2003), employing a multifactor checklist to create a scale from definitive psychopathy to likely psychopathy to unlikely psychopathy.

63. See R. Veit et al., "Aberrant Social and Cerebral Responding in a Competitive Reaction Time Paradigm in Criminal Psychopaths," *NeuroImage* 49 (2010): 3365–72; K. A. Kiehl, "A Cognitive Neuroscience Perspective on Psychopathy: Evidence for Paralimbic System Dysfunction," *Psychiatry Research* 142 (2006): 107–28.

64. Ishikawa et al., "Autonomic Stress Reactivity and Executive Functions in Successful and Unsuccessful Criminal Psychopaths from the Community," *Journal of Abnormal Psychology* 110 (2001): 423, 425, identifying successful psychopaths as those with personality traits central to psychopathy but lacking a criminal record. See generally James Fallon, *The Psychopath Inside: A Neuroscientist's Personal Journey into the Dark Side of the Brain* (New York: Current, 2013), describing life as law-abiding neuroscientist and professor with personality traits and brain structure of a psychopath. See also A. Raine et al., "Increased Executive Functioning, Attention, and Cortical Thickness in White-Collar Criminals," *Human Brain Mapping* 33 (2012): 2932–40: "These findings . . . are broadly consistent with the idea that white-collar criminals engage in a careful and rational calculation of both the costs and benefits of offending" (2939).

65. *In re* Martenies, 350 N.W.2d 470 (Minn. Ct. App. 1984).

66. Mark B Dunnell, *Dunnell Minnesota Digest* (New York: LexisNexis, 2014), § 2.07.

67. See Kent A. Kiehl, et al., "Limbic Abnormalities in Affective Processing by Criminal Psychopaths as Revealed by Functional Magnetic Resonance Imaging," *Biological Psychiatry* 50 (2001): 677–84. Kiehl administered the *Hare PCL* to determine where certain prisoner-participants fell on the psychopathy continuum. He then used fMRI to record the participants' neurological emotional processing during performance of an affective memory task. The study showed that criminal psychopaths exhibit less limbic activation when processing affective stimuli than criminal nonpsychopaths and noncriminal control participants.

68. Jon Ronson, *The Psychopath Test: A Journey through the Madness Industry* (New York: Riverhead, 2012), 58.

69. Robert Hare, "This Charming Psychopath: How to Spot Social Predators before They Attack," *Psychology Today*, January 1, 1994.

70. Yaling Yang et al., "Localization of Deformations within the Amygdala in Individuals with Psychopathy," *Archives of General Psychiatry* 66, no. 9 (2009): 986–94. MRI scans show "significant volume reductions and regional morphological alterations in bilateral amygdala in psychopathic individuals" (993).

71. See Philip K. Dick, *Collected Stories of Phillip K. Dick*, vol. 4, *The Minority Report* (Burton, MI: Subterranean, 2013); *Minority Report* (Paramount Pictures, 2002). If a crime is highly likely to occur, why should we wait for it to occur?

72. See Paul H. Robinson, "Punishing Dangerousness: Cloaking Preventative Detention as Criminal Justice," *Harvard Law Review* 114 (2001): 1429–56.

73. See, e.g., Gideon Yaffe, "Criminal Attempts," *Yale Law Journal* 124 (2014): 95; Stephen J. Morse, "Reason, Results, and Criminal Responsibility," *University of Illinois Law Review* (2004): 363, 387–95; George P. Fletcher, *Rethinking Criminal Law* (Oxford, UK: Oxford University Press, 2000), 166–97.

74. For useful distinctions, see, e.g., Michael S. Moore, *Act and Crime: The Philosophy of Action and Its Implications for Criminal Law* (Oxford, UK: Oxford

University Press, 1993); Husak, "Rethinking the Act Requirement." For a discussion of why an analysis of the act element of attempt should be recharacterized as an inquiry into intent, see Yaffe, "Criminal Attempts."

75. See Robinson v. California, 370 U.S. 660, 667 (1962), holding a California statute unconstitutional under the Eighth Amendment because it criminalized the status of "be[ing] addicted to the use of narcotics." Powell v. Texas, 392 U.S. 514, 536–37 (1968), holding that an inebriate was not being prosecuted for his status as an alcoholic, but for committing act of being drunk in public; Jones v. Los Angeles, 444 F.3d 1118, 1138 (9th Cir. 2006) (*vacated by* 505 F.3d 1006 (9th Cir. 2007)) (opinion vacated by agreed-upon settlement), holding a Los Angeles ordinance unconstitutional under the Eighth Amendment because prohibiting "unavoidable act of sitting, lying, or sleeping at night while being involuntarily homeless" criminalized the status of homelessness.

76. See, e.g., 42 U.S.C.A. § 264(d)(1), permitting apprehension and examination of individuals with certain communicable diseases; 42 U.S.C.A. § 265, empowering the surgeon general to prohibit "the introduction of persons and property" into the United States from countries with known communicable diseases.

77. See Judith Rich Harris, *No Two Alike: Human Nature and Human Individuality* (New York: W. W. Norton, 2007); Giorgio A. Ascoli, *Trees of the Brain, Roots of the Mind* (Cambridge, MA: MIT Press, 2015).

78. See, e.g., Adam J. Kolber, "The Subjective Experience of Punishment," *Columbia Law Review* 109 (2009): 189.

79. See, e.g., Geoffrey M. Cooper, *The Cell: A Molecular Approach; The Molecular Composition of Cells*, 2nd ed. (Sunderland, UK: Sinauer Associates, 2000).

80. See ibid.

81. See, e.g., Burton E. Tropp, *Molecular Biology: Genes to Proteins*, 4th ed. (Burlington, VT: Jones and Bartlett Learning, 2012).

82. See, e.g., Y. Wang et al., "A Genetic Susceptibility Mechanism for Major Depression: Combinations of Polymorphisms Defined the Risk of Major Depression and Subpopulations," *Medicine* 94 (2015), e778–e782.

83. Kirti A. Gautam et al., "c.29C>T Polymorphism in the Transforming Growth Factor-β1 (TGFB-1) Gene Correlates with Increased Risk of Urinary Bladder Cancer," *Cytokine* 75 (2015): 344–48.

84. Ibid.

85. See, e.g., J. D. Boardman et al., "Genes in the Dopaminergic System and Delinquent Behaviors across the Life Course: The Role of Social Controls and Risks," *Criminal Justice and Behavior* 41 (2014): 713–31.

86. A. Caspi et al., "Role of Genotype in the Cycle of Violence in Maltreated Children," *Science* 297 (2002): 851–54.

87. See J. A. Tehrani et al, "Mental Illness and Criminal Violence," *Social Psychiatry and Psychiatric Epidemiology* 33 (1998): S81.

88. See G. Guo, M. E. Roettger, and J. C. Shih, "Contributions of the DAT1

and DRD2 Genes to Serious and Violent Delinquency among Adolescents and Young Adults," *Human Genetics* 121 (2007): 125–36.

89. *Similar* and *same* are used here as normatively relative terms, indicating actual shared environments or environmental factors so alike that, in effect, *A* and *B* may be said to have something akin to a shared environment. Of course, environmental factors are so variable (even twins in the same household may encounter different treatment at school or even from parents) that research focused on heritability can never entirely control for environment, but controlling for similar environments best enables us to approach that goal. See, e.g., Thomas J. Bouchard and Matt McGue, "Genetic and Environmental Influences on Human Psychological Differences," *Developmental Neurobiology* 54 (2003): 4–45.

90. Ibid., 35–36.

91. See, e.g., G. J. Myers et al., "Prenatal Methylmercury Exposure from Ocean Fish Consumption in the Seychelles Child Development Study," *Lancet* 361 (2003): 1686.

92. See J. R. Hibbeln et al., "Maternal Seafood Consumption in Pregnancy and Neurodevelopmental Outcomes in Childhood (ALSPAC study): An Observational Cohort Study," *Lancet* 369 (2007): 578–85; R. K. McNamara and S. E. Carlson, "Role of Omega-3 Fatty Acids in Brain Development and Function: Potential Implications for the Pathogenesis and Prevention of Psychopathology," *Prostaglandins, Leukotrienes and Essential Fatty Acids* 75 (2006): 329–49; C. Iribarren et al., "Dietary Intake of N-3, N-6 Fatty Acids and Fish: Relationship with Hostility in Young Adults—the CARDIA study," *European Journal of Clinical Nutrition* 58 (2004): 24–31.

93. See A. Caspi et al., "Role of Genotype in the Cycle of Violence in Maltreated Children," *Science* 297 (2002): 851–54.

94. See Exod. 20:5–6 (King James); Num. 14:18.

95. See, e.g., R. Toro et al., "Prenatal Exposure to Maternal Cigarette Smoking and the Adolescent Cerebral Cortex," *Neuropsychopharmacology* 33 (2008): 1019–27; V. W. Swayze et al., "Magnetic Resonance Imaging of Brain Anomalies in Fetal Alcohol Syndrome," *Pediatrics* 99 (2006): 232–40.

96. See E. D. Levin et al., "Prenatal Nicotine Effects on Memory in Rats: Pharmacological and Behavioral Challenges," *Developmental Brain Research* 97 (1996): 207–15, finding even a low level of nicotine may disrupt neurotransmitter development. See also Raja A. S. Mukherjee et al., "Low Level Alcohol Consumption and the Fetus," *British Medical Journal* 330 (2005): 375–76, arguing that pregnant women should abstain from alcohol because, despite popular anecdotal evidence about benefits of drinking red wine, scientific literature has not found a conclusively safe dose that does not subject unborn child to risks of fetal alcohol syndrome.

97. See B. Maughan et al., "Prenatal Smoking and Early Childhood Conduct Problems," *Archives of General Psychiatry* 61 (2004): 842, finding that "(1) prenatal

smoking is not simply a proxy indexing genetic risk for antisocial behavior but that (2) it is also unlikely to be a unique cause of early childhood behavior problems."

98. Children may be exposed to lead-contaminated house dust (tainted by disintegrating lead paint), lead-contaminated soil tracked into the house, or to lead contained in old dishware (manufactured prior to lead content regulations). See, e.g., H. W. Mielke and S. Zahran, "The Urban Rise and Fall of Air Lead (Pb) and the Latent Surge and Retreat of Societal Violence," *Environment International* 43 (2012): 48–55; Bruce P. Lanphear and Klaus J. Roghmann, "Pathways of Lead Exposure in Urban Children," *Environmental Law Research* 74 (1997): 67–73.

99. See Moffitt, "Adolescence-Limited and Life-Course Persistent Antisocial Behavior," 677.

100. Adrian Raine, P. A. Brennan, and D. P. Farrington, *Biosocial Bases of Violence: Conceptual and Theoretical Issues* (New York: Plenum Press, 1997); Adrian Raine and P. H. Venables, "Classical Conditioning and Socialization—A Biosocial Interaction," *Personality and Individual Differences* 2 (1981): 273–83; H. J. Eysenck, *Crime and Personality*, 3rd ed. (London: Routledge and Kegan Paul, 1977).

101. See Moffitt, "Adolescence-Limited and Life-Course Persistent Antisocial Behavior," 684.

102. Compare Mark A. Cohen, "The Monetary Value of Saving a High-Risk Youth," *Journal of Quantitative Criminology* 14 (1998): 27: "The present value of saving a high-risk youth is estimated to be $1.7 to $2.3 million"; Mark A. Cohen and Alex R. Piquero, "New Evidence on the Monetary Value of Saving a High-Risk Youth," *Journal of Quantitative Criminology* 25 (2009): 46, providing a new estimate of $3.2–$5.8 million, discounted to present value at age fourteen.

103. See, e.g., Stephen C. Richards, ed., *The Marion Experiment: Long-Term Solitary Confinement and the Supermax Movement* (Carbondale: Southern Illinois University Press, 2015); Lisa Guenther, *Solitary Confinement: Social Death and Its Afterlives* (Minneapolis: University of Minnesota Press, 2013); Sharon Shalev, *Supermax: Controlling Risk through Solitary Confinement* (Berkeley: University of California Press, 2011).

104. See Ann L. Webber and Joanne Wood, "Amblyopia: Prevalence, Natural History, Functional Effects and Treatment," *Clinical and Experimental Optometry* 88 (2005): 365, 368, 371.

105. See, e.g., Paul Gendreau et al., "Changes in EEG Alpha Frequency and Evoked Response Latency during Solitary Confinement," *Journal of Abnormal Psychology* 79, no. 1 (1972): 54–59, finding significant decline in brain-wave activity in prisoners in solitary confinement for only one week; A. Vrca et al., "Visual Evoked Potentials in Relation to Factors of Imprisonment in Detention Camps," *International Journal of Legal Medicine* 109, no. 3 (1996): 114–17, revealing brains of prisoners subjected to solitary confinement were among most severely impaired.

106. "Alone and Afraid: Children Held in Solitary Confinement and Isolation in Juvenile Detention and Correctional Facilities," *American Civil Liberties Union*,

June 2014, 3–4, explaining the distinction between adolescent and adult brains, effects of solitary confinement on adolescent brain development, and long-term harm of those effects.

107. Phineas Gage was a railroad worker responsible for tamping explosive charges. One day, Gage accidentally ignited the charge, propelling the tamping iron through his skull and orbitofrontal cortex. Amazingly, Gage survived, but the brain damage caused a dramatic change in behavior and personality. For further discussion of Phineas Gage, see Malcolm Macmillan, *An Odd Kind of Fame: Stories of Phineas Gage* (Cambridge, MA: MIT Press, 2002).

108. See, e.g., State v. Brown, 907 So.2d 1 (La. 2005), addressing the claim that execution of defendant with serious brain injury is unconstitutional; State v. Stanko, 741 S.E.2d 708 (S.C. 2013), considering expert testimony "that the damaged lobe of Appellant's brain played an important role in impulse control, judgment, and empathy."

109. See, e.g., Ky. Rev. Stat. Ann. § 504.020: "A person is not responsible for criminal conduct if at the time of such conduct, as a result of mental illness or intellectual disability, he lacks substantial capacity either to appreciate the criminality of his conduct or to conform his conduct to the requirements of law"; Me. Rev. Stat. Ann. tit. 17A, § 38: "Evidence of an abnormal condition of the mind may raise a reasonable doubt as to the existence of a required culpable state of mind"; Missoula v. Paffhausen, 289 P.3d 141, 148 (Mont. 2012): "Automatism refers to behavior performed in a state of unconsciousness or semi-consciousness such that the behavior cannot be deemed volitional."

110. Nita A. Farahany, "Cruel and Unequal Punishments," *Washington University Law Review* 86 (2009): 859–915.

111. 536 U.S. 304 (2002).

112. The enactments Farahany referred to are La. Code Crim. Proc. Ann. art. 905.5.1(H) (2008); Nev. Rev. Stat. Ann. § 174.098 (West Supp. 2008); Va. Code Ann. § 19.2-264.3:1.1 (A) (2008); Cal. Penal Code § 1376 (West Supp. 2007); Del. Code Ann. tit. 11, § 4209(d)(3)(d) (2007); 725 Ill. Comp. Stat. 5/114-15(d) (2006); Idaho Code Ann. § 19-2515A(1)(a) (2004); Utah Code Ann. §§ 77-15a-101 to 102 (2003). (Note added.)

113. Farahany, "Cruel and Unequal Punishments," 915.

114. State v. Brown, 907 So.2d 1 (La. 2005).

115. Compare Markus Christen and Sabine Müller, "Effects of Brain Lesions on Moral Agency: Ethical Dilemmas in Investigating Moral Behavior," *Current Topics in Behavioral Neuroscience* 19 (2014): 159–88. The authors concluded that in order to estimate an offender's blameworthiness, his capacity to distinguish between right and wrong must be assessed. When brain injury results in a very low IQ, it is intuitive that the offender may be unable to distinguish rationally between right and wrong in a given situation. The connection between intelligence and moral judgment appears to us linear and straightforward (even if it is less so in reality). On

the other hand, when a brain injury does not affect an offender's intelligence, but instead compromises emotional capacities, it is less intuitive to imagine that that offender may be less able to distinguish between right and wrong. The path from brain area to particular emotional aptitude, and then from emotional aptitude to moral judgment, is complex and presents idiosyncratically in case studies.

116. *Brown*, 907 So.2d at 31.

117. La. Code Crim. Proc. Ann. Art. 905.5.1(H)(2).

118. State v. Anderson, 996 So.2d 973 (La. 2008).

119. U.S. v. Candelario-Santana, 916 F. Supp. 2d 191 (D.P.R. 2013).

120. Ibid., 196.

121. Ibid., 197.

122. Ibid., 207. But see Hall v. Florida, 134 S.Ct. 1986, 1994 (2013), ruling that scoring above 70 on IQ test, especially within test's margin of error, does not preclude finding that defendant suffers from intellectual disability, which would make him ineligible for death penalty under *Atkins*; Brumfield v. Cain, 135 S.Ct. 2269, 2272 (2015), holding the state postconviction court's determination that defendant's score of 75 on IQ test demonstrated that defendant did not possess "subaverage intelligence" was unreasonable.

123. State v. Stanko, 741 S.E.2d 708 (S.C. 2013).

124. Ibid., 713.

125. Ibid., 726.

126. *State ex rel.* Clayton v. Griffith, 457 S.W.3d 735 (Mo. 2015).

127. Ibid., 737.

128. Ibid., 753.

129. Roper v. Simmons, 543 U.S. 551 (2005).

130. 497 U.S. 417 (1990).

131. *Roper*, 543 U.S. at 617–18 (Scalia, J., dissenting).

132. Laurence Steinberg et al., "Are Adolescents Less Mature Than Adults? Minors' Access to Abortion, the Juvenile Death Penalty, and the Alleged APA 'Flip-Flop,'" *American Psychologist* 64 (2009): 583, 593.

133. See Adrian Raine, *The Anatomy of Violence: The Biological Roots of Crime* (New York: Pantheon, 2013), 91; Adrian Raine and Yaling Yang, "Neural Foundations to Moral Reasoning and Antisocial Behavior," *Social Cognitive and Affective Neuroscience* 1 (2006): 203–13.

134. M. Koenigs et al., "Damage to the Prefrontal Cortex Increases Utilitarian Moral Judgments," *Nature* 446 (2007): 908–11.

135. See *M*A*S*H: Goodbye, Farewell, and Amen* (Twentieth Century Fox television broadcast, February 28, 1983), for a dramatic presentation of the dilemma.

136. In high-conflict personal dilemmas, during which participants must decide whether they would commit a morally reprehensible action in order to achieve a utilitarian goal (e.g., pushing one person off a bridge to save five people from dying), participants with VMPC damage were twice as likely to choose the utilitarian

option than the control groups (Koenigs et al., "Damage to the Prefrontal Cortex," 908–10).

137. "In the absence of an emotional reaction to harm of others in personal moral dilemmas, VMPC patients may rely on explicit norms endorsing the maximization of aggregate welfare and prohibiting the harming of others" (ibid., 910).

138. Raine and Yang, "Neural Foundations to Moral Reasoning," 203.

139. The Standard Issue Moral Judgment Interview scores participants on their level of moral judgment, according to Lawrence Kohlberg's six-stage theory of moral development. Ann Colby and Lawrence Kohlberg, *The Measurement of Moral Judgment*, vol. 2, *Standard Issue Scoring Manual* (New York: Cambridge University Press, 1987). Compare Steven W. Anderson et al., "Impairment of Social and Moral Behavior Related to Early Damage in Human Prefrontal Cortex," *Nature Neuroscience* 2 (1999): 1032, 1033–34, finding that two patients with early-onset prefrontal cortex damage scored in preconventional stage, characterized by "egocentric perspective with decisions based on avoidance of punishment," with Jefferey L. Saver and Antonio Damasio, "Preserved Access and Processing of Social Knowledge in a Patient with Acquired Sociopathy Due to Ventromedial Frontal Damage," *Neuropsychologia* 29 (1991): 1241, 1244–45, recounting that patient who had his VMPC removed in the course of meningioma treatment at age thirty-five scored in late-conventional, early postconventional stage.

Chapter Three

1. Though, in the event that the car overheated, we might let it cool down, and that could be an efficacious repair strategy. Compare Neil Levy, *Neuroethics: Challenges for the 21st Century* (Cambridge: Cambridge University Press, 2007), 222, recounting an amusing take on agency confusion.

2. Stephen J. Morse, "New Neuroscience, Old Problems," in *Neuroscience and the Law: Brain, Mind, and the Scales of Justice*, ed. Brent Garland (New York: Dana, 2004), distinguished human beings from other physical objects and sentient beings by noting that humans are the only fully intentional creatures, thus human action is guided by reason and not by neural structure and mechanisms. This capacity to reason, argued Morse, makes human beings responsible agents deserving of (and even entitled to) punishment for wrongs they choose to commit.

3. Raphael Van Riel and Robert Van Gulick, "Scientific Reduction," in *The Stanford Encyclopedia of Philosophy*, ed. Edward N. Zalta, http://plato.stanford.edu/archives/spr2016/entries/scientific-reduction (citing Herbert Feigl, *The "Mental" and the "Physical"* [Minneapolis: University of Minnesota Press, 1967], 77).

4. Francis Crick, *The Astonishing Hypothesis: The Scientific Search for the Soul* (New York: Scribner's, 1994), 3.

5. Van Riel and Van Gulick, "Scientific Reduction."

6. See Anne Ruth Mackor, "What Can Neuroscience Say about Responsibility?," in *Neuroscience and Legal Responsibility*, ed. Nicole A. Vincent (New York: Oxford University Press, 2013), 53 (citing Richard Dawkins, "Straf is wetenschappelijk achterhaald," *NRC-Handelsblad*, January 14, 2006, 14: "lawsuits about the guilt or the diminished responsibility of human beings are just as absurd as lawsuits against cars, or so Dawkins claims.").

7. Stephen J. Morse, "Common Criminal Law Compatibilism," in *Neuroscience and Legal Responsibility*, ed. Nicole A. Vincent (New York: Oxford University Press, 2013), 27.

8. A point that is true despite Morse's parenthetical cynicism about cognitive neuroscience's mechanistic view. Morse ("Common Criminal Law Compatibilism," 45), stating that his being "convinced" of cognitive neuroscience's view on intent would itself be illusion by cognitive neuroscience's standards and knowing anything, as state of mind, would too be illusion.

9. See Morse ("Common Criminal Law Compatibilism," 38). Perhaps the two most pertinent affirmative defenses are duress and insanity, yet the criteria for establishing those defenses reflect the state of the art (science) surrounding the notion of free will at the time of their promulgation.

10. Ibid., 42, noting that "ordinary people are intuitive compatibilists."

11. Ibid., 44, explaining how consent in contract is assailable if contract becomes simply the outcome of various "neuronal circumstances."

12. See Ronald N. Giere, *Scientific Perspectivism* (Chicago: University of Chicago Press, 2006), 20. Humans are capable of seeing only a very small region of the electromagnetic spectrum, which is labeled as visible light. Light exists along a much wider spectrum, however, with shorter-waved gamma rays at one end and longer radio waves at the other (ibid., 17).

13. See Paul S. Davies, "Skepticism concerning Human Agency," in *Neuroscience and Legal Responsibility*, ed. Nicole A. Vincent (New York: Oxford Univ. Press, 2013), explaining the importance to human evolution of belief in some shared form of morality.

14. Paul M. Churchland, "Eliminative Materialism and the Propositional Attitudes," *Journal of Philosophy* 78 (1981): 75: Folk psychology's "explanatory impotence and long stagnation inspire little faith that its categories will find themselves neatly reflected in the framework of neuroscience. On the contrary, one is reminded of how alchemy must have looked as elemental chemistry was taking form."

15. Ibid., 72.

16. Ibid.

17. Ibid.

18. See ibid., 81: "Alchemy is a terrible theory, well-deserving of its complete elimination, and the . . . [four fundamental spirit defenses of it] is reactionary, obfuscatory, retrograde, and wrong. But in historical context, that defense might have seemed wholly sensible, even to reasonable people."

19. Ibid., 76–78 (citing Hillary Putnam, "Robots: Machines or Artificially Created Life?," *Journal of Philosophy* 61 [1964]: 675). Churchland divided contemporary functionalism into two threads: (1) The normative character of folk psychology, i.e., no descriptive theory of neural mechanisms can replace folk psychology as a normative characterization of rationality, beliefs, and desires (see Daniel Dennett, "Intentional Systems," *Journal of Philosophy* 68 [1971]: 87; Karl R. Popper, *Objective Knowledge* [Oxford, UK: Clarendon, 1972]; Joseph Margolis, *Persons and Minds* [Boston: D. Reidel, 1978]); and (2) The abstract nature of folk psychology, i.e., folk psychology allows for broader, more efficient description of internal states that may differ in their neurological manifestations but are of the same nature (see Jerry A. Fodor, *Psychological Explanation* [New York: Random House, 1968], 116).

20. Paul Churchland, "Eliminative Materialism," 82.

21. Ibid., 84–89, sketching scenarios of potential neuroscientific advances into the neural mechanics of knowledge and learning to show folk psychology's vulnerability and analytical weaknesses.

22. Ibid., 84.

23. Joshua Greene and Jonathan Cohen, "For the Law, Neuroscience Changes Nothing and Everything," *Philosophical Transactions of the Royal Society of London B: Biological Science* 359 (2004): 1776. See also Andrew Scull, "Mind, Brain, Law and Culture," *Brain* 130 (2007): 590: "We are, as Greene and Cohen would have it, mere puppets, helplessly acting out in a purely mechanical way our preprogrammed pathways."

24. See, e.g., Benjamin Libet et al., "Time of Conscious Intention to Act in Relation to Onset of Cerebral Activity (Readiness Potential): The Unconscious Initiation of a Freely Voluntary Act," *Brain* 106 (1983): 623–42.

25. See, e.g., Elizabeth R. Sowell et al., "*In Vivo* Evidence for Post-Adolescent Brain Maturation in Frontal and Striatal Regions," *Nature Neuroscience* 2 (1999): 859–61, finding maturation in frontal and striatal regions in postadolescent brains through MRI scans.

26. Antoine Bechara, "The Role of Emotion in Decision-Making: Evidence from Neurological Patients with Orbitofrontal Damage," *Brain and Cognition* 55 (2004): 30–40, finding that orbitofrontal cortex damage inhibits decision making and impulse control and may lead to antisocial behavior.

27. Roper v. Simmons, 543 U.S. 551 (2005). See chapter 2.

28. Greene and Cohen, "For the Law," 1778–79.

29. See also Andrew E. Lelling, "Eliminative Materialism, Neuroscience and the Criminal Law," *University of Pennsylvania Law Review* 141 (1993): 1475–76, in a note discussing eliminative materialism's attack on fundamental assumptions underlying folk psychology.

30. See Stuart Grassian, "Psychiatric Effects of Solitary Confinement," *Washington University Journal of Law and Policy* 22 (2006): 331: "even a few days of solitary

confinement will predictably shift the electroencephalogram (EEG) pattern toward an abnormal pattern characteristic of stupor and delirium"; Manabu Makinodan et al., "A Critical Period for Social Experience-Dependent Oligodendrocyte Maturation and Myelination," *Science* 337 (2012): 1357–60, suggesting that social isolation leads to a decrease in myelination in prefrontal cortex, thus showing the disproportionate effect such isolation has on adolescent neurological development.

31. See Daniel Reisel, "Towards a Neuroscience of Restorative Justice," in *The Psychology of Restorative Justice*, ed. Theo Gavrielides (New York: Routledge, 2015), 49.

32. See Adrian Raine, *The Anatomy of Violence: The Biological Roots of Crime* (New York: Pantheon, 2013).

33. Jeffery Rosen, "The Brain on the Stand," *New York Times*, March 11, 2007 (quoting Robert Sapolsky): "You can have a horrendously damaged brain where someone knows the difference between right and wrong but nonetheless can't control their behavior. . . . At that point, you're dealing with a broken machine, and concepts like punishment and evil and sin become utterly irrelevant. Does that mean the person should be dumped back on the street? Absolutely not. You have a car with the brakes not working, and it shouldn't be allowed to be near anyone it can hurt."

34. Van Riel and Van Gulick, "Scientific Reduction" (citing Joseph Levine, "On Leaving Out What It's Like," in *Consciousness: Psychological and Philosophical Essays*, ed. Martin Davies and Glyn W. Humphreys [Oxford, UK: Blackwell, 1993]; Levine, "Conceivability and the Metaphysics of Mind," *Noûs* 32 [1998]: 449–80).

35. Greene and Cohen, "For the Law," 1781.

36. Michael S. Moore, "Responsible Choices, Desert-Based Legal Institutions, and the Challenges of Contemporary Neuroscience," in *New Essays in Political and Social Philosophy*, ed. Ellen Frankel Paul, Fred D. Miller Jr., and Jeffrey Paul (Cambridge: Cambridge University Press, 2012), 259.

37. See Eric Kandel and J. H. Schwartz, "Molecular Biology of Learning," *Science* 218 (1982): 433–43. In both lower forms of life and humans, neurons learn through intracellular recordings of electrical activity that results from conditioning and other forms of learning.

38. See Stephen J. Morse, "Psychopathy and Criminal Responsibility," *Neuroethics* 1 (2008): 205–12, arguing that law should be reformed to excuse those with severe psychopathy from blame because they lack neural function (empathy), but that such psychopaths should still be subject to some legal repercussions such as civil confinement. See also Thomas Nadelhoffer and Walter P. Sinnott-Armstrong, "Is Psychopathy a Mental Disease?," in *Neuroscience and Legal Responsibility*, ed. Nicole A. Vincent (New York: Oxford University Press, 2013), 229, citing empirical evidence that psychopathy, because of its neurological manifestations, should support insanity defense as mental disease or defect despite current practice in

various jurisdictions of excluding psychopathy from requisite classification of "disease of the mind" or "mental disease or defect."

39. See Virginia Hughes, "Science in Court: Head Case," *Nature* 464 (2010): 342 (quoting Morse).

40. Morse, "Common Criminal Law Compatibilism," 31.

41. See M. R. Bennett and P. M. S. Hacker, *History of Cognitive Neuroscience* (Malden, MA: Wiley-Blackwell, 2008); Michael S. Pardo and Dennis Patterson, "Philosophical Foundations of Law and Neuroscience," *University of Illinois Law Review*, no. 4 (2010): 1211; Brian Knutson et al., "Neural Predictors of Purchases," *Neuron* 53 (2007): 147.

42. Eric R. Scerri and Lee McIntyre, "The Case for the Philosophy of Chemistry," *Synthese* 111 (1997): 213–32; Michael Esfeld and Christian Sachse, "Theory Reduction by Means of Functional Sub-types," *International Studies in the Philosophy of Science* 21 (2007): 1–17. (Citations in original.)

43. J. A. Fodor, "Special Sciences (Or, The Disunity of Science as a Working Hypothesis)," *Synthese* 28 (1974): 97–115; Fodor, "Special Sciences: Still Autonomous after All These Years," *Noûs Supplement: Philosophical Perspectives* 31 (1997): 149–63; Hillary Putnam, *Meaning and the Moral Sciences* (London: Routledge and Kegan Paul, 1978); Robert Van Gulick, "Nonreductive Materialism and the Nature of Intertheoretical Constraint," in *Emergence or Reduction? Essays on the Prospects of Nonreductive Physicalism*, ed. A. Beckermann, H. Flohr, and J. Kim (Berlin: Walter de Gruyter, 1992). (Citations in original.)

44. Van Riel and Van Gulick, "Scientific Reduction."

45. Shakespeare, *Hamlet*, act 1, sc. 5.

46. Benjamin Libet, "Unconscious Cerebral Initiative and the Role of Conscious Will in Voluntary Action," *Behavioral and Brain Sciences* 8 (1985): 529–39. Libet's experiment required subjects, attached to an EEG, to flex their fingers at their discretion and to watch a clock to register the moment of their decision. Libet showed that the subjects' awareness of their intent to flex their finger occurred after a shift of electrical potential in their brains, suggesting that neural processes for voluntary movements precede reported awareness of the decision to make such movements.

47. Daniel M. Wegner, "Précis of the Illusion of Conscious Will," *Behavioral and Brain Sciences* 27 (2004): 649, 682 (emphasis added). See also Wegner, *The Illusion of Conscious Will* (Cambridge, MA: MIT Press, 2002), noting that the broad range of neural phenomena demonstrate that our brains make decisions before we are even conscious that decisions are to be made.

48. Michael S. Moore, "Responsible Choices," 255.

49. Ibid., 269.

50. Ibid., 273.

51. Some scholars refute Libet's findings on the basis of subject bias, inaccuracies in the timing of reporting, and its reliance on backward causation. See Max Velmans, *Understanding Consciousness* (New York: Routledge, 2000), 235–37, crit-

icizing Libet's use of subject self-reporting; Alexander Batthyany, "Mental Causation and Free Will after Libet and Soon: Reclaiming Conscious Agency," in *Irreducibly Conscious*, ed. Alexander Batthyany and Avshalom Elitzur (Heidelberg,
Ger.: Universitätsverlag Winter, 2009), 135, noting inconsistency with Libet's asking
his subjects to allow "urge" to move to appear without any preplanning when those
subjects are consciously bringing about that urge; Daniel Dennett, "The Self as a
Responding—and Responsible—Artifact," *Annals of the New York Academy of
Science* 1001 (2003): 42–43, highlighting ambiguities in Libet's use of electrodes in
measuring readiness potential but self-reporting from a clock to determine when
the conscious decision was made; Edoardo Bisiach, "The (Haunted) Brain and Consciousness," in *Consciousness in Contemporary Science*, ed. Anthony J. Marcel and
Edoardo Bisiach (New York: Oxford University Press, 1988), 101–20, arguing that
Libet's subjects made an illusory judgment to antedate the moment sensation began to the moment of initial neuronal activity.

52. I am grateful to my colleague Professor Paul S. Davies for working this
through with me.

53. See, e.g., Zihong Jiang et al., "Social Isolation Exacerbates Schizophrenia-
like Phenotypes via Oxidative Stress in Cortical Interneurons," *Biological Psychiatry* 73 (2013): 1024.

54. See Stephen J. Morse, introduction to *A Primer on Criminal Law and Neuroscience*, ed. Stephen J. Morse and Adina L. Roskies (New York: Oxford University Press 2013), xxi: "we already knew from common-sense observation and rigorous behavioral studies that juveniles are on average less rational than adults";
Richard A. Posner, "Justices Should Use More than Their Gut and 'Brain Science'
to Decide Cases," *Slate*, June 26, 2012: "The court has learned from brain science
that teenagers are immature! But we knew that." See also Roper v. Simmons, 543
U.S. 551, 618–20 (2005) (Scalia, J. dissenting), recounting the premeditation and
"callous[ness]" of Christopher Simmons's murder of Shirley Cook and referencing
amici briefs by various states detailing brutal murders committed by individuals
under eighteen years of age.

55. Selim Berker, "The Normative Insignificance of Neuroscience," *Philosophy
and Public Affairs* 37 (2009): 293. The article was selected by *The Philosopher's
Annual* as one of the ten best philosophy papers published in 2009.

56. See Joshua D. Greene et al., "An fMRI Investigation of Emotional Engage-
ment in Moral Judgment," *Science* 293 (2001): 2105; Greene et al., "The Neural
Bases of Cognitive Conflict and Control in Moral Judgment," *Neuron* 44 (2004):
389; Greene et al., "Cognitive Load Selectively Interferes with Utilitarian Moral
Judgment," *Cognition* 107 (2008): 1144; Greene, "From Neural 'Is' to Moral
'Ought': What Are the Moral Implications of Neuroscientific Moral Psychology?,"
Nature Reviews Neuroscience 4 (2003): 846; Greene, "The Secret Joke of Kant's
Soul," *Moral Psychology* 3 (2008): 35.

57. Peter Singer, "Ethics and Intuitions," *Journal of Ethics* 9 (2005): 331–52,

arguing that research in neuroscience should lead us to reconsider role of intuition in normative ethics because these intuitions are biological vestiges of our evolutionary history.

58. Greene, "fMRI Investigation."

59. Greene defined characteristically deontological judgments as the type of judgment that tends to be reached on the basis of emotional responses Greene, "Secret Joke," 39. Berker criticized Greene's definition as being based on typical deontological judgments and found it to be inconsistent with Greene's definition of characteristically utilitarian moral judgments, based on typical utilitarian principles.

60. Greene, "fMRI Investigation"; Greene et al., "Neural Bases of Cognitive Conflict"; Greene, "Cognitive Load."

61. See Philippa Foot, "The Problem of Abortion and the Doctrine of Double Effect," *Oxford Review* 5 (1967): 5, first posing the trolley dilemma, which juxtaposes pushing one man onto trolley tracks to save five others with throwing a switch that kills one person to save five others.

62. In certain circumstances, inaction because of selfish motives is perceived as even more morally impermissible than action motivated by neutral ones when both result in the same harm (ibid.).

63. See Joshua Greene, *Moral Tribes: Emotion, Reason, and the Gap between Us and Them* (New York: Penguin, 2013). Though it can be made more difficult for the consequentialist by positing the hypothetical in terms of taking the life of one to harvest her organs to save five others, and the consequentialist needs to explain why the two hypothetical scenarios are normatively different.

64. "[N]early all deontologists judge that it is permissible to divert the trolley in the trolley driver dilemma" (Berker, "Normative Insignificance of Neuroscience," 299). Berker did not, however, cite to those deontologists.

65. Berker, "Normative Insignificance of Neuroscience," pointed to three methodological problems that do not quite undermine the empirical findings but render those findings unstable. First, Greene found that at least one brain region associated with emotion was found to be correlated with consequentialist judgment, which calls into question Greene's characterization of deontological judgments as driven by purely emotional processes and consequentialist judgments as driven by cognitive processes (ibid., 307–8). Second, Berker noted that, when interpreted correctly, response-time data collected by Greene et al. do not, in fact, confirm the prediction about comparative response times (ibid., 308). Finally, Greene's "me hurt you" criteria when forming dilemmas that give rise to deontological judgments essentially claim that characteristically deontological judgments only concern bodily harms. Deontology, however, prohibits lying, promise-breaking, coercion, and the like (ibid., 311–12).

66. Ibid., 329; James Woodward and John Allman, "Moral Intuition: Its Neural Substrates and Normative Significance," *Journal of Physiology-Paris* 101 (2007):

195; John Allman and James Woodward, "What Are Moral Intuitions and Why Should We Care about Them? A Neurobiological Perspective," *Philosophical Issues* 18 (2008): 164. See generally Gerd Gigerenzer, "Moral Intuition = Fast and Frugal Heuristics?," in *Moral Psychology*, vol. 2, ed. Walter Sinnott-Armstrong (Cambridge, MA: MIT Press, 2008), arguing that moral judgment and actions can be influenced by simple heuristics.

67. See Amos Tversky and Daniel Kahneman, "Judgment under Uncertainty: Heuristics and Biases," *Science* 185 (1974): 1124–31. Salience bias is a mental shortcut through which a person's judgments are unconsciously informed by the most readily observable phenomena.

68. See Berker, "Normative Insignificance of Neuroscience," 317; Cass R. Sunstein, "Moral Heuristics," *Behavioral and Brain Sciences* 28 (2005): 531–73.

69. See Morse, "Psychopathy and Criminal Responsibility," 205–12. Even Morse would be able to find utility in neuroscientific insights about psychopathy and has argued that the law should be reformed accordingly to allow psychopathy as an affirmative defense. His conclusion that psychopathy should be an excusing incapacity could not, of course, rely solely on behavioral evidence because an element of (successful?) psychopathy is the ability to behave as though one does *not* lack empathy. The most successful psychopaths may be quite charming. See Richard W. Larsen, *Bundy: The Deliberate Stranger* (Englewood Cliffs, NJ: Prentice Hall, 1980).

70. Berker, "Normative Insignificance of Neuroscience," 318.

71. Adina Roskies and Walter Sinnott-Armstrong, "Between a Rock and a Hard Place: Thinking about Morality," *Scientific American*, July 29, 2008, http://www.scientificamerican.com/article/thinking-about-morality/.

72. Berker, "Normative Insignificance of Neuroscience."

73. Though deontology can come in many different flavors, such as agent-centered, patient-centered, and contractualist deontological theories, the object here is to focus on the paradigmatic formulation, which should, at least sufficiently for present purposes, demonstrate the differences between instrumentalism and noninstrumentalism insofar as punishment decisions are concerned. See Larry Alexander and Michael Moore, "Deontological Ethics," *Stanford Encyclopedia of Philosophy*, ed. Edward N. Zalta. https://plato.stanford.edu/entries/ethics-deontological/.

74. See generally Immanuel Kant, *Groundwork of the Metaphysics of Morals*, 2nd ed. (New York: Cambridge University Press, 2011), introducing the categorical imperative. See also Michael S. Moore, *Placing Blame: A Theory of the Criminal Law* (Oxford, UK: Oxford University Press, 1997), 28: "By seeking to achieve other goods through punishment, we necessarily lessen our ability to achieve the good of retribution."

75. See Herbert Morris, "Persons and Punishment," *Monist* 52 (1968): 479–80. One retributivist justification for punishment is that those who break the rules reap the benefit of not restraining themselves when others have properly restrained

themselves, thus violating the mutual social advantage of agreeing to be bound by the rules. Retributive punishment restores balance to society by punishing those who break the rules.

76. See, e.g., H. L. A. Hart, ed., *Punishment and Responsibility* (Oxford, UK: Oxford University Press, 1968); David Dolinko, "Three Mistakes of Retributivism," *UCLA Law Review* 39 (1992): 1623–58; Dolinko, "Some Thoughts about Retributivism," *Ethics* 101 (1991): 537–39; Michael S. Moore, "Moral Reality," *Wisconsin Law Review* (1982): 1061–1156; Moore, "Moral Reality Revisited," *Michigan Law Review* 90 (1992): 2424–2533; J. L. Mackie, "Retribution: A Test Case for Ethical Objectivity," in *Philosophy of Law*, ed. Joel Feinberg and Hyman Gross (Belmont, CA: Wadsworth, 1991); Mackie, "Morality and the Retributive Emotions," *Criminal Justice Ethics* 1 (1982): 3–10; Herbert Fingarette, "Punishment and Suffering," *Proceedings and Addresses of the American Philosophical Association* 50 (1977): 499; John Rawls, "Two Concepts of Rules," *Philosophical Review* 64 (1955): 3–32; A. M. Quinton, "On Punishment," *Analysis* 14 (1954): 133–42.

77. Those who defend torture typically limit their support to singular acts of torture by state actors in emergency situations, the classic example being torturing a known terrorist in order to determine the location of a ticking bomb. Such torture is not punishment but arguably morally justifiable in a utilitarian sense because it avoids a greater harm. See Mirko Bagaric and Julie Clarke, *Torture: When the Unthinkable Is Permissible* (Albany: State University of New York, 2007), 29.

78. Adam J. Kolber, "The Subjective Experience of Punishment," *Columbia Law Review* 109 (2009): 182.

79. Michael S. Moore, *Placing Blame: A Theory of the Criminal Law* (Oxford, UK: Oxford University Press, 2010), 139–52. Moore dismissed the primary argument against retributive justice stemming from its grounding in "ressentiment" by arguing instead its grounding in guilt.

80. Ibid., 159: "only when harsh treatment is imposed on offenders in order to give them their just deserts does such harsh treatment constitute *punishment*" (emphasis in original).

81. Ibid., 164.

82. Ibid., 116 (note omitted).

83. Daniel Kahneman, Paul Slovic, and Amos Tversky, *Judgment under Uncertainty: Heuristics and Biases* (Cambridge: Cambridge University Press, 1982), 1131. Heuristics like representativeness, availability of instances or scenarios, and adjustment from an anchor or starting point "are employed in making judgments under uncertainty. . . . These heuristics are highly economical and usually effective, but they lead to systematic and predictable errors."

84. Ibid.

85. Ibid.

86. See chapter 2, Conclusion.

87. See, e.g., Stephen J. Morse, "Psychopathy and Criminal Responsibility"; Morse, "Lost in Translation? An Essay on Law and Neuroscience," in *Law and*

Neuroscience, Current Legal Issues 13, ed. Michael Freeman (Oxford, UK: Oxford University Press, 2011), 537–38, noting that neuroscientific insights into neural structures and functions associated with legally relevant capacities, such as capacity for rationality and control, may inform retrospective evaluations of criminal responsibility.

88. See Kolber, "Subjective Experience." Even determinism could be sufficient; if all we care about is coincidence of act and mental state, it would not matter what generated that coincidence.

89. See Morse, "Common Criminal Law Compatibilism," 44, stating that though the present science does not support everything that neuroscience advocates believe will occur, the possibility for such affirmation is possible in the future as the science develops; though Morse seems dubious; Morse, "Lost in Translation?," 534, conceding that law will be "fundamentally challenged" if neuroscience conclusively shows that folk psychology is wrong.

90. Morse, "Psychopathy and Criminal Responsibility," 208.

91. Ibid.

92. See Scott O. Lilienfeld, Ashley L. Watts, and Sarah Francis Smith, "Successful Psychopathy," *Current Directions in Psychological Science* 24 (2015): 298–303, examining the current state of successful psychopath research, addressing controversies about successful psychopathy, and providing evidence for competing models of successful psychopathy.

93. Morse, "Psychopathy and Criminal Responsibility," 209 (citing Paul Litton, "Responsibility Status of the Psychopath," *Rutgers Law Journal* 39 [2008]: 349–92).

94. Morse, "Psychopathy and Criminal Responsibility," 209.

95. Psychopathic behavior is correlated with low IQ scores. Hanna Heinzen et al., "Psychopathy, Intelligence and Conviction History," *International Journal of Law and Psychiatry* 34, no. 5 (2011): 336–40.

96. Consider cues to distinguish emotional states from same somatic effect, e.g., arousal and fear.

97. See Daniel Kahneman, *Thinking, Fast and Slow* (New York: Farrar, Straus and Giroux, 2011), 12–13, 20–22.

98. Morse, "Psychopathy and Criminal Responsibility," 209.

99. Ibid.

100. Ibid., 210.

101. For consideration of the different senses of responsibility, see Katrina L. Sifferd, "Translating Scientific Evidence into the Language of the 'Folk': Executive Function as Capacity-Responsibility," in *Neuroscience and Legal Responsibility*, ed. Nicole A. Vincent (New York: Oxford University Press, 2013), 183–204; H. L. A. Hart, "Legal Responsibility and Excuses," in Hart, ed., *Punishment and Responsibility*.

102. Stephen J. Morse, "Deprivation and Desert," in *From Social Justice to Criminal Justice: Poverty and the Administration of Criminal Law*, ed. William C. Heffernan and John Kleinig (New York: Oxford University Press, 2000), 114.

103. Ibid., 115 (emphasis added).

104. Ibid., 117.

105. Ibid., 120. Morse's guiding light—appropriate emotions—shines about as dimly as Moore's imagined guilt as felt by the virtuous person.

106. See Michael S. Moore, "A Natural Law Theory of Interpretation," *Southern California Law Review* 58 (1985): 286: "there is a right answer to moral questions, a moral reality if you like. . . . [R]eal morals . . . have a necessary place in the interpretation of any legal text"; Moore, "Moral Reality," canvassing views of skeptics of moral realism; Moore, "Moral Reality Revisited," describing a moral realist thesis and making a positive case for moral realism.

107. Morse, "Deprivation and Desert," 140.

108. Ibid., 127.

109. Daniel Dennett, "Book Review: Against Moral Responsibility," *Naturalism* (2012), http://www.naturalism.org/resources/book-reviews/dennett-review-of -against-moral-responsibility.

110. Victoria McGeer, "Co-reactive Attitudes and the Making of Moral Community," in *Emotions, Imagination, and Moral Reasoning*, ed. Robyn Langdon and Catriona Mackenzie (New York: Psychology Press, 2012), 299; Philip Pettit, "Responsibility Incorporated," *Ethics* 117 (2007): 171–201; Peter F. Strawson, "Freedom and Resentment," in *Freedom and Resentment and Other Essays* (London: Harper and Row, 1974).

111. See Lawrence v. Texas, 539 U.S. 558, 603 (2003) (J. Scalia dissenting): "Social perceptions of sexual and other morality change over time, and every group has the right to persuade its fellow citizens that its view of such matters is the best."

112. Bruce N. Waller, *Against Moral Responsibility* (Cambridge, MA: MIT Press, 2011), 1.

113. Ibid., 8.

114. Ibid.

115. Ibid., 10.

116. Ibid., 41.

117. Succinctly: "Under the influence of moral responsibility . . . we 'find the person in whom the decisive junction of causes lies,' assign the sanctions to that individual, and the issue is closed. . . . We overestimate the importance of individual character to behavioral outcome, and underestimate the impact of situational/environmental influences. . . . Under the baleful influence of moral responsibility, we get the fundamental attribution error on steroids" (ibid., 144).

118. Ibid., 168: "The moral responsibility system is in crisis, as scientific research encroaches on those areas where grounds for moral responsibility were taking refuge. Efforts to save that system take on many different shapes, but none seem successful."

119. Ibid., 114. See also ibid., 131: "one thing that makes the moral responsibility system undesirable is its systematic blocking of deeper inquiry into the causes that shape our values and our behavior."

120. Ibid., 180 (citing Derk Pereboom, *Living without Free Will* [Cambridge: Cambridge University Press, 2001], 212; Jonathan Bennett, "Accountability," in *Philosophical Subjects*, ed. Zak van Stratten [Oxford, UK: Oxford University Press, 1980], 31).

121. Waller, *Against Moral Responsibility*, 302.

Chapter Four

1. John Stuart Mill, *On Liberty* (New York: D. Appleton, 1863), 23: "the only purpose for which power can be rightfully exercised over any member of a civilized community, against his will, is to prevent harm to others."

2. See Dan B. Dobbs, *The Law of Torts* (St. Paul, MN: West Group, 2000), 277: "The duty owed by all people generally . . . is the duty to exercise the care that would be exercised by a reasonable and prudent person under the same or similar circumstances to avoid or minimize risks of harm to others" (citations omitted); see also Vaughan v. Menlove, 132 E.R. 490 (1837), generally regarded as the source of the reasonable person standard.

3. Breunig v. Am. Family Ins. Co., 173 N.W.2d 619, 622–23 (Wis. 1970).

4. American Law Institute, *Restatement (Third) of Torts: Physical and Emotional Harm* (St. Paul, MN: American Law Institute, 2012), § 3.

5. American Law Institute, *Restatement (Second) of Torts* (St. Paul, MN: American Law Institute, 1965), § 283: "Unless the actor is a child, the standard of conduct to which he must conform to avoid being negligent is that of a reasonable man under like circumstances."

6. American Law Institute, *Restatement (Third) of Torts*, § 11(c); see, e.g., Shapiro v. Tchernowitz, 155 N.Y.S.2d 1011, 1014 (Sup. Ct. 1956), holding that insane defendants are liable for all torts, except those in which malice is necessary ingredient; Sforza v. Green Bus Lines, 268 N.Y.S. 446, 447 (Mun. Ct. 1934): "an insane person is civilly liable for his torts, and whether the tort be one of nonfeasance or misfeasance does not seem to affect the liability."

7. American Law Institute, *Restatement (Third) of Torts*, § 10(b).

8. Ibid., § 11 cmt. e (citing Kenneth S. Abraham, *The Forms and Functions of Tort Law*, 3rd ed. [Eagan, MN: Foundation Press, 2007], 59–60: "mental infirmities are invisible, hard to measure, and incompletely verifiable").

9. Dobbs, *Law of Torts*, 286. See also David E. Seidelson, "Reasonable Expectations and Subjective Standards in Negligence Law: The Minor, the Mentally Impaired, and the Mentally Incompetent," *George Washington Law Review* 50 (1981): 17, 19: "In the overwhelming majority of negligence actions, whatever else may be said of defendant's conduct, it is probably accurate to say that he acted in a manner consonant with his own best judgment in the circumstances."

10. The foregoing list is compiled from Dobbs, *Law of Torts*, 287–88.

11. American Law Institute, *Restatement (Third) of Torts*, § 11 cmt. e.

12. See Dobbs, *Law of Torts*, 285 (citing Mochen v. State, 43 A.D.2d 484 (N.Y. 1974); Stacy v. Jedco Construction, Inc., 457 S.E.2d 875 (N.C. 1995); Feldman v. Howard, 214 N.E.2d (Ohio App. 1966), rev'd on other grounds 226 N.E.2d 564 (Ohio 1967); Birkner v. Salt Lake County, 771 P.2d 1053 (Utah 1989)).

13. See Sherry v. Asing, 531 P.2d 648, 661 (Haw. 1975): "Because of varying degrees of intelligence and capacities possessed by children in the same age group, it has been necessary to except children from the objective standards of care"; Young v. Grant, 290 So.2d 706, 710 (La. Ct. App. 1974), holding that in considering contributory negligence, child's mental illness can be taken into account.

14. Richard A. Posner, "Justices Should Use More than Their Gut and 'Brain Science' to Decide Cases," *Slate*, June 26, 2012, http://www.slate.com/articles/news_and_politics/the_breakfast_table/features/2012/_supreme_court_year_in_review/supreme_court_year_in_review_the_justices_should_use_more_than_their_emotions_to_decide_how_to_rule_.html.

15. Dobbs, *Law of Torts*, 297–98.

16. American Law Institute, *Restatement (Third) of Torts*, § 10(c): Children are held to the adult standard when engaging in "dangerous activity that is characteristically undertaken by adults."

17. Ibid., § 10(a): "A child's conduct is negligent if it does not conform to that of a reasonable careful person of the same age, intelligence, and experience."

18. Ibid., § 10 cmt. a.

19. Jason Chein et al., "Peers Increase Adolescent Risk Taking by Enhancing Activity in the Brain's Reward Circuitry," *Developmental Science* 14 (2011): F1–F10.

20. Ibid., F2–F5.

21. Ibid., F7.

22. See Beatriz Luna et al., "Maturation of Widely Distributed Brain Function Subserves Cognitive Development," *NeuroImage* 13 (2001): 786–93, detailing brain function of children, adolescents, and adults during an antisaccade task.

23. Ibid., 791: "Synaptic pruning and myelination during childhood and adolescence are important for enhancing widely distributed brain functions by refining synaptic connections and enhancing the transfer of information throughout the brain in a rapid manner."

24. Catherine Label and Christian Beaulieu, "Longitudinal Development of Human Brain Wiring Continues from Childhood into Adulthood," *Journal of Neuroscience* 31 (2011): 10939.

25. See Elizabeth R. Sowell et al., "*In Vivo* Evidence for Post-Adolescent Brain Maturation in Frontal and Striatal Regions," *Nature Neuroscience* 2 (1999): 859–61.

26. A. Rae Simpson, "Brain Changes," *MIT Young Adult Development Project*, http://hrweb.mit.edu/worklife/youngadult/brain.html.

27. See generally Oliver Sacks, *Hallucinations* (New York: Vintage Books, 2013); Sacks, *The Man Who Mistook His Wife for a Hat* (New York: Summit Books, 1985). Sacks has detailed numerous examples: amorous tendencies brought on by

neurosyphilis (ibid., 102–3); uncharacteristically facetious and indifferent attitude caused by orbitofrontal tumor (ibid., 116–18); autoscopic phenomena caused by parietal or temporal lobe damage (Sacks, *Hallucinations*, 289).

28. Arachnoid cysts, for example, while visually ominous on an MRI scan, may never cause symptoms. See National Institute of Neurological Disorders and Stroke, *NINDS Arachnoid Cyst Information Page*, http://www.ninds.nih.gov/disorders/all -disorders/arachnoid-cysts-information-page.

29. Dave Zirin, "Jovan Belcher's Murder-Suicide: Did the Kansas City Chiefs Pull the Trigger?," *The Nation*, http://www.thenation.com/blog/177787/jovan-belchers -murder-suicide-did-kansas-city-chiefs-pull-trigger#.

30. Ibid.

31. American Law Institute, *Restatement (Third) of Torts*, §§ 29–36.

32. See W. Keeton et al., *Prosser and Keeton on the Law of Torts § 65*, 5th ed. (Eagan, MN: West Group, 1984), 451–52. Alabama, Maryland, North Carolina, and Virginia still use contributory negligence, in which the plaintiff's negligence, in any amount, bars the plaintiff's action completely, unless the last clear chance doctrine applies (Dobbs, *Law of Torts*, 504).

33. See Keeton et al., *Prosser and Keeton*, 470; Dobbs, *Law of Torts*, 503. Comparative negligence systems take two forms. Pure, or complete, comparative negligence applies comparative fault to all parties, and no plaintiff is barred by her contributory negligence (Dobbs, *Law of Torts*, 505). Modified, or incomplete, comparative negligence systems bar recovery when a plaintiff's contributory negligence reaches a certain threshold (ibid.). In the "greater than" subcategory of comparative negligence, the plaintiff's recovery is barred if the plaintiff's fault "exceeds that of the defendant" (ibid.). In the "equal to" subcategory of comparative negligence, the plaintiff's recovery is barred if the plaintiff's fault is equal to the defendant's (ibid.).

34. Engle v. Liggett Group, Inc., 945 So.2d 1246 (Fla. 2006).

35. U.S. v. Phillip Morris USA, Inc., 449 F. Supp. 2d 1, 881 (D.D.C. 2006).

36. Ibid., 833–34.

37. Ibid., 833.

38. Ibid., 835.

39. Ibid., 836–37.

40. Ibid., 840. See Grinnell v. American Tobacco Co., 883 S.W.2d 791, 799 (Tex. App. 1994), defining addictive substances as "neurologically active," or resulting in "particular exhortation of certain receptors in the central nervous system." See also Brian M. Lowe, *Emerging Moral Vocabularies: The Creation and Establishment of New Forms of Moral and Ethical Meanings* (Lanham, MD: Lexington, 2006), 152 (quoting a 1980 Tobacco Institute document: "the entire matter of addiction is the most potent weapon a prosecuting attorney can have in a lung cancer/cigarette case. We can't defend continued smoking as 'free choice' if the person was 'addicted.'").

41. Athina Markou, "Neurobiology of Nicotine Dependence," *Philosophical Transactions of the Royal Society of London B: Biological Sciences* 363 (2008): 3159–68.

42. See Eric J. Nestler and Robert C. Malenka, "The Addicted Brain," *Scientific American* 290 (2004): 81.

43. See B. Douglas Bernheim and Antonio Rangel, "Addiction and Cue-Triggered Decision Processes," *American Economic Review* 94 (2004): 1558–90.

44. Ibid.

45. See U.S. v. Phillip Morris USA, 449 F. Supp. 2d at 1366–1763. See generally Gregory N. Connolly et al., "Trends in Nicotine Yield in Smoke and Its Relationship with Design Characteristics among Popular US Cigarette Brands, 1997–2005," *Tobacco Control* 16 (2007), e5; Richard D. Hurt and Channing R. Robertson, "Prying Open the Door to the Tobacco Industry's Secrets about Nicotine: The Minnesota Tobacco Trial," *Journal of the American Medical Association* 280 (1998): 1173–81; Barry Meier, "U.S. Brings First Charges in Inquiry on Tobacco Companies," *New York Times*, January 8, 1998, A16; David A. Kessler, "The Control and Manipulation of Nicotine in Cigarettes," *Tobacco Control* 3 (1994): 362, 368.

46. Phillip Morris USA, Inc. v. Douglas, 110 So.3d 419 (Fla. 2013).

47. The general phase I findings for the *Engle* class included (1) smoking cigarettes causes a various schedule of medical complications, diseases, and cancers, (2) nicotine in cigarettes is addictive, (3) tobacco companies placed cigarettes on the market that were defective and unreasonably dangerous, (4) tobacco companies concealed or omitted material information or failed to disclose a material fact about the health effects or addictive nature of smoking cigarettes, (5) smokers and the public relied on this misinformation to their detriment, (6) all of the tobacco companies sold or supplied cigarettes that were defective, (7) all the tobacco companies sold or supplied cigarettes that did not conform to representations of fact made by the companies, and (8) all of the tobacco companies were negligent. *Engle*, 945 So.2d at 1276–77. (Note added.)

48. *Douglas*, 110 So.3d at 431–32.

49. David L. Wallace, "Addiction Postulates and Legal Causation, or Who's in Charge, Person or Brain?," *Journal of the American Academy of Psychiatry and the Law* 41 (2013): 92–97. Wallace was a litigation partner at Herbert Smith Freehills in New York. David Wallace Company Profile, *Herbert Smith Freehills*, http://www.herbertsmithfreehills.com/people/david-l-wallace. He now is Wallace Law PLLC.

50. Ibid., 93.

51. Stuart Fox, "Laws Might Change as the Science of Violence Is Explained," *Live Science*, June 7, 2010, http://www.livescience.com/6535-laws-change-science-violence-explained.html (quoting Stephen Morse).

52. M. R. Bennett and P. M. S. Hacker, *Philosophical Foundations of Neuroscience* (Malden, MA: Blackwell, 2003), 3: Neuroscience's "discoveries in no way affect the conceptual truth that these powers and their exercise in perception,

thought, and feelings are *attributes of human beings*, not of their parts—in particular, *not of their brain*."

53. Michael S. Pardo and Dennis Patterson, *Minds, Brains, and Law: The Conceptual Foundations of Law and Neuroscience* (New York: Oxford University Press, 2013), 22: "[P]ersons are not their brains."

54. See Stephen J. Morse, "Lost in Translation? An Essay on Law and Neuroscience," in *Law and Neuroscience*, Current Legal Issues 13, ed. Michael Freeman (Oxford, UK: Oxford University Press, 2011), 532: "I do not mean to imply dualism here. I am simply accepting the folk-psychological view that mental states . . . play a genuinely causal role in explaining human behaviour"; M. R. Bennett and P. M. S. Hacker, *History of Cognitive Neuroscience* (Malden, MA: Wiley-Blackwell, 2008), 241: "The central theme of our book was to demonstrate the incoherence of brain-body dualism"; Pardo and Patterson, introduction to *Minds, Brains, and Law*, xiii: "We are not dualists."

55. Pardo and Patterson, "Philosophical Foundations," 1249.

56. Ibid.

57. Walter Glannon, "Brain, Behavior, and Knowledge," *Neuroethics* 4 (2011): 191–94.

58. Wallace, "Addiction Postulates," 93.

59. Harold Kalant, "What Neurobiology Cannot Tell Us about Addiction," *Addiction* 105 (2010): 780–89.

60. Alan I. Leshner, "Addiction Is a Brain Disease, and It Matters," *Science* 278 (1997): 45.

61. Ibid., 46.

62. Ibid.

63. Wallace, "Addiction Postulates," 94 (quoting *DSM-IV*).

64. Kalant, "What Neurobiology Cannot Tell Us," 782.

65. Ibid. (citing William Renthal et al., "ΔFosB Mediates Epigenetic Desensitization of the c-fos Gene after Chronic Amphetamine Exposure," *Journal of Neuroscience* 28 [2008]: 7344–49; Deanna L. Wallace et al., "The Influence of ΔFosB in the Nucleus Accumbens on Natural Reward-Related Behavior," *Journal of Neuroscience* 28 [2008]: 10272–77).

66. Kalant, "What Neurobiology Cannot Tell Us," 784.

67. G. R. Breese et al., "Stress Enhancement of Craving during Sobriety: A Risk for Relapse," *Alcoholism: Clinical and Experimental Research* 29 (2005): 185–95. (Note in original.)

68. Kalant, "What Neurobiology Cannot Tell Us," 785.

69. An analogous context warrants comparison: Manufacturers and prescribers of selective serotonin reuptake inhibitors (SSRIs) could not escape liability for the higher risk of suicidality in children and young adults who took antidepressants, even though suicidality also is caused by several endogenous and exogenous factors. See Floyd ex rel. Ray v. U.S., 2010 WL 4905010 (M.D. Ga. 2010), in which the

court found that a recent breakup with the boyfriend of a teenager who committed suicide was not enough to dissuade the court from finding SSRIs to be the general cause of suicidality in the teenager; Mason v. SmithKline Beecham Corp., 2010 WL 2697173 (C.D. Ill. 2010), in which federal preemption of state tort law did not apply in a suit surrounding the suicide of a twenty-two-year-old two days after being prescribed Paxil; Tucker v. SmithKline Beecham Corp., 701 F. Supp. 2d 1040 (S.D. Ind. 2010), in which the court denied the SSRI manufacturer's motion of summary judgment founded in claims of ambiguous causation of the suicide of a Catholic priest.

70. See Bob Herbert, "In America; Tobacco Dollars," *New York Times*, November 28, 1993, http://www.nytimes.com/1993/11/28/opinion/in-america-tobacco -dollars.html (quoting R. J. Reynolds executive's response to why he did not smoke: "We reserve that right for the poor, the young, the black and the stupid"); Dorie E. Apollonio and Ruth E. Malone, "Marketing to the Marginalised: Tobacco Industry Targeting of the Homeless and Mentally Ill," *Tobacco Control* 14 (2005): 409, discussing how tobacco companies target homeless and mentally ill; Centers for Disease Control, "Cigarette Smoking among Adults and Trends in Smoking Cessation—United States," *Morbidity and Mortality Weekly Report* 58 (2009): 1227, noting that adults aged twenty-five years and older with "low educational attainment" had the highest prevalence of smoking; Sarah S. Lochlann Jain, " 'Come Up to the Kool Taste': African American Upward Mobility and the Semiotics of Smoking Menthols," *Public Culture* 15 (2003): 295–322, examining tactics that tobacco companies used to target African Americans in inner cities.

71. Marc N. Potenza, "What Integrated Interdisciplinary and Translational Research May Tell Us about Addiction," *Addiction* 105 (2010): 792–96. There is something of a vicious cycle here: "pre-clinical studies indicate that not only does substance-naïve impulsive decision-making predict substance self-administration, but also that substance intake, probably in a developmentally sensitive fashion, influences decision-making" (792). So those with some neural predisposition to addictive behavior will be induced by the addictive substance itself to continue the addictive behavior. The same propensities that caused addiction in the first place are emphasized by the addictive substance in ways that result in continuation of the addictive and, in the case of nicotine, destructive behavior, perhaps long after the addict "knows better." That would suggest the commercial desirability of getting new smokers while they are young. See Pamela M. Ling and Stanton A. Glantz, "Why and How the Tobacco Industry Sells Cigarettes to Young Adults: Evidence from Industry Documents," *American Journal of Public Health* 92 (2002): 908–16; K. Michael Cummings et al., "Marketing to America's Youth: Evidence from Corporate Documents," *Tobacco Control* 11 (2002).

72. The economic burden of smoking was determined, at least roughly, in the Master Settlement Agreement of 1998 between the major tobacco manufacturers and the attorney generals of forty-six states, which allocated more than $200 billion to recoup state Medicaid expenditures related to smoking-induced disease. "Master Settlement Agreement," National Association of Attorneys General (1998),

http://publichealthlawcenter.org/sites/default/files/resources/master-settlement
-agreement.pdf.

73. Ellen Wertheimer, "The Smoke Gets in Their Eyes: Product Category Liability and Alternative Feasible Designs in the Third Restatement," *Tennessee Law
Review* 61 (1994): 1429, 1445, concluding that tobacco would surely be per se defective if not for its place in our history and economy.

74. See Mark Hyman, "Why Kids under 14 Should Not Play Tackle Football,"
Time, November 6, 2012, http://ideas.time.com/2012/11/06/why-kids-under-14-should
-not-play-tackle-football/: "A teenager entering high school can make a judgment
about the ups and downs of playing tackle football. . . . The same isn't true of a
6-year-old." Children are allowed to play youth tackle football as early as age five
and when as light as thirty-five pounds (Pop Warner League, *Ages and Weights*,
http://www.popwarner.com/football/footballstructure.htm).

75. For discussions of the cumulative effects of concussions, see Meheroz H.
Rabadi and Barry D. Jordan, "The Cumulative Effect of Repetitive Concussion in
Sports," *Clinical Journal of Sport Medicine* 11 (2001): 194–98; Kevin M. Guskiewicz
et al., "Cumulative Effects Associated with Recurrent Concussion in Collegiate
Football Players," *Journal of the American Medical Association* 290 (2003): 2549–
55; Rosemarie S. Moser et al., "Prolonged Effects of Concussion in High School
Athletes," *Neurosurgery* 57 (2005): 300–306.

76. See Esteban Toledo et al., "The Young Brain and Concussion: Imaging as a
Biomarker for Diagnosis and Prognosis," *Neuroscience and Biobehavioral Reviews*
36 (2012): 1526, highlighting the difficulty in currently determining the effects of
youth concussions: "A current paradox exists in which many studies show the extent
and severity of the consequences with an increased incidence and prevalence of
acute as well as chronic problems, while statistics indicate that the disease burden
seems to decrease over time."

77. Ann C. McKee et al., "Chronic Traumatic Encephalopathy in Athletes: Progressive Tauopathy after Repetitive Head Injury," *Journal of Neuropathology and
Experimental Neurology* 68 (2009): 709, 711.

78. See Boston University CTE Center, *What Is CTE?*, http://www.bu.edu/cte
/about/what-is-cte/; see also Bennet I. Omalu et al., "Chronic Traumatic Encephalopathy in a National Football League Player," *Neurosurgery* 57 (2005): 129; Elliot J.
Pellman, "Background on the National Football League's Research on Concussion
in Professional Football," *Neurosurgery* 53 (2003): 797–98.

79. For evidence of brain-based modulation of addictive urges, see Nasir H.
Naqvi et al., "Damage to the Insula Disrupts Addiction to Cigarette Smoking," abstract, *Science* 315 (2007): 531: "We found that smokers with brain damage involving the insula, a region implicated in conscious urges, were more likely than smokers with brain damage not involving the insula to undergo a disruption of smoking
addiction, characterized by the ability to quit smoking easily, immediately, without
relapse, and without persistence of the urge to smoke."

80. See American Law Institute, *Restatement (Third) of Torts*, 8, Scope Note,

noting that the inability to encourage mitigation has historically been a problem with allowing recovery for pure emotional injury; American Law Institute, *Restatement (Second) of Torts*, § 918: "one injured by the tort of another is not entitled to recover damages for any harm that he could have avoided by the use of reasonable effort or expenditure after the commission of the tort"; American Law Institute, *Restatement (Third) of Torts*, § 3, retaining the mitigation principle.

81. Steve Fainaru and Mark Fainaru-Wada, "Youth Football Participation Drops," *ESPN.com*, November 14, 2013, http://espn.go.com/espn/otl/story/_/page/popwarner /pop-warner-youth-football-participation-drops-nfl-concussion-crisis-seen-causal-factor.

82. Mary H. Kathol et al., "Magnetic Resonance Imaging of Athletic Soft Tissue Injuries," *Iowa Orthopaedic Journal* 9 (1989), discussing improvements that MRI offers to reveal previously invisible soft tissue injury.

83. See Yvette I. Sheline et al., "Hippocampal Atrophy in Recurrent Major Depression," *Proceedings of the National Academy of Sciences* 93 (1996): 3908–13, explaining how those diagnosed with depression suffer from reduction in hippocampus volume; Luca Santarelli et al., "Requirement of Hippocampal Neurogenesis for the Behavioral Effects of Antidepressants," *Science* 301 (2003): 805–9, explaining that interference with antidepressant-modulated neurogenesis in the hippocampus erases the effects of an antidepressant in mice.

84. See, e.g., M. H. Teicher et al., "The Neurobiological Consequences of Early Stress and Childhood Maltreatment," *Neuroscience and Biobehavioral Review* 27 (2003): 33: "The major structural consequences of early stress include reduced size of the mid-portions of the corpus callosum and attenuated development of the left neocortex, hippocampus, and amygdala." An MRI reveals structural differences between maltreated subjects and healthy controls, implicating certain localities as markers for those who have experienced trauma.

85. See American Law Institute, *Restatement (Third) of Torts*, § 2 cmt. b: "While a showing of negligence generally suffices for compensatory damages, the standard for awarding punitive damages commonly refers to the defendant's reckless conduct—or reckless indifference to risk, or reckless disregard for risk."

86. American Law Institute, *Restatement (Second) of Torts*, § 436A, cmt a. to §§ 312 and 313, explains that "the rule stated in this section does not give protection to mental and emotional tranquility in itself"—which reiterates the language of 436A. Sections 312 and 313 state that when emotional disturbance causes bodily harm (section 312 takes up intentionally and unreasonably caused harm and 313 discusses unintentionally and negligently caused harm), the defendant is liable for that harm, even though it was emotional disturbance that caused the bodily harm.

87. Ibid., § 436A cmt. b.

88. See Omalu et al., "Chronic Traumatic Encephalopathy," 128–34; Pellman, "Background on the National Football League's Research," 797–98, describing behavioral and psychological symptoms of CTE, including forgetfulness and moodiness, and how those symptoms result from CTE.

89. See Paul Bach-y-Rita, "Brain Plasticity as a Basis for Recovery of Function in Humans," *Neuropsychologia* 28 (1990): 547–54; Michael M. Merzenich and William M. Jenkins, "Reorganization of Cortical Representations of the Hand Following Alterations of Skin Inputs Induced by Nerve Injury, Skin Island Transfers, and Experience," *Journal of Hand Therapy* 6 (1993): 89–104.

90. American Law Institute, *Restatement (Third) of Torts*, § 47.

91. Ibid., § 47 cmt. a.

92. Ibid., § 47(b).

93. Ibid., § 48.

94. Ibid., § 48 cmt. a. The Second Restatement categorically denies recovery for negligently caused emotional injury without accompanying bodily harm. See American Law Institute, *Restatement (Second) of Torts*, § 436A. Section 47 of the Third Restatement adds two exceptions: danger from immediate bodily harm and certain categories of activities that lack clear boundaries. See American Law Institute, *Restatement (Third) of Torts*, § 47. The reorganization underlying section 48 of the Third Restatement shifts the language orienting physical harm with emotional harm to chapter 2 of the Third Restatement. See American Law Institute, *Restatement (Third) of Torts*, § 4–6. Section 48 of the Third Restatement contains the essence of section 436(3) of the Second Restatement. See American Law Institute, *Restatement (Second) of Torts*, § 436(3).

95. American Law Institute, *Restatement (Third) of Torts*, § 47 cmt. e.

96. Ibid., § 47 cmt. k.

97. Ibid.

98. Ibid.

99. Ibid.

100. Ibid.

101. Ibid., § 47 cmt. k, illus. 2.

102. Ibid., § 47 cmt. d: "Some of the policies that might dictate a withdrawal of liability are . . . reflected in limitations on liability in this section, such as the requirement that the emotional harm suffered be serious."

103. Ibid., § 47, Reporter's Note at 194–95.

104. Though there is disagreement about the extent to which mirror neurons give rise to empathy, it is generally agreed that these cells underlie imitation, other-regarding, and theory of mind—allowing humans to share in the experiences of others. See Sandra Blakeslee, "Cells That Read Minds," *New York Times*, January 10, 2006, http://www.nytimes.com/2006/01/10/science/10mirr.html?pagewanted=all; see also Jonas T. Kaplan and Marco Iacoboni, "Getting a Grip on Other Minds: Mirror Neurons, Intention Understanding, and Cognitive Empathy," *Social Neuroscience* 1 (2006): 175–83.

105. Perrotti v. Gonicberg, 877 A.2d 631 (2005) (cited in American Law Institute, *Restatement (Third) of Torts*, § 47, Reporter's Note at 194–95).

106. Gagnon v. Rhode Island Co., 101 A. 104 (R.I. 1917).

107. Sadegh Nabavi et al., "Engineering a Memory with LTD and LTP," *Nature* 511 (2014): 348–52.

108. Ibid.

109. Ibid.

110. American Law Institute, *Restatement (Third) of Torts*, § 48.

111. A psychopathic family member, presumably, would not be entitled to recovery. See R. J. Blair, "Facial Expressions, Their Communicatory Functions and Neuro-cognitive Substrates," *Philosophical Transactions of the Royal Society of London B: Biological Sciences* 358 (2003): 561–72, describing psychopaths' lack of empathy.

112. American Law Institute, *Restatement (Third) of Torts*, § 48 cmt. d.

113. Comment g recognizes that the requirements of the section may be "described or criticized as arbitrary." But the commentary defends the lines drawn by the black letter: "[T]hese rules are 'arbitrary' in the same way that an age limit of 16 to obtain a driver's license is arbitrary. Some rule is necessary, and the one selected will be excessively restrictive for some and too lenient for others. . . . Limits are required for emotional harm because of its ubiquity, and an alternative to workable and effective limits for such liability could be a rule of no liability." American Law Institute, *Restatement (Third) of Torts*, § 48 cmt. d.

114. American Law Institute, *Restatement (Third) of Torts*, 48 cmt. g.

115. Siri Leknes and Irene Tracey, "A Common Neurobiology for Pain and Pleasure," *Nature Neuroscience* 9 (2008): 314–20.

116. Donald D. Price, "Psychological and Neural Mechanisms of the Affective Dimension of Pain," *Science* 288 (2000): 1769–72.

117. C. S. Wheatley Jr., "Future Pain and Suffering as Elements of Damages for Physical Injury," *American Law Reports* 81 (1932): 423, annotation.

118. McDougald v. Garber, 536 N.E.2d 372, 375 (N.Y. 1989).

119. See Petro v. Town of W. Warwick ex rel. Moore, 889 F. Supp. 2d 292, 345 (D.R.I. 2012): "Under the familiar eggshell plaintiff rule, which is applicable to [civil action for deprivation of rights] suits, defendants are liable for the full extent of Jackson's pain and suffering, regardless of whether it was exacerbated by Jackson's mental health issues."

120. Tor D. Wager et al., "An fMRI-Based Neurologic Signature of Physical Pain," *New England Journal of Medicine* 368 (2013): 1388, fig. 2B.

121. A. Vania Apkarian, "A Brain Signature for Acute Pain," *Trends in Cognitive Science* 17 (2013): 310. One commentator described the value of the study as, first, identifying a general neural pain signature that would appear to be accurate across different subjects performing different tasks, and, second, quite provocatively as far as the law is concerned, beginning "to address what perception itself is." Other commentators reminded that the results were limited to the experimental design and did not take into account the many circumstances that accompany pain. See Yi Lu, Geraldine T. Klein, and Michael Y. Wang, "Can Pain Be Measured Objectively?,"

Neurosurgery 73 (2013), N24: "we need to remember that the pain signature described in this study were the result of moderate thermal pain applied to the forearm of young healthy volunteers."

122. See Michael E. Robinson et al., "Pain Measurement and Brain Activity: Will Neuroimages Replace Pain Ratings?," *Journal of Pain* 14 (2013): 323–27.

123. Adam J. Kolber, "Pain Detection and the Privacy of Subjective Experience," *American Journal of Law and Medicine* 33 (2007): 441.

124. Ibid.

125. Eric Eich et al., "Questions concerning Pain," in *Well-Being: The Foundations of Hedonic Psychology*, ed. Daniel Kahneman, Edward Deiner, and Norbert Schwarz (New York: Russell Sage Foundation, 1999), 160. (Note in original.)

126. Ibid. (Note in original.)

127. Chris J. Main, "The Nature of Chronic Pain," in *Malingering and Illness Deception*, ed. Peter W. Halligan, Christopher Bass, and David A. Oakley (New York: Oxford University Press, 2003), 171, 172 (note in original). Kolber, "Pain Detection," 437.

128. For an enumeration of pain and suffering compensation schemes, see Jack H. Olender, "Showing Pain and Suffering," *American Jurisprudence Trials*, vol. 5, ed. Thompson West (Rochester, MN: Lawyers Co-operative, 1966), § 30.

Chapter Five

1. Guido Calabresi, *The Costs of Accidents: A Legal and Economic Analysis* (New Haven, CT: Yale University Press, 1970), 26–31, positing "cheapest cost avoider," which seeks a point of deterrence where costs of accident and accident prevention are lowest.

2. Ibid.

3. That is particularly true with regard to liability for design and failure to warn defects, which is grounded in negligence (American Law Institute, *Restatement (Third) of Torts: Product Liability* [St. Paul, MN: American Law Institute, 2012], §§ 2(b)–2(c)).

4. Ibid., § 2(a).

5. For a discussion of some of the empirical problems with instrumentalist theories of tort law, see G. Edward White, *Tort Law in America: An Intellectual History*, exp. ed. (New York: Oxford University Press, 2003), attacking fundamental assumptions of instrumental theories of tort.

6. See Ernest J. Weinrib, *Corrective Justice* (New York: Oxford University Press, 2012), 72–75. Weinrib has defended a pure form of corrective justice in which the analysis examines only the tortfeasor and the injured party. See Ernest J. Weinrib, "Correlativity, Personality, and the Emerging Consensus on Corrective Justice," *Theoretical Inquiries in Law* 2 (2001): 107, 108. Central to Weinrib's theory is the

principle of correlativity whereby only the tortfeasor, not a third party source of compensation (i.e., insurance), must pay the entire damage amount to the injured party.

7. Jules L. Coleman, *Risks and Wrongs* (Oxford, UK: Oxford University Press, 2002), 311–15. Coleman adopted a similar view of corrective justice as the central objective of tort law and argued that tortfeasors have a moral duty to correct the harms for which they are responsible. But Coleman did not equate grounds of redress with mode of redress. In other words, "even if the injurer has the duty to repair injustice, it does not follow that justice requires that the duty be discharged by the injurer." Coleman, "The Mixed Conception of Corrective Justice," *Iowa Law Review* 77 (1991): 427, 443. Thus, Coleman left room for third-party sources of compensation.

8. For particularly rich contemporary commentary not treated here, see, e.g., Stephen Perry, "The Moral Foundations of Tort Law," *Iowa Law Review* 77 (1992): 449; Perry, "Political Authority and Political Obligation," in *Oxford Studies in the Philosophy of Law 2*, ed. Leslie Green and Brian Leiter (New York: Oxford University Press, 2012); Perry, "Torts, Rights, and Risks," in *Philosophical Foundations of the Law of Torts*, ed. John Oberdiek (New York: Oxford University Press, 2014). There are noninstrumental, moral responsibility–based, theories of tort, such as Perry's, that clearly contemplate a fit among morality, conceptions of blameworthiness, and tort in ways that would be inconsistent with neuroscientific insights insofar as those insights undermine blameworthiness and desert generally. The focus here, though, is on the corrective justice theory of tort, the dominant noninstrumental perspective. Tort law's goal, "at least according to those (like me) who are not utilitarians, is to achieve corrective justice—the restoration of the *ex ante* status quo, or its moral equivalent." Heidi M. Hurd, "The Innocence of Negligence," *Contemporary Readings in Law and Social Justice* 8, no. 2 (2016): 78.

9. Aristotle, *Nicomachean Ethics*, ed. Roger Crisp (Cambridge: Cambridge University Press, 2000).

10. Aristotle, *Nicomachean Ethics*, 75: "One kind [of justice] is that which is manifested in distributions of honour or money or the other things that fall to be divided among those who have a share in the constitution."

11. Aristotle, *Nicomachean Ethics*, 75–77: "One [other type of justice] is that which plays a rectifying part in transactions between man and man. . . . [T]herefore, corrective justice will be the intermediate between loss and gain."

12. Compare Jules Coleman, Scott Hershovitz, and Gabriel Mendlow, "Theories of the Common Law of Torts," in *Stanford Encyclopedia of Philosophy*, ed. Edward M. Zalta, http://plato.stanford.edu/entries/tort-theories/#CorJus, explaining that corrective justice places a duty on a wrongdoer to repair a wrong committed and resulting in loss suffered by the victim, whatever that loss may be; what actually constitutes harm is not strictly a matter of corrective justice.

13. Stephen Perry, "Comment on Coleman: Corrective Justice," *Indiana Law Journal* 67 (1992): 381, 397. Perry appeared to at least kick the tires of such a ques-

tion when discussing Coleman's concept of corrective justice as restoring an initial just distribution: "If the primary responsibility to annul wrongful losses has come to rest with the state, then regardless of whether that responsibility is original or mandated, Coleman is best regarded, for reasons we have already considered, as talking not about corrective but about distributive justice."

14. See Weinrib, "Corrective Justice," *Iowa Law Review* 77 (1992): 403–9.

15. Palsgraf v. Long Island Railroad Co., 162 N.E. 99 (N.Y. 1928).

16. Ibid., 99.

17. Ibid., 100.

18. Ibid. (citing Frederick Pollock, *The Law of Torts*, 11th ed. [London: Stevens and Sons, 1920], 455; Martin v. Herzog, 228 N. Y. 164, 170 (1920); compare John W. Salmond, *Salmond on the Law of Torts* [London, Sweet and Maxwell, 1973], 24).

19. Ernest J. Weinrib, *The Idea of Private Law* (Oxford, UK: Oxford University Press, 2012), 159–60.

20. Palsgraf, 162 N.E. 99 at 100.

21. Which is perhaps even more curious in a society with social welfare systems including universal health care.

22. Weinrib, *Idea of Private Law*, 160.

23. Ibid., 161.

24. There is, after all, nothing in the black letter of the doctrine that so constrains the proximate causation or duty inquiries. See American Law Institute, *Restatement (Second) of Torts* (St. Paul, MN: American Law Institute, 1965), § 281, stating elements of cause of action for negligence; American Law Institute, *Restatement (First) of Torts* (St. Paul, MN: American Law Institute, 1934), § 281: same.

25. See Larry Alexander, "Theory's a What Comes Natcherly," *San Diego Law Review* 37 (2000): 777, 779, denying the possibility of grand theory of tort law because of the complexity evolved from judicial doctrine and statutory interpretations over time.

26. Weinrib, *Idea of Private Law*, 84.

27. Ibid. (emphasis added).

28. Ibid.

29. Ibid., 86.

30. Ibid., 89 (quoting Immanuel Kant, *Critique of Pure Reason*, trans. F. Max Muller [New York: Macmillan, 1922], emphasis in original).

31. Ibid.

32. Ibid.

33. Ibid., 90.

34. Ibid.

35. Ibid., 91.

36. See, e.g., Arthur Schopenhauer, *On the Basis of Morality* (London: Swan Sonnenschein, 1903, part 2, 21); Sally S. Sedgwick, "Hegel's Critique of the Subjective Idealism of Kant's Ethics," *Journal of the History of Philosophy* 26 (1998): 89–105.

37. Weinrib, *Idea of Private Law*, 91.

38. Gerry Greenstone, "The History of Bloodletting," *British Columbia Medical Journal* 52 (2010): 12.

39. Coleman, *Risks and Wrongs*, 12: "the difference between rational choice liberalism and the market paradigm is that the latter [*sic*: "former"] sees all law entirely in terms of its role in promoting efficiency whereas the latter claims that certain parts of the law help to create and sustain the conditions under which markets can flourish and contribute to stability."

40. Coleman, *Risks and Wrongs*, 3–4.

41. Ibid., 285.

42. Ibid.

43. Ibid.

44. Ibid., 306. Coleman had originally defined the first part of his theory of corrective justice, "the annulment theory," rather simply as the belief that "the point of corrective justice is to eliminate, rectify, or annul wrongful (or unjust) losses. . . . Wrongful gains and losses cannot be annulled to create other wrongful gains or losses." After receiving strong criticism from Weinrib that the annulment theory failed to account for the duty a tortfeasor owes a victim (see Weinrib 1983, 37, 39), Coleman abandoned the annulment theory for a mixed conception of corrective justice, which adopts Weinrib's duty conception of corrective justice (ibid., 311–24).

45. Ibid., 314 (emphasis added); the "responsibility" here would seem to be moral, at least after a fashion.

46. Coleman's conception may described as mixed in that it kept annulment theory's premise that the wrongfulness of a loss created a right in the victim to be made whole by the tortfeasor, but it built in the recognition that tortfeasors owe a special duty to those they injure, thus mixing the relational and annulment conceptions of corrective justice (ibid., 318–19).

47. Ibid., 326 (emphasis added).

48. Coleman et al., "Theories of the Common Law of Torts."

49. Weinrib, *Idea of Private Law*, 91.

50. Ibid.

51. It would seem unlikely that either Kant or Weinrib intended free will in a strict compatibilist sense, though nothing much changes in the foregoing if they did.

52. Ibid., 93.

53. Ibid., 94.

54. Coleman, *Risks and Wrongs*, 333–34.

55. Ibid., 334.

56. Ibid. (note omitted).

57. Ibid.

58. Ibid. (emphasis added).

59. Coleman et al., "Theories of the Common Law of Torts."

60. Coleman, *Risks and Wrongs*, 346.

61. Ibid. (note omitted).

62. Ibid.

63. Ibid., 349.

64. Ibid., 352.

65. Ibid. This too would seem to be consistent with Coleman's argument that morality operates where markets fail.

66. Ibid.

67. See, e.g., Stephen Perry, "Responsibility for Outcomes, Risk, and the Law of Torts," in *Philosophy and Tort Law*, ed. Gerald Postema (Cambridge: Cambridge University Press, 2001), 119–20, in which Perry described his outcome-responsibility approach to tort law and noted that "Fault will only exist in those cases where the defendant acted so as to increase the risk substantially, or else could have reduced or eliminated the risk at a relatively low cost to himself"; John Goldberg and Benjamin Zipursky, *Tort Law: Responsibilities and Redress* (New York; Aspen, 2004), xix, stating that torts "imposes on members of society a set of legal obligations— i.e. *responsibilities*—to avoid injuring others"; Jason M. Solomon, "Civil Recourse as Social Equity," *Florida State University Law Review* 39 (2011): 244–45, making a case that civil recourse theory offers the notion of equality that is normative foundation of tort law.

68. See Goldberg and Zipursky, *The Oxford Introductions to U.S. Law: Torts* (New York: Oxford University Press, 2010), 68–69, establishing some similarities between civil recourse and corrective justice theorists but distinguishing the two approaches.

69. Benjamin Zipursky, "*Palsgraf*, Punitive Damages, and Preemption," *Harvard Law Review* 125 (2012): 1757–58.

70. See Goldberg and Zipursky, "Torts as Wrongs," *Texas Law Review* 88 (2010): 917–86.

71. See ibid., 950.

72. See ibid., 951.

73. Some of those loss-based constructions of tort liability include Perry's concept of outcome responsibility, his conception of tort liability as dependent on the existence of standard patterns of interaction rather than always applicable strict liability, and his reworking of Coleman's concepts of primary and secondary rights and rights invasions. See, e.g., Perry, "Comment on Coleman"; Perry, "The Distributive Turn: Mischief, Misfortune and Tort Law," in *Analyzing Law: New Essays in Legal Theory*, ed. Brian Bix, 141, 157–61 (New York: Oxford University Press, 1998); Perry, "Responsibility for Outcomes."

74. Coleman, *Risks and Wrongs*.

75. They explained: "[W]hile we conceive of torts as private wrongs, we also concede that government is central to the tort system's operation in a manner that many scholars have overlooked and that a challenge for tort theory is to explain what is distinctively 'private' about tort, given the state's role. In short, there is a need for a cogent and doctrinally grounded account of two distinct concepts and

the connection between them: tortious wrongdoing ('wrongs') and civil recourse ('recourse')." See Goldberg and Zipursky, "Torts as Wrongs," 919.

76. Ibid., 943–44 (notes omitted).

77. Ibid., 926 (citing Coleman, *Risks and Wrongs*, 222–24, 314–18, 330–32).

78. Coleman, *Risks and Wrongs*, 320.

79. See Goldberg and Zipursky, "Torts as Wrongs," 932–34, arguing for a focus on wrong rather than loss, as civil recourse entails; additionally noting that Weinrib holds a similar view of the prominence of wrong that suggests civil recourse's greater affinity with at least Weinrib's corrective justice.

80. *Blameworthiness* here is redolent of Moore's guilt basis and measure of retribution. See chapter 3, "Blame, Desert, and Culpability."

81. Goldberg and Zipursky, "Torts as Wrongs," 944 (emphasis added).

82. For a discussion of Third Restatement examples of such exceptional circumstances, see the section "Negligent Infliction of Direct Emotional Harm" in chapter 4: plaintiff as a passenger in apparently doomed aircraft or one who is exposed to HIV as examples of satisfying immediacy requirement; PTSD as example of satisfying severity requirement; use of fetal heart monitoring to allay mother's fear that her baby was injured in car accident as being appropriate deviation from objectivity requirement (citing American Law Institute, *Restatement (Third) of Torts*, § 47 cmts. k, d, Reporter's Notes at 194–95).

83. Goldberg and Zipursky, "Torts as Wrongs," 956.

84. See chapter 4, "Ultimate Inscrutability?," highlighting PTSD as example of emotional injury that manifests itself in distinctive brain structures and patterns.

85. See Bruce Waller, *Against Moral Responsibility* (Cambridge, MA: MIT Press, 2011).

86. See Galen Strawson, *Freedom and Belief* (New York: Oxford University Press, 2010).

87. See Neil Levy, *Hard Luck: How Luck Undermines Free Will and Moral Responsibility* (New York: Oxford University Press, 2011).

88. For some examples of no-fault statutes, see Fla. Stat. 627.737 (2015), concerning tort exemption, limitation on rights to damages, and punitive damages; Minn. Stat. 65B.51 (2015), detailing deduction of collateral benefits from tort recovery, limitation on right to recover damages; N.J. Rev. Stat. 17:28-1.9 (2016), describing immunity from liability for certain auto insurance providers.

89. See Benjamin Zipursky, "Civil Recourse, Not Corrective Justice," *Georgetown Law Journal* 91 (2003): 695.

90. Ibid., 700 (notes omitted).

91. Goldberg and Zipursky, "The Moral of *MacPherson*," *University of Pennsylvania Law Review* 146 (1998): 1830–31 (notes omitted).

92. See chapter 4, "Compensable Injury," discussing required evidence of physical manifestation for compensable emotional injury.

93. See ibid., "Mental versus Physical Disability," discussing administrative and

policy arguments for why an adult with the mental capacity of a five-year-old is treated differently legally from the way an actual five-year-old is treated.

94. See ibid., Addiction, discussing the use of contributory and comparative negligence to deny sickened smokers' recovery when cigarette manufacturers knew that a certain percentage would be victimized and even knew and took advantage of characteristics of those victims.

95. Coleman et al., "Theories of the Common Law of Torts." It would seem that Coleman, then, would have to claim some moral substance for his perspective, corrective justice, if he criticizes civil recourse for having none. So Goldberg and Zipursky may indeed have been correct to conclude that corrective justice is not amoral.

96. Goldberg and Zipursky, "The Myths of *MacPherson*," *Journal of Tort Law* 9 (forthcoming), Fordham Law Legal Studies Research Paper No. 2770725.

97. MacPherson v. Buick Motor Company, 217 N.Y. 382 (N.Y. 1916).

98. "One of the wheels was made of defective wood, and its spokes crumbled into fragments. The wheel was not made by the defendant; it was bought from another manufacturer. There is evidence, however, that its defects could have been discovered by reasonable inspection, and that inspection was omitted" (ibid., 385).

99. William L. Prosser, *Handbook of The Law of Torts* § 31, 180. Greenman v. Yuba Power Prods. Inc., 377 P.2d 897 (Cal. 1963); Escola v. Coca Cola Bottling Co. of Fresno, 150 P.2d 436, 465 (Cal. 1944) (Traynor, J., concurring).

100. Goldberg and Zipursky point to *Rowland*, *Dillon*, and *Tarasoff* as specific decisions indirectly resulting from Posner's instrumentalist interpretation of *MacPherson*. Rowland v. Christian, 69 Cal. 2d 108 (1968), eliminating plaintiff-status categories in premises liability in favor of general duty of reasonable care; Dillon v. Legg, 68 Cal. 2d 728 (1968), granting action for negligent infliction of emotional distress to certain "bystanders"; Tarasoff v. Regents of University of California, 17 Cal. 3d 425 (Cal. 1976), recognizing therapists' duty to warn certain third parties endangered by their patients.

101. Goldberg and Zipursky, "The Myths of *MacPherson*," 13.

102. Ibid., 18 (emphasis added).

103. Ibid., 19 (emphasis added).

104. "Needless to say, we think the partial shift of de facto tort lawmaking to legislatures has been something of a disaster" (ibid., 23).

105. Ibid., 24 (emphasis added).

106. Hurd, "Innocence of Negligence," 48–95.

107. Ibid., 49.

108. Ibid., 51.

109. Peter Westen, "Individualizing the Reasonable Person in Criminal Law," *Criminal Law and Philosophy* 2 (2008): 143 (quoted in Hurd, "Innocence of Negligence," 50).

110. Hurd, "Innocence of Negligence," 51.

111. Ibid., 52.

112. Hurd engaged this line of reasoning when writing "Of course, both utilitarian and desert-oriented tort theorists *do* advocate penalties for those who cause harm accidently. The question is how they justify the imposition of liability on those who have not chosen their harmful actions under any description that triggered their appreciation that their actions might be harmful" (ibid., 54).

113. Ibid., 71.

114. Ibid.

115. Ibid., 71–72.

116. Sarah Rose Cavanagh, "Female Sexual Desire: An Evolutionary Biology Perspective," *Psychology Today*, June 19, 2013, https://www.psychologytoday.com /blog/once-more-feeling/201306/female-sexual-desire-evolutionary-biology-per spective: "Females have a long history of choice, such that they developed neural mechanisms to evaluate the quality of their partners and to adjust their level of desire accordingly."

117. Stephen Jay Gould, introduction to *Dance of the Tiger: A Novel of the Ice Age*, ed. Björn Kurtén (New York: Random House, 1980), xvii–xviii. "Just-so stories," as defined by Gould, are those arguments or presumptions that "rely on the fallacious assumption that everything exists for a purpose." Indeed, it may turn out that heroic behavior is *not* actually adaptive, for the hero at least. It would seem likely that a gene or predisposition for cowardice would prevail insofar as cowards live to fight (or run) another day and heroes often (or, at least more frequently) do not. Communities, of course, would have no reason to praise cowardice; it is much more efficacious for the group to celebrate (and sing anthems about) courage.

118. Hurd, "Innocence of Negligence," 72.

119. Hurd defined supererogatory actions as those that "go beyond the call of duty," or are heroic in nature. She contrasted suberogatory acts as deeds that "abuse rights" (ibid., 75). (Note added.)

120. Ibid., 75–76 (citing Neera Kapur Badhwar, "Friendship, Justice and Supererogation," *American Philosophical Quarterly* 22 [1985]: 123–31).

121. Bruce N. Waller, *The Stubborn System of Moral Responsibility* (Cambridge, MA: MIT Press, 2014). "Moral responsibility is like masturbation: it leads to blindness" (233). Waller emphasized the power that moral reasoning has to shape human agents as a sort of myopia, one that allows those making moral condemnations to look only at the acts of the one being morally condemned, not the possible root causes of those actions. Waller also suggested why moral responsibility acts in this way: It does so as a sort of survival tactic, for further inquiries into the root causes of immorality may make the inquiring party question the reasons for moral responsibility itself.

122. Hurd, "Innocence of Negligence," 77.

123. Daniel Wegner, *The Illusion of Conscious Will* (Cambridge, MA: MIT Press, 2002).

124. Ibid., ix.

125. Ibid., 11.

126. Ibid., 13.

127. Ibid.

128. Ibid. (quoting David Hume, *A Treatise of Human Nature*, ed. L. A. Selby-Bigge [London: Oxford University Press, 1896], 400–401).

129. Wegner, *Illusion of Conscious Will*, 14.

130. Ibid., 15.

131. Ibid. (emphasis in original).

132. Ibid., 21.

133. Hurd, "Innocence of Negligence," 78.

134. As seen in chapter 2, some "bad character" status such as alcoholism, narcotic addiction, and homelessness have been criminalized in some measure in various US jurisdictions; chapter 2, note 79.

135. Hurd, "Innocence of Negligence," 79.

136. Ibid., 81, 83. Hurd pointed out that badness of character does not enter into the damages calculus, that some inadvertence that seems character-based is plainly not the product of bad character, that attaching badness of character to criminal sentencing creates evidentiary difficulties, and that such concerns about bad character are almost impossible to make consistent with our notions of causation when it comes to concepts like contributory negligence.

137. Ibid., 86.

138. Ibid., 86–87.

139. Levy, *Hard Luck*.

140. See Waller, *Against Moral Responsibility*.

141. See Strawson, *Freedom and Belief*.

142. For a helpful resource for tracking the rise of free will skepticism, see Chandler Klebs and George Ortega, "Exploring the Illusion of Free Will," *Casual Consciousness*, http://causalconsciousness.com/index.html (cited in Gregg Caruso, introduction to *Exploring the Illusion of Free Will and Moral Responsibility* [Lanham, MD: Lexington, 2013], 2).

143. See T. M. Scanlon, *What We Owe to Each Other* (Cambridge, MA: Harvard University Press, 1998), 286.

144. Levy, *Hard Luck*, 205.

145. For representative views see, e.g., Nomy Arpaly, *Unprincipled Virtue* (New York: Oxford University Press, 2002), 91; Angela Smith, "Responsibility for Attitudes: Activity and Passivity in Mental Life," *Ethics* 115 (2005): 236; Smith, "Control, Responsibility, and Moral Assessment," *Philosophical Studies* 138 (2008): 368; Susan Wolf, *Freedom within Reason* (Oxford, UK: Oxford University Press, 1990). (Note added.)

146. This "indifference" would seem to track negligence well for present purposes. (Note added.)

147. Levy, *Hard Luck*, 205 (emphasis in original).

Chapter Six

1. Peter A. Alces, "Guerrilla Terms," *Emory Law Journal* 56 (2007): 1511, 1512.

2. See Chun Siong Soon et al., "Unconscious Determinants of Free Decisions in the Human Brain," *Nature Neuroscience* 11 (2008): 543, finding an outcome of a decision to be encoded in prefrontal and parietal cortex activity up to ten seconds before entering awareness; Antoine Bechara et al., "Deciding Advantageously before Knowing the Advantageous Strategy," *Science* 275 (1997): 1293–95, explaining that nonconscious biases guide behavior before conscious knowledge does; Benjamin Libet et al., "Time of Conscious Intention to Act in Relation to Onset of Cerebral Activity (Readiness Potential)," *Brain* 106 (1983): 623–42, discovering that recorded cognitive activity (readiness potential) preceded reported time of conscious intention to act by several hundred milliseconds.

3. In fact, neuroscience may even be able to predict our decisions before we make them. See Brian Knutson et al., "Neural Predictors of Purchases," *Neuron* 53 (2007): 147–56, suggesting that activation of nucleus accumbens precedes and supports consumers' purchasing decisions.

4. See Yannis Bakos, Florencia Marotta-Wurgler, and David R. Trossen, "Does Anyone Read the Fine Print? Consumer Attention to Standard-Form Contracts," *Journal of Legal Studies* 43 (2014): 1–45, finding that fewer than two of every thousand online software consumers accessed end user license agreements, and that those who did, read only small portions of agreement.

5. American Law Institute, *Restatement (Second) of Contracts* (St. Paul, MN: American Law Institute, 1981).

6. *Contract* is defined as a "promise or set of promises for the breach of which the law gives a remedy, or the performance of which the law in some way recognizes as a duty" (American Law Institute, *Restatement (Second) of Contracts*, § 1). The restatement defines *promise* as the "manifestation of intention to act or refrain from acting in a specified way" (ibid., § 2(1)). *Agreement* is defined as the "manifestation of mutual assent," and *bargain* is defined as "an agreement to exchange promises or to exchange a promise for a performance or to exchange performances" (ibid., § 3). See also Peter A. Alces, "Contract Reconceived," *Northwestern University Law Review* 96 (2001): 39–97, discussing the interplay among contract, promise, bargain, and agreement.

7. See Larry T. Garvin, "Small Business and the False Dichotomies of Contract Law," *Wake Forest Law Review* 40 (2005): 295–388, exploring how contractual pressures to which consumers are subject often similarly affect small businesses.

8. See generally Margaret J. Radin, *Boilerplate: The Fine Print, Vanishing Rights, and the Rule of Law* (Princeton, NJ: Princeton University Press, 2013); Omri Ben-Shahar, foreword to "'Boilerplate': Foundations of Market Contracts Symposium," *Michigan Law Review* 104 (2006): 821–26; Tess Wilkinson-Ryan, "A Psychological Account of Consent to Fine Print," *Iowa Law Review* 99 (2014): 1745–84; Florencia

Marotta-Wurgler and Robert Taylor, "Set in Stone? Change and Innovation in Consumer Standard-Form Contracts," *New York University Law Review* 88 (2013): 240–85.

9. The American Law Institute has undertaken various initiatives to bridge this gap. For example, the *Principles of the Law of Software Contracts* (St. Paul, MN: American Law Institute, 2010) addressed the concern that consumers do not read the terms of electronic standard form contracts. With this in mind, the *Principles* suggest that (1) continuing the current legal direction, (2) adopting more specific rules about which terms become part of the agreement, and (3) focusing on best practices of disclosure could increase readership. In addition, the *Principles'* influence can be seen in the proposed third restatement on consumer contracts, which lists adequate notice and opportunity to review as integral to consumers' consent (American Law Institute, *Restatement of the Law, Consumer Contracts: Preliminary Draft No. 2* [St. Paul, MN: American Law Institute, 2015]). Thus, under the proposed new restatement, increased disclosure is offered as the answer to the challenges proposed by modern contracting.

10. For example, stop-loss orders in day trading of stocks can be sent and received without human intervention. For further discussions of the implications that contracts formed by electronic agents present for the doctrine, see Anthony J. Bellia Jr., "Contracting with Electronic Agents," *Emory Law Journal* 50 (2001): 1047–94; Tom Allen and Robin Widdison, "Can Computers Make Contracts?," *Harvard Journal of Law and Technology* 9 (1996): 25–52.

11. See Uniform Commercial Code (U.C.C.), May 1949 Draft, § 2-207 (1949).

12. As chief reporter, Llewellyn was the principal architect of the U.C.C., particularly article 2, governing the sale of goods. William Twining noted that Llewellyn spent the greater portion of fifteen years (1937–52) on the code, and "his contributions represent one of his greatest achievements" (William Twining, *Karl Llewellyn and the Realist Movement*, 2nd ed. [Cambridge: Cambridge University Press, 2012], 270). In addition to drafting article 2, Llewellyn "expended an enormous amount of time on the first drafts of the Sales Comments during the period of 1943–5" (ibid., 328). Moreover, Llewellyn's article greatly influenced the Second Restatement of Contracts. Allan Farnsworth, Reporter for the Second Restatement, confirmed that the effects of article 2 can been seen in the restatement's treatment of impracticability, frustration, mistake, misrepresentation, cure, resale, cover, foreseeability, liquidated damages, unconscionability, and repudiation (E. Allan Farnsworth, "Ingredients in the Redaction of the 'Restatement (Second) of Contracts,'" *Columbia Law Review* 81 [1981], 1–12).

13. In fact, the only time we could be certain, in the case of form contracts between commercial entities, that both parties consent to the terms of the writings would be when the parties have negotiated and agreed on a single writing that accurately reflects their intent. If the writing is the product of negotiation and reflects the parties' authentic consent, then we do not have a battle of the forms and do

not need special rules to discern (or supply) the requisite consent. We could then rely on conceptions of *contract*, *agreement*, and *bargain* under the U.C.C. The code defines *agreement* as the "bargain of the parties in fact, as found in their language or inferred from other circumstances, including course of performance, course of dealing, or usage of trade" U.C.C., art. 1 (2001): § 1-201(b)(3). *Contract* is defined as the "total legal obligation that results from the parties' agreement" (ibid., § 1-201(b)(12)).

14. When a contract is based on performance, we look to the code hierarchy to determine what the terms are. See U.C.C., art. 2 (2002), § 2-207(3). The code hierarchy consists first of express terms on which the parties agree (ibid.). The hierarchy then looks to course of performance of the same contract, course of dealing from prior contracts between the parties, and usage of trade within the industry U.C.C., art. 1, § 1-303. Finally, the hierarchy supplements any missing terms with code gap-fillers, which include terms for the construction of price, delivery, payment, and warranties. See generally U.C.C., art. 2, part 3.

15. Portions of this and the following section are developed from Peter A. Alces, "The Death of Consent?," in *Comparative Contract Law: British and American Perspectives*, ed. Larry A. DiMatteo and Martin Hogg (Oxford, UK: Oxford University Press, 2016).

16. Carnival Cruise Lines, Inc. v. Shute, 499 U.S. 585 (1991).

17. It would seem that at best such a proffer might amount to only an attempt at modification. Compare U.C.C., art. 2, § 2-207(2): "The additional terms are to be construed as proposals for addition to the contract."

18. Carnival Cruise, 499 U.S. at 594.

19. Williams v. Walker-Thomas Furniture Co., 350 F.2d 445 (D.C. Cir. 1965).

20. Carnival Cruise, 499 U.S. at 600–601.

21. ProCD, Inc. v. Zeidenberg, 86 F.3d 1447 (7th Cir. 1996).

22. Hill v. Gateway 2000, 105 F.3d 1147 (7th Cir. 1997).

23. Easterbrook failed to appreciate that Zeidenberg was a merchant and not a consumer. Qualification as a "merchant" transactor under the code does not depend on where (i.e., at what store) the sale occurred. See U.C.C., art. 2, § 2-104.

24. See, e.g., John E. Murray Jr., "The Dubious Status of the Rolling Contract Formation Theory," *Duquesne Law Review* 50 (2012): 80: "There is no need to continue the deliberate misconstructions of statutes or precedent that [rolling contract] theory requires"; Roger C. Bern, "Terms Later Contracting: Bad Economics, Bad Morals, and a Bad Idea for a Uniform Law, Judge Easterbrook Notwithstanding," *Journal of Law and Policy* 12 (2004): 642–43: *ProCD* "and its initial progeny, *Hill v. Gateway 2000, Inc.*, however, have been deservedly and widely criticized, variously described as a 'swashbuckling tour de force that dangerously misinterprets legislation and precedent,' a 'real howler' that is 'dead wrong' on its interpretation of section 2-207 of the UCC, a decision that 'flies in the face of UCC policy and precedent,' a 'detour from traditional U.C.C. analysis' 'contrary to public policy,' with analysis that 'gets an "F" as a law exam'" (notes omitted); William H.

Lawrence, "Rolling Contracts Rolling over Contract Law," *San Diego Law Review* 41 (2004): 1099, 1100: Judge Easterbrook's "analysis is replete with distortion and avoidance of the relevant contract principles." But see Eric A. Posner, "ProCD v. Zeidenberg and Cognitive Overload in Contractual Bargaining," *University of Chicago Law Review* 77 (2010): 1194: "Time, then, has vindicated *ProCD*, which will be remembered as a masterpiece of realist judging, one of the great opinions in the canon of contract law cases." For a sampling of cases following the decisions, see M.A. Mortenson Co. v. Timberline Software Corp., 998 P.2d 305 (Wash. 2000) and DeFontes v. Dell, Inc., 984 A.2d 1061 (R.I. 2009). Acceptance of Judge Easterbrook's analysis, however, has not been universal. See Klocek v. Gateway, Inc., 104 F. Supp. 2d 1332 (D. Kan. 2000) and Wachter Mgmt. Co. v. Dexter & Chaney, Inc., 144 P.3d 747 (Kan. 2006). For a case that perhaps stakes out a middle ground between the two sides, see Brower v. Gateway 2000, Inc., 676 N.Y.S.2d 569 (N.Y. App. Div. 1998), accepting product-now-terms-later model, but ultimately concluding that part of arbitration agreement was unenforceable because it was unconscionable. A recent case has pointed out the importance of the parties' manifestation of assent even in contracts of adhesion. Berkson v. Gogo LLC, 97 F. Supp. 3d 359 (E.D.N.Y. 2015), holding that the buyer did not consent to the online transaction because he was unaware that he was binding himself to additional terms.

25. See ProCD, 86 F.3d at 1451.

26. For a discussion claiming that "pay now, terms later" contracts present terms no more odious than those presented before the time of purchase, see Florencia Marotta-Wurgler, "Are 'Pay Now, Terms Later' Contracts Worse for Buyers? Evidence from Software License Agreements," *Journal of Legal Studies* 38 (2009): 309–43.

27. See ProCD, 86 F.3d at 1451.

28. U.C.C., art. 2, § 2-207(2).

29. Compare American Law Institute, *Restatement (Second) of Contracts*, § 22(2), stating that a "manifestation of mutual assent may be made even though neither offer nor acceptance can be identified and even though the moment of formation cannot be determined."

30. ProCD, 86 F.3d at 1452.

31. See Posner, "ProCD v. Zeidenberg and Cognitive Overload," 1181, 1187, pointing out that the ProCD box did not include such an acceptance provision, and so ProCD's offer could have been construed differently.

32. ProCD, 86 F.3d at 1452.

33. Hill, 105 F.3d 1147.

34. See Hill, 105 F.3d at 1148.

35. Hill, 105 F.3d at 1149.

36. See American Law Institute, *Principles of the Law of Software Contracts: Summary Overview to Topic 2* (St. Paul, MN: American Law Institute, 2010), 117: "To ensure enforcement of their standard form, software transferors should disclose

terms on their website prior to a transaction and should give reasonable notice of and access to the terms upon initiation of the transfer, whether initiation is by telephone, Internet, or selection in a store."

37. The recently enacted Credit Card Act of 2009 requires the card issuer to "apply amounts in excess of the minimum payment amount first to the card balance bearing the highest rate of interest, and then to each successive balance bearing the next highest rate of interest, until the payment is exhausted" (15 U.S.C.A. § 1666c(b)(1)).

38. Arguments advanced in this section are derived in part from Alces, "Guerrilla Terms," 1511–62.

39. "[E]conomists seem virtually unanimous in assuming that people are preference-driven choosers (that is, dispositionists)" (Jon Hanson and David Yosifon, "The Situational Character: A Critical Realist Perspective on the Human Animal," *Georgetown Law Journal* 93 [2004], 8).

40. For a thoughtful and comprehensive perspective on cost-benefit analysis, see Matthew D. Adler and Eric A. Posner, "Implementing Cost-Benefit Analysis When Preferences Are Distorted," in *Cost-Benefit Analysis: Legal, Economic, and Philosophical Perspectives*, ed. Matthew D. Adler and Eric A. Posner (Chicago: University of Chicago Press, 2000).

41. See, e.g., Hanson and Yosifon "Situational Character," 129, 154, highlighting social psychological studies to argue that we should view people not as dispositional, rational actors but as "situational characters."

42. See Jon D. Hanson and Douglas A. Kysar, "Taking Behavioralism Seriously: The Problem of Market Manipulation," *New York University Law Review* 74 (1999): 630–749; Hanson and Kysar, "Taking Behavioralism Seriously: Some Evidence of Market Manipulation," *Harvard Law Review* 112 (1999): 1420, 1438.

43. See Alces, "Guerrilla Terms," 1523–33.

44. See, e.g., Werner Güth, Rolf Schmittberger, and Bernd Schwarze, "An Experimental Analysis of Ultimatum Bargaining," *Journal of Economic Behavior and Organization* 3 (1982): 367, 381–82, tables 9 and 10, showing situations in which players rejected rational solution, that is, selected an outcome that implied lower payoffs for both players.

45. See, e.g., Russell B. Korobkin and Thomas S. Ulen, "Law and Behavioral Science: Removing the Rationality Assumption from Law and Economics," *California Law Review* 88 (2000): 1055, asserting that "There is simply too much credible experimental evidence that individuals frequently act in ways that are incompatible with the assumptions of rational choice theory."

46. See Laurie R. Santos and Michael L. Platt, "Evolutionary Anthropological Insights into Neuroeconomics," in *Neuroeconomics: Decision Making and the Brain*, ed. Paul W. Glimcher and Ernst Fehr (Cambridge, MA: Academic Press, 2013), 109, 110: "the biases that pervade human choice may be more deeply imbedded in our nervous system than [once] thought"; Joseph W. Kable, "Valuation, Intertemporal Choice, and Self-Control," in Glimcher and Fehr, *Neuroeconomics*,

173–89, discussing neural bases for choice in decisions with immediate versus future consequences.

47. See generally Cass R. Sunstein, ed., *Behavioral Law and Economics* (New York: Cambridge University Press, 2000).

48. See Hanson and Yosifon, "Situational Character"; Hanson and Yosifon, "The Situation: An Introduction to the Situational Character, Critical Realism, Power Economics, and Deep Capture," *University of Pennsylvania Law Review* 152 (2003); Adam Benforado, Jon Hanson, and David Yosifon, "Broken Scales: Obesity and Justice in America," *Emory Law Journal* 53 (2004): 1645–1806; Ronald Chen and Jon Hanson, "The Illusion of Law: The Legitimating Schemas of Modern Policy and Corporate Law," *Michigan Law Review* 103 (2004): 1–149; Chen and Hanson, "Categorically Biased: The Influence of Knowledge Structures on Law and Legal Theory," *Southern California Law Review* 77 (2004): 1103–1254.

49. See Hanson and Yosifon, "The Situation," 154: "We are, in essence, not rational actors, but 'situational characters'"; Hanson and Yosifon, "Situational Character," 6, arguing that situationism provides "a more realistic depiction of the human animal" than dispositionism.

50. A typical dispositionist assumption would be "that by their very nature humans enjoy the freedom to order their actions as they see fit" (Hanson and Yosifon, "Situational Character," 10).

51. Hanson and Yosifon, "The Situation," 263. For a discussion of the various ways in which advertisers manipulate situation, see Piotr Winkielman, Kent C. Berridge, and Julia L. Wilbarger, "Unconscious Affective Reactions to Masked Happy versus Angry Faces Influence Consumption Behavior and Judgments of Value," *Personality and Social Psychology Bulletin* 31 (2005): 121–35, explaining how affective reactions from even subconscious exposure to happy or angry faces can modulate behavior and consumption; Margo Wilson and Martin Daly, "Do Pretty Women Inspire Men to Discount the Future?," *Proceedings of the Royal Society of London B: Biological Sciences* 271 (2004): 177–79, finding that temporal discounting increased significantly in men who saw photographs of attractive women; Hilke Plassmann et al., "Marketing Actions Can Modulate Neural Representations of Experienced Pleasantness," *Proceedings of the National Academy of Sciences* 105 (2008): 1050–54, finding that manipulations of the price of wine resulted in an increase in activity in the medial orbitofrontal cortex, an area of the brain thought to encode experienced pleasantness; Benedetto De Martino et al., "Frames, Biases, and Rational Decision-Making in the Human Brain," *Science* 313 (2006): 684–87, finding that framing effects associated with amygdala activity contribute to deviations in standard economic accounts of human rationality.

52. Hanson and Yosifon, "Situational Character," 10–13.

53. Ibid., 8–10.

54. Ibid., 13–20.

55. Ibid., 13.

56. See ibid., 13–15.

57. Ibid., 16, referring to the tort "reasonable person" standard.

58. See Hanson and Yosifon, "The Situation," 178.

59. See Adam Benforado and Jon Hanson, "The Great Attributional Divide: How Divergent Views of Human Behavior Are Shaping Legal Policy," *Emory Law Journal* 57 (2008): 311–408, discussing the attributional divide between situationism and dispositionism that shapes most policy debates; see also Matthew A. Edwards, "The Virtue of Mandatory Disclosure," *Notre Dame Journal of Law, Ethics and Public Policy* 28 (2014): 68, questioning whether law can be used to foster "stable dispositions" in human beings; Trenton G. Smith and Attila Tasnádi, "The Economics of Information, Deep Capture and the Obesity Debate," *American Journal of Agricultural Economics* 96 (2014): 533–41, discussing role of public perception on obesity epidemic.

60. See Hanson and Yosifon, "Situational Character," 25–33.

61. Daniel M. Wegner, *The Illusion of Conscious Will* (Cambridge, MA: MIT Press, 2002), 40, describing a case of phantom limb.

62. Ibid., 96.

63. Hanson and Yosifon, "Situational Character," 132–33 (emphasis in original).

64. Xavier Gabaix and David Laibson, "Shrouded Attributes, Consumer Myopia, and Information Suppression in Competitive Markets," *Quarterly Journal of Economics* 121 (2006): 505–40.

65. Ibid., 509.

66. See Hal S. Scott, "The Risk Fixers," *Harvard Law Review* 91 (1978): 737, 759–62.

67. Gabaix and Laibson, "Shrouded Attributes," 505 (emphasis added).

68. Ibid., 506. Succinctly, a shrouded attribute "is a product attribute that is hidden by a firm, even though the attribute could be nearly costlessly revealed" (ibid., 512; citation omitted); see also Oren Bar-Gill, *Seduction by Contract: Law, Economics, and Psychology in Consumer Markets* (Oxford, UK: Oxford University Press, 2012), 2, explaining that sellers respond to consumer biases by "design[ing] products, contracts, and pricing schemes to maximize not the *true* (net) benefit from their product, but the (net) benefit as *perceived by the imperfectly rational consumer*" (emphasis in original).

69. Gabaix and Laibson, "Shrouded Attributes," 528 (citing Brad M. Barber, Terrance Odean, and Lu Zheng, "Out of Sight, Out of Mind: The Effects of Expenses on Mutual Fund Flows," *Journal of Business* 78 [2005]: 2095–2120).

70. Gabaix and Laibson, "Shrouded Attributes," 509.

71. Ibid., 511.

72. See Michael Kosfeld and Ulrich Schüwer, "Add-On Pricing in Retail Financial Markets and the Fallacies of Consumer Education," *Review of Finance* 21 (2017): 1189–1216, discussing how price discrimination can lead firms to shroud high add-on prices for myopes and unshroud low add-on prices for sophisticates; Chun-Hui Miao, "Consumer Myopia, Standardization and Aftermarket Monop-

olization," *European Economic Review* 54 (2010): 931–46, describing how the possibility of market segregation leads to even more pervasive disincentive to educate myopes.

73. See Alces, "Guerrilla Terms," 1517nn12–15 and accompanying text, presenting an objectivist approach to contract.

74. Gabaix and Laibson, "Shrouded Attributes," 510.

75. Ibid., 519–20. New sophisticates are then capable of educating the remaining myopes, snowballing the curse of debiasing. As Douglas Baird noted, "When the market works effectively . . . [the typical buyer] benefits from the presence of other, more sophisticated buyers. A seller in a mass market often cannot distinguish among her buyers. To make a profit, she cannot focus exclusively on the unsophisticated. As ignorant of computers as I am, I can always see whether the more knowledgeable are buying a particular model" ("The Boilerplate Puzzle," *Michigan Law Review* 104 [2006], 936; citation omitted). See also George L. Priest, "A Theory of the Consumer Product Warranty," *Yale Law Journal* 90 (1981): 1297, 1299–1302, detailing the general tendency of sellers to exploit myopic buyers.

76. The connection is not difficult: Add-ons and hidden (guerrilla) contract terms are, "for economic purposes[,] . . . both just features of the product" (Margaret Jane Radin, "Boilerplate Today: The Rise of Modularity and the Waning of Consent," *Michigan Law Review* 104 [2006], 1223, 1229). Radin posited that the "collapse of the contract-product distinction is a trope that has become very prominent in contract theory" (ibid., citing Arthur Allen Leff, "Contract as Thing," *American University Law Review* 19 [1970]: 131, 144–51, 155). Gabaix and Laibson, "Shrouded Attributes," 509n11, implicitly made the product-contract connection in their discussion of credit-card terms and conditions.

77. In the credit-card context, for example, "It is typical for major issuers to amend their agreements in important respects with remarkable frequency" (Ronald J. Mann, *Charging Ahead: The Growth and Regulation of Payment Card Markets* [Cambridge: Cambridge University Press, 2006], 132); see also Omri Ben-Shahar and Carl E. Schneider, *More Than You Wanted to Know: The Failure of Mandated Disclosure* (Princeton, NJ: Princeton University Press, 2014), 103–4, concluding that even most sophisticated consumers among us may find it difficult to keep up with the changing terms; see also Peter A. Alces and Michael M. Greenfield, "They Can Do What!? Limitations on the Use of Change-of-Terms Clauses," *Georgia State Law Review* 26 (2010), surveying examples of contracts with unilateral adjustment rights in one party.

78. See American Law Institute, *Restatement of the Law, Consumer Contracts: Preliminary Draft No. 2* (St. Paul, MN: American Law Institute, 2015), § 2, Reporters' Notes: "credible empirical evidence . . . suggests that consumers rarely read standard contract terms no matter how these terms are disclosed."

79. Ibid.

80. See ibid., § 2(b).

81. See Ben-Shahar and Schneider, *More Than You Wanted to Know*, 67: "The internet transactions of disclosees are easily tracked, so we *know* that nobody reads the terms (like the iTunes contract) they agree to"; ibid., 12: "*Much* that is disclosed people sensibly ignore. They rightly calculate that reading an end-users license agreement won't change their minds" (emphasis in original); Yannis Bakos, Florencia Marotta-Wurgler, and David R. Trossen, "Does Anyone Read the Fine Print? Consumer Attention to Standard-Form Contracts," *Journal of Legal Studies* 43, no. 1 (2014): 2: "only one or two in 1,000 shoppers access a product's [end-user license agreement] for at least 1 second"; Florencia Marotta-Wurgler, "Will Increased Disclosure Help? Evaluating the Recommendations of the ALI's 'Principles of the Law of Software Contracts,'" *University of Chicago Law Review* 78 (2011): 168, noting extremely low readership in end-user license agreements.

82. See Omri Ben-Shahar and Carl E. Schneider, "The Failure of Mandated Disclosure," *University of Pennsylvania Law Review* 159 (2011): 671: "some direct as well as indirect evidence suggests that almost no consumers read boilerplate, even when it is fully and conspicuously disclosed"; Ben-Shahar and Schneider, *More Than You Wanted to Know*, 43: "readership rates of privacy statements and end-user license agreements are virtually zero"; Marotta-Wurgler, "Will Increased Disclosure Help?," 183: "Depending on the methodology, I estimate that moving from browsewraps to clickwraps would increase shoppers' readership rates by 0.04 percent to 1.32 percent relative to a baseline readership rate of around 0 percent. An average estimate of the effect across six methodologies is 0.36 percent"; Bar-Gill, *Seduction by Contract*, 1: "That no one reads the fine print is old news."

83. Ben-Shahar and Schneider, *More Than You Wanted to Know*, 45: "Even in ideal circumstances informed-consent disclosures fail"; ibid., 12: "Mandated disclosure is a fundamental failure that cannot be fundamentally fixed"; Bar-Gill, *Seduction by Contract*, 2–3: "The prevalence of contracts and prices that cannot be fully explained within a rational-choice framework proves the robustness of the biases and misperceptions driving the behavioral economics theory"; ibid., 3: "imagine an imperfectly rational consumer trying to choose among several . . . complex, multidimensional contracts. The task is a daunting one. Many consumers will simply avoid it. Markets don't work well when consumers do not shop for the best deal."

84. Ben-Shahar and Schneider, *More Than You Wanted to Know*, 7.

85. Though those are real problems too. (Note added.)

86. Ben-Shahar and Schneider, *More Than You Wanted to Know*, 8.

87. Ibid., 7: "mandated disclosure is ill suited to its ends"; ibid., 10: "mandated disclosure seems plausible only on logically reasonable but humanly false assumptions"; ibid., 12: "not only does the empirical evidence show that mandated disclosure regularly fails, failure is inherent in it"; ibid.: "Mandated disclosure is a fundamental problem that cannot be fundamentally fixed."

88. In situations where "the exchange of money precedes the communication of detailed terms," the contract is not formed until after the consumer has the opportunity to read the terms "for the first time in the comfort of home" (ProCD, Inc. v. Zeidenberg, 86 F.3d 1447, 1451 (7th Cir. 1996)). Rolling contracts are not formed "when the consumer orders and pays for the goods and the seller ships them," but when "the prescribed 'accept or return' time expires"; Robert A. Hillman, "Rolling Contracts," *Fordham Law Review* 71, no. 3 (2002): 744.

89. Robert A. Hillman and Jeffrey J. Rachlinski, "Standard-Form Contracting in the Electronic Age," *New York University Law Review* 77 (2002): 469–70, pointing out that the Internet makes product and corporate reviews more accessible to consumers, and so e-businesses and brick-and-mortar stores both have greater incentives to maintain good reputations and may be less likely to offer or enforce onerous terms.

90. Gabaix and Laibson, "Shrouded Attributes," 506–7.

91. For a discussion of the Truth in Lending Act's attempt to use annual percentage rate (APR) disclosures as a "normalized measure of the total cost of credit," see Bar-Gill, *Seduction by Contract*, 174–76. Bar-Gill has noted that, despite its promise, APR disclosure has fallen short of its goals for three reasons: the APR is typically not disclosed early enough, it fails to capture the total cost of the loan by ignoring certain price dimensions, and it "fails to account for the pre-payment option" (ibid., 177–78).

92. See Nathaniel D. Daw and Philippe N. Tobler, "Value Learning through Reinforcement: The Basics of Dopamine and Reinforcement Learning," in Glimcher and Fehr, *Neuroeconomics*, 283; Nathaniel D. Daw, "Advanced Reinforcement Learning," in Glimcher and Fehr, *Neuroeconomics*, 299.

93. See, e.g., U.C.C., art. 2, §§ 2-313 (express warranties), 2-314 (implied warranty of merchantability), 2-315 (implied warranty of fitness for particular purpose).

94. See ibid., § 2-316(3)(a): "all implied warranties are excluded by expressions like 'as is,' 'with all faults,' or other language that in common understanding calls the buyer's attention to the exclusion of warranties and makes plain that there is no implied warranty."

95. Ibid., § 2-316(2) states that, in order to exclude or modify the implied warranties of merchantability or fitness, the exclusion must be "conspicuous." "Conspicuous" is defined as a term that is written, displayed, or presented in such a way that "a reasonable person against which it is to operate ought to have noticed it" (U.C.C., art. 1, § 1-201(b)(10)). For example, conspicuous terms include headings or language in capital letters, larger fonts, or contrasting type (ibid., § 1-201(b)(10)(A)–(B)).

96. For example, stores could release oxytocin, a neurohypophysial hormone important for feelings of intimacy and sexual reproduction, to modulate consumer behavior. See Thomas Baumgartner et al., "Oxytocin Shapes the Neural Circuitry of Trust and Trust Adaptation in Humans," *Neuron* 58, no. 4 (2008): 639–50, finding that subjects who were administered oxytocin showed no decrease in trusting

behavior even after trust had been repeatedly breached by a reduction in the activation of the amygdala, midbrain regions, and dorsal striatum; Michael Kosfeld et al., "Oxytocin Increases Trust in Humans," *Nature* 435 (2005): 673–76, finding that administration of oxytocin made subjects more trusting and more willing to accept social risk in interpersonal interactions.

97. See Daniel Kahneman, *Thinking, Fast and Slow* (New York: Farrar, Straus and Giroux, 2011), 20–21: "*System* 1 operates automatically and quickly, with little or no effort and no sense of voluntary control. *System* 2 allocates attention to the effortful mental activities that demand it, including complex computations. The operations of *System* 2 are often associated with the subjective experience of agency, choice, and concentration." For an overview of the literature recognizing this process in different but essentially similar terms, see Jonathan St. B. T. Evans, "Dual-Processing Accounts of Reasoning, Judgment, and Social Cognition," *Annual Review of Psychology* 59 (2008): 255–78.

98. See Joshua Greene et al., "The Neural Bases of Cognitive Conflict and Control in Moral Judgment," *Neuron* 44 (2004): 392–96, noting that dlPFC shows an increase in activity during utilitarian moral judgments; see also Greene, *Moral Tribes: Emotion, Reason, and the Gap between Us and Them* (New York: Penguin Press, 2013), 136–37: "Reasoning . . . depends critically on the [dlPFC]. This is not to say that reasoning occurs exclusively in the DLPFC. On the contrary, the DLPFC is more like the conductor of an orchestra than a solo musician."

99. See Paul J. Whalen, "Fear, Vigilance, and Ambiguity: Initial Neuroimaging Studies of the Human Amygdala," *Current Directions in Psychological Science* 7, no. 6 (1998) 177, reviewing brain-imaging studies (using fMRI or PET) that reveal an increase in activity in the amygdala when participants view pictures expressing facial emotion, particularly fear; see also Greene, *Moral Tribes*, 122, noting the amygdala as the "brain region known for its role in emotion."

100. One study, considered at length in chapter 3, used fMRI to measure neural responses as test subjects were faced with personal and impersonal moral dilemmas in the classic trolley case. The personal moral dilemma asked subjects whether they would personally push one large man onto the trolley tracks to divert a trolley and save five workers farther down the track. The impersonal dilemma asked subjects whether they would flip a switch to divert a trolley headed for five workers to a separate track heading toward only one worker. From a utilitarian perspective, the personal and impersonal dilemma pose the same question: Would you sacrifice one to save five? Most of us will make the utilitarian judgment in impersonal cases. In a more personal setting, we are more likely to be guided by the amygdala's emotional response. See Greene et al., *Neural Bases*, 391–92, finding bilateral increase in amygdala during personal moral dilemma.

101. See Green et al., *Neural Bases*, 392–96; see also Greene, *Moral Tribes*, 122: "the 'impersonal' dilemmas, the ones like the *switch* case, elicited increased activity in the dlPFC" (emphasis in original).

102. Kahneman, *Thinking, Fast and Slow*, noted that our fast thinking system "is rarely indifferent to emotional words" (367). Even physicians, perhaps known for methodical decision making, are influenced by emotional characterizations of problems. In one study, physicians were asked to choose between radiation and surgery for their patient. When the medical literature described the outcome of surgery as a 90 percent *survival* rate, a much larger percentage of physicians chose surgery than when the outcome was described as a 10 percent *mortality* rate.

103. Joshua D. Greene, "The Cognitive Neuroscience of Moral Judgment," in *The Cognitive Neurosciences IV*, ed. Michael S. Gazzaniga (Cambridge, MA: MIT Press, 2009), 991–93. Greene's dual-process theory of moral judgment argued that personal dilemma decisions (e.g., push the man) are made through an emotional appraisal process whereas impersonal dilemma decisions (e.g., pull the switch) are made through a more utilitarian appraisal process. Thus, moral judgment becomes an interaction of various neural systems pitting emotional reaction against controlled cognitive processing. Selim Berker, however, in "The Normative Insignificance of Neuroscience," *Philosophy and Public Affairs* 37 (2009): 293–329, offered criticisms of Greene's methods to undermine these empirical conclusions. Berker's criticisms are discussed at length in chapter 3.

104. Karolina M. Lempert and Elizabeth A. Phelps, "Neuroeconomics of Emotion and Decision Making," in Glimcher and Fehr, *Neuroeconomics*, 233: "Although a two-system approach to describing the relation between emotion and decisions has been useful and provides a simple explanation for some decision-related behaviors, it is clearly not sufficient to capture the range of means by which affect influences choices."

105. See Joshua I. Gold and Michael N. Shadlen, "The Neural Basis of Decision Making," *Annual Review of Neuroscience* 30 (2007): 535–74, reviewing advances in understanding neural mechanism of decision making in neurophysiological studies involving monkeys making categorical choices based on sensory stimulus; Yang and Shadlen (2007), showing that monkeys were able to combine probabilistic information from shape combinations in choosing the target that furnished the highest probability of reward; Roozbeh Kiani and Michael N. Shadlen, "Representation of Confidence Associated with a Decision by Neurons in the Parietal Cortex," *Science* 324, no. 5928 (2009): 759–64, arguing that firing rates of neurons in monkeys' lateral intraparietal cortex represent choice certainty.

106. See Samuel M. McClure et al., "Separate Neural Systems Value Immediate and Delayed Monetary Rewards," *Science* 306, no. 5695 (2004): 503–7, finding that decisions involving immediately available rewards activated parts of the limbic system associated with the midbrain dopamine system; see also M. Marsel Mesulam, *Principles of Behavioral and Cognitive Neurology*, 2nd ed. (New York: Oxford University Press, 2000), 1–12, explaining behavioral neuroanatomy generally and how limbic system's structure links emotion to cognition and behavior.

107. The somatic marker theory proposed that an emotional defect significantly

impairs decision making. Particularly, the theory proposed that somatic marker signals received from the body's interaction with the external world are regulated in the ventromedial prefrontal cortex (the part of the brain regulating emotion) to help regulate decision making in complex and uncertain situations. See Antoine Bechara et al., "Emotion, Decision Making and the Orbitofrontal Cortex," *Cerebral Cortex* 10, no. 3 (2000): 295–307; Barnaby D. Dunn, Tim Dalgleish, and Andrew D. Lawrence, "The Somatic Marker Hypothesis: A Critical Evaluation," *Neuroscience and Biobehavioral Reviews* 30 (2006): 239–71. But see E. T. Rolls et al., "The Orbitofrontal Cortex," *Philosophical Transactions of the Royal Society of London B: Biological Sciences* 351 (1996): 1433–43, arguing that somatic marker theory rests on biological inefficiency and that a more efficient explanation is to connect execution of behavior directly to higher functioning areas of brain; J. Panksepp, review of *Looking for Spinoza: Joy, Sorrow, and the Feeling Brain*, by Antonio Damasio, *Consciousness and Emotion* 4 (2003): 111, 126–27, criticizing somatic marker theory for being an extreme view in arguing that mental states are not composed merely of different bodily states but of somatosensory feedback in other aspects of emotional processing.

108. Antoine Bechara et al., "Deciding Advantageously," 1293–95, finding that normal participants in gambling task were able to choose advantageous decks of cards before they consciously understood the best strategy, whereas patients with damage to prefrontal portions of their brains associated with emotions continued to choose disadvantageously even after they knew best strategy.

109. See Daw and Tobler, "Value Learning through Reinforcement," 287–89, discussing the role of dopamine neurons in predicting error-based learning models.

110. Tania Singer et al., "Empathic Neural Responses Are Modulated by the Perceived Fairness of Others," *Nature* 439 (2006): 466–69, showing that men have lower empathy-related responses than women to the administration of pain to a participant who acted unfairly. For sources detailing the study's findings, see Paul W. Glimcher and Ernst Fehr, eds., *Neuroeconomics: Decision Making and the Brain* (Cambridge, MA: Academic, 2013), 205, 490, 520; Stephen Pinker, *The Better Angels of Our Nature: Why Violence Has Declined* (New York: Viking, 2012), 531–32.

111. Lempert and Phelps, "Neuroeconomics of Emotion," 231, discussing cognitive emotion regulation studies that show emotion affects calculation of subjective value with regard to decision making and memory reconsolidation studies that show the updating function of memory may be used to recondition reaction to certain stimulus in order to elicit different emotional responses and, therefore, different choices.

112. American Law Institute *Restatement (Second) of Contracts*, § 12(2): "A natural person who manifests assent to a transaction has full legal capacity to incur contractual duties thereby unless he is (a) under guardianship, or (b) an infant, or (c) mentally ill or defective, or (d) intoxicated."

113. For the contract doctrine's unconscionability provisions, see American Law Institute, *Restatement (Second) of Contracts*, § 208; U.C.C., art. 2, § 2-302.

114. See Debra Pogrund Stark, "Ineffective in Any Form: How Confirmation Bias and Distractions Undermine Improved Home-Loan Disclosures," *Yale Law Journal Online* 122 (2013): 377–400, highlighting experiments that demonstrate disclosure language often is either not read entirely or not appreciated because of distracting conversation.

115. The Federal Trade Commission, for example, utilizes a number of laws to protect consumers from payday lenders: Section 5 of the Federal Trade Commission Act, the Truth in Lending Act, the Electronic Fund Transfer Act, and the Credit Practices Rule. See Federal Trade Commission, *Payday Lending* [web page], https://www.ftc.gov/news-events/media-resources/consumer-finance/payday-lending.

Chapter Seven

1. See generally Margaret Jane Radin, *Boilerplate: The Fine Print, Vanishing Rights, and the Rule of Law* (Princeton, NJ: Princeton University Press, 2013), considering the extent to which firms should be allowed to create their own legal environment through boilerplate language; Douglas G. Baird, "The Boilerplate Puzzle," *Michigan Law Review* 104 (2006): 933, 934: "Boilerplate, while not a vice itself, is frequently the symptom of a problem that the law should appropriately address."

2. Actually in the sense of a "horrible or unreal creature of the imagination." *Random House Unabridged Dictionary* (New York: Random House, 2001).

3. See chapter 4, "Addiction."

4. The theories considered in this chapter are those that have the most currency and offer treatments of human agency typical of the genre. But the survey is not, could not be, exhaustive: The effort to find "contract as [something]" is a cottage industry, as though untold fame and fortune await the legal scholar who discovers the Holy Grail. We might best conclude, however, that contract is ultimately "morally impossible." See Peter A. Alces, "The Moral Impossibility of Contract," *William and Mary Law Review* 48 (2007): 1647–72.

5. See Nathan B. Oman, "The Failure of Economic Interpretations of the Law of Contract Damages," *Washington and Lee Law Review* 64 (2007): 875 notes that without a unifying theory of contract, "contract doctrine represents little more than the random final product of a long chain of historical accidents."

6. Much as many other now disproven medical techniques, bleeding, for example, probably seemed to make sense and worked occasionally as a matter of happenstance. See Benjamin Rush, "A Defence of Blood-Letting, as a Remedy for Certain Diseases," in *Medical Inquires and Observations* (Philadelphia: Kimber and Richardson, 1812), 4, providing anecdotes describing successes of bloodletting.

7. See Virginia Hughes, "Science in Court: Head Case," *Nature* 464 (2010): 342 (quoting Stephen Morse). See also Stephen J. Morse, "Law, Responsibility, and the Sciences of the Brain/Mind," in *Oxford Handbook of Law, Regulation, and*

Technology, ed. Roger Brownsword, Eloise Scotford, and Karen Yeung (Oxford, UK: Oxford University Press, 2016), 11 ("Brains don't convince each other; people do"); Morse, "Brain Overclaim Syndrome and Criminal Responsibility: A Diagnostic Note," *Ohio State Journal of Criminal Law* 3 (2006), 397 ("Brains do not commit crimes; people commit crimes.").

8. See Michael Moore, *Placing Blame: A Theory of the Criminal Law* (Oxford, UK: Oxford University Press, 2010), 147–48: "to *feel* guilty causes the judgement that we *are* guilty, in the sense that we are morally responsible. . . . [G]uilt feelings typically engender the judgement that we deserve punishment . . . not only in the weak sense of desert—that it would not be unfair to be punished—but also and more important in the strong sense that *we ought* to be punished."

9. Something like this is suggested in Richard Joyce, *The Evolution of Morality* (Cambridge, MA: MIT Press, 2006), 116–17.

10. See Peter A. Alces, *A Theory of Contract Law: Empirical Insights and Moral Psychology* (New York: Oxford University Press, 2011), 268.

11. Radin has opined that "the EU's disapproval of onerous clauses seems on the rise . . . while at the same time the US is allowing more exculpatory clauses and restrictions of remedy to be enforced." Radin, *Boilerplate*, 234–40.

12. See Stephen Wolfram, *A New Kind of Science* (Champaign, IL: Wolfram Media, 2002), 551: "It is inevitable that there will be situations where one cannot recognize the origins of behavior that one sees—even when this behavior is in fact produced by very simple rules."

13. Compare Judith Rich Harris, *No Two Alike: Human Nature and Human Individuality* (New York: W. W. Norton, 2007), 84, discussing findings of behavioral geneticists on the effect of even small initial differences in home environment on ultimate differences in adult personality.

14. Daniel C. Dennett, *Elbow Room: The Varieties of Free Will Worth Wanting* (Cambridge, MA: MIT Press, 1984), 96, has argued that small initial variations among people average out in the end and we all make choices on a level playing field. Others disagree. See Bruce Waller, *Against Moral Responsibility* (Cambridge, MA: MIT Press, 2011), 117–18: "the supposition that those with disadvantaged starting points will be compensated by later good fortune—'after all, luck averages out in the long run'—is absurd" (quoting Dennett, *Elbow Room*, 95); Neil Levy, *Hard Luck: How Luck Undermines Free Will and Moral Responsibility* (New York: Oxford University Press, 2011), 199: "Dennett is wrong. Constitutive luck, good and bad tends to ramify, not even out. . . . Chance events that are genuinely lucky and that actually compensate for constitutive luck are rare and extraordinary."

15. Peevyhouse v. Garland Coal & Mining Co., 382 P.2d 109 (Okla. 1962). The Peevyhouses sued Garland Coal Company on a lease agreement that allowed Garland to strip-mine coal on the Peevyhouses' property and under which Garland agreed to, but failed to, restore the property once the mining was complete. In finding that Garland breached the contract, the Oklahoma Supreme Court awarded the Peevyhouses the difference in land value resulting from the breach

($300) rather than cost of restoring the land ($25,000). Such an outcome allowed Garland to receive the benefit of the bargain twice since the Peevyhouses presumably would have charged Garland the additional $25,000 upon entering the lease had they known Garland would not effect that repair at the end of the lease term.

16. Jacob & Youngs, Inc. v. Kent, 230 N.Y. 239 (1921). In building a house for Kent, Jacob & Youngs used Cohoes pipe instead of the contracted-for Reading pipe. Kent demanded that the work be done over again, which would have required expensive demolition. In finding for Jacob & Youngs in the suit to collect the remaining balance, Justice Cardozo reasoned that the proper measure of damages was not the cost to replace the Cohoes pipe with Reading pipe, but the difference in value between the two brands of pipe, which was zero because the two brands or pipe were of the same type and quality.

17. Uniform Commercial Code (U.C.C.), art. 1, § 1-304 (2001); American Law Institute, *Restatement (Second) of Contracts* (St. Paul, MN: American Law Institute, 1981), § 205.

18. U.C.C., art. 2, § 2-302 (2002); American Law Institute, *Restatement (Second) of Contracts*, § 208.

19. Specifically fraud, duress, and illegality. American Law Institute, *Restatement (Second) of Contracts*, § 164 (misrepresentation makes a contract voidable); ibid., § 175 (duress by threat makes a contract voidable); ibid., § 178 (term unenforceable on grounds of public policy, including unlawful or illegal behavior).

20. Benjamin E. Hermalin et al., "Contract Law," in *Handbook of Law and Economics*, ed. A. Mitchell Polinsky and Steven Shavell (London: Elsevier, 2007), 15: "most economic analysis of contract law is aimed at filling the gaps in incomplete agreements and setting default rules that operate when the parties have expressed no preference regarding a particular issue" (citing Richard Craswell, "Contract Law, Default Rules, and the Philosophy of Promising," *Michigan Law Review* 88 [1989]: 489; Steven Shavell, "An Economic Analysis of Altruism and Deferred Gifts," *Journal of Legal Studies* 20 [1991]: 401–22).

21. Or, at least, certainly not "willfully," whatever that might mean. See Richard A. Posner, "Let Us Never Blame a Contract Breaker," *Michigan Law Review* 107 (2009): 1353–64, arguing that it encourages inefficient conduct to deem certain types of breach—those motived by a desire to avoid the opportunity costs of forgoing the sale of contracted-for goods at higher prices to third parties—as "willful" breach supported by the risk of incurring punitive damages.

22. See American Law Institute, *Restatement (Second) of Contracts*, § 12: mentally ill and minors lack the capacity to contract.

23. In general, contract damages aim to place the promisee in the position she would have been had the contract been performed (ibid., § 347). Tort law damages generally aim to restore the injured party to the substantially equivalent position had no tort been committed (American Law Institute, *Restatement (Second) of Torts* [St. Paul, MN: American Law Institute, 1965], § 903 cmt. a).

24. As through the operation of procedural unconscionability principles. The

unconscionability doctrine allows a court to rule directly on the enforceability of a contract or clause to avoid troublesome results (American Law Institute, *Restatement (Second) of Contracts*, § 208, cmt. a; Reporter's Notes, cmt. d), citing cases finding unconscionability; see also U.C.C., art. 2, § 2-303, cmt 1. In making such a determination, the court may consider the relative bargaining positions of the parties, including level of education, stress of the bargaining position, experience of the parties, and other factors that may affect capacity (U.C.C., art. 2, § 2-303, cmt. d).

25. See chapter 6.

26. See, e.g., U.C.C., art. 9, offering specific provisions for consumers who are party to secured transactions. Secured parties may not collect collateral in partial satisfaction of an obligation in consumer transactions (U.C.C., art. 9, § 9-620(g)). Debtors may waive the right to redeem collateral except in consumer-goods transactions (U.C.C., art. 9, § 9-624(c)). Consumers are entitled to a description of any liability for deficiency assessed against them (U.C.C., art. 9, § 9-614(1)(B)).

27. See, e.g., Magnuson-Moss Warranty Act, 15 U.S.C. § 2302(a), requiring "full and conspicuous disclosure of [warranty] terms and conditions."

28. See, e.g., U.C.C., art. 2, § 2-719(3): "Limitation of consequential damages for injury to the person in the case of consumer goods is prima facie unconscionable."

29. Tort damages may include pain and suffering. American Law Institute, *Restatement (Second) of Torts*, § 904, cmt. c: "If bodily harm of any kind is alleged, physical pain and suffering resulting from it can be shown without any specific allegation. In other words, it is regarded as general damages."

30. See Timothy J. Sullivan, "Punitive Damages in the Law of Contract: The Reality and the Illusion of Legal Change," *Minnesota Law Review* 61 (1977): 207–52, analyzing the growing willingness of courts to award punitive damages in contract cases: "By characterizing the plaintiff's action as one for tortious breach of a public duty, the courts were able to permit [punitive damage] recoveries without seeming insult to the general rule that such awards were not appropriate in contract actions"; Brown v. Coates, 253 F.2d 36, 40 (D.C. Cir. 1958), awarding punitive damages for breach of contract for sale of home because a real estate agent assumes fiduciary obligations toward his client; Romero v. Mervyn's, 109 N.M. 249, 258 (1989), justifying the award of punitive damages when a store breached a promise to pay a customer's medical bills because a jury could have inferred that the store acted with malice; Gruenberg v. Aetna Ins. Co., 9 Cal. 3d 566, 573 (1973), awarding damages for breach of "implied covenant of good faith and fair dealing in every contract" although not acknowledging damages as being punitive.

31. American Law Institute, *Restatement (Second) of Contracts*, § 90.

32. Robert A. Hillman, *Principles of Contract Law* (St. Paul, MN: West Academic, 2004), 17–39, 91–102.

33. "A contract is a promise or a set of promises for the breach of which the law gives a remedy, or the performance of which the law in some way recognizes as a duty" (American Law Institute, *Restatement (Second) of Contracts*, § 1).

34. Ibid., § 2 cmt. (a).

35. See Helga Varden, "Kant and Lying to the Murderer at the Door . . . One More Time: Kant's Legal Philosophy and Lies to Murderers and Nazis," *Journal of Social Philosophy* 41 (2010): 403: "Does Kant really mean to say that people hiding Jews in their homes should have told the truth to the Nazis, and that if they did lie, they became co-responsible for the heinous acts committed against those Jews who, like Anne Frank, were caught anyway?"

36. Promissory estoppel originated as a doctrine of contract law, but many theorists argue that it has evolved into a separate cause of action, independent of contract. See Michael B. Metzger and Michael J. Phillips, "Emergence of Promissory Estoppel as an Independent Theory of Recovery," *Rutgers Law Review* 35 (1983): 472–557; Eric Mills Holmes, "The Four Phases of Promissory Estoppel," *Seattle University Law Review* 20 (1996): 45–79.

37. See, e.g., Lord Wright, "Ought the Doctrine of Consideration to Be Abolished from the Common Law?," *Harvard Law Review* 49 (1936): 1225–53; Ernest G. Lorenzen, "Causa and Consideration in the Law of Contracts," *Yale Law Journal* 28 (1919): 646; James D. Gordon III, "Dialogue about the Doctrine of Consideration," *Cornell Law Review* 75 (1990): 987–1006; Mark B. Wessman, "Should We Fire the Gatekeeper? An Examination of the Doctrine of Consideration," *University of Miami Law Review* 43 (1993): 45–117.

38. Robert E. Scott, "The Rise and Fall of Article 2," *Louisiana Law Review* 62 (2002): 1035–36, describing how article 2 accounted for reputational concerns through its remedial scheme for broken contracts; see also William Twining, *Karl Llewellyn and the Realist Movement*, 2nd ed. (Cambridge: Cambridge University Press, 2012), 336: "commercial self-interest spurs most businessmen to act within widely recognized leeways of decency and honesty: gross abuses tend to be self-defeating."

39. iCloud Terms and Conditions, *Apple*, http://www.apple.com/legal/internet-services/icloud/en/terms.html.

40. See chapter 6, Consumer Consent.

41. Traditionally, a "meeting of the minds" has occurred when two parties contracting with one another have thought they were contracting to the same agreement and reached an agreement about that same subject matter—when "a harmonious and full mental accord" has been reached (Costigan 1920). For the evolution of this subjective test of whether two contracting parties actually shared a state of mind, see Richmond & Alleghany R.R. Co. v. R.A. Patterson Tobacco Co., 169 U.S. 311, 312 (1898): "The contract is the concrete result of the meeting of the minds of the contracting parties. The evidence thereof is but the instrument by which the fact that the will of the parties did meet is shown"; Douglas v. Smulski, 131 A.2d 225 (Conn. C.P. 1957), disparaging concepts of "quasi-contracts" as betraying the will theory behind the "meeting of the minds" conception of contracts; Spring Lake NC, LLC v. Holloway, 110 So.3d 916 (Fla. Dist. Ct. App. 2013), forgoing a truly subjective test of contract by admitting most modern contracts do not truly represent a "meeting of the minds."

42. See chapter 6, "Making Sense of Situation."

43. See generally E. Allan Farnsworth, *Changing Your Mind: The Law of Regretted Decisions* (New Haven, CT: Yale University Press, 1998); Peter A. Alces, "Regret and Contract 'Science,'" *Georgetown Law Journal* 89 (2000): 143–70.

44. See American Law Institute, *Restatement (Second) of Contracts*, §§ 151–53; Howard O. Hunter, *Modern Law of Contracts* (St. Paul, MN: Thomson Reuters, 2016), § 19:1; Peter A. Alces, *A Theory of Contract Law*, 85–88.

45. See, e.g., Linda S. Mullenix, "Gaming the System: Protecting Consumers from Unconscionable Contractual Forum-Selection and Arbitration Clauses," *Hastings Law Journal* 66 (2015): 755–56; Michael Terasaki, "Do End User License Agreements Bind Normal People?," *Western State University Law Review* 41 (2014): 467, 487.

46. See Oren Bar-Gill, *Seduction by Contract: Law, Economics, and Psychology in Consumer Markets* (Oxford, UK: Oxford University Press, 2012), 34, suggesting that the dominant party should have an obligation to disclose to the consumer what the dominant party knows about the consumer that consumer himself does not know.

47. See Randy E. Barnett, "Contract Is Not Promise; Contract Is Consent," in *Philosophical Foundations of Contract Law*, ed. Gregory Klass, George Letsas, and Prince Saprai (Oxford, UK: Oxford University Press, 2014), 45.

48. American Law Institute, *Restatement (Second) of Contracts*, §§ 18–19.

49. See generally Peter A. Alces, "Guerrilla Terms," *Emory Law Journal* 56 (2007): 1511–62; see also chapter 6, "Making Sense of Situation," discussing Gabaix and Laibson's finding that sellers have no incentive to increase buyer sophistication.

50. Carnival Cruise Lines, Inc. v. Shute, 499 U.S. 585 (1991).

51. ProCD, Inc. v. Zeidenberg, 86 F.3d 1447 (7th Cir. 1996).

52. Hill v. Gateway 2000, Inc., 105 F.3d 1147 (7th Cir. 1997).

53. Oren Bar-Gill and Omri Ben-Shahar, Reporter's Notes to *Restatement (Third) Consumer Contracts*, 6, http://www.ballardspahr.com/~/media/files/alerts/2012-12 -06-outline.pdf.

54. In explaining the tension found in *Hill* and *ProCD* between freedom to contract and Easterbrook's confidence in market forces, Posner has maintained that "the best explanation is that Judge Easterbrook is foremost a common law judge— not an ideologue or theorist—who in these two cases smells a rat (or several rats): the plaintiffs" (Eric A. Posner, "ProCD v. Zeidenberg and Cognitive Overload in Contractual Bargaining," *University of Chicago Law Review* 77 [2010]: 1188).

55. Bar-Gill and Ben-Shahar, Reporter's Notes to *Restatement (Third) Consumer Contracts*, 3, referring to American Law Institute, *Principles of the Law of Software Contracts* (St. Paul, MN: American Law Institute, 2010).

56. See, e.g., Robin Bradley Kar, "Contract as Empowerment," *University of Chicago Law Review* 83 (2016): 101–74, endorsing a contractualist approach to legal obligation.

57. For an exemplary list of noninstrumentalist theories, see Gregory Klass, Introduction to *Philosophical Foundations of Contract Law*, ed. Gregory Klass, George Letsas, and Prince Saprai (Oxford, UK: Oxford University Press, 2014), 4–5.

58. Charles Fried, *Contract as Promise: A Theory of Contractual Obligation* (Cambridge, MA: Harvard University Press, 1981), 14–17.

59. Ibid., 3.

60. Ibid., 4.

61. Ibid., 8.

62. See Lawrence v. Texas, 539 U.S. 558, 603 (2003) (J. Scalia dissenting): "Social perceptions of sexual and other morality change over time, and every group has the right to persuade its fellow citizens that its view of such matters is the best."

63. Fried, *Contract as Promise*, 13.

64. Ibid.

65. Ibid., 14 (emphasis added).

66. See chapter 6, "Making Sense of Situation."

67. Fried, *Contract as Promise*, 16.

68. Barnett, "Contract Is Not Promise."

69. Fried, *Contract as Promise*, 20–21.

70. See chapter 2.

71. Dori Kimel, *From Promise to Contract: Towards a Liberal Theory of Contract* (Portland: Hart, 2003), 31.

72. Kimel, *From Promise to Contract*, 97–98.

73. Fried, *Contract as Promise*, 14. Kimel replied to that speculation in "Personal Autonomy and Change of Mind in Promise and in Contract," in *Philosophical Foundations of Contract Law*, ed. Gregory Klass, George Letsas, and Prince Saprai (Oxford, UK: Oxford University Press, 2014), 100.

74. See Kimel, "Personal Autonomy," 100.

75. See ibid., 100–101 (emphasis in original).

76. Seana Valentine Shiffrin, "The Divergence of Contract and Promise," *Harvard Law Review* 120 (2007): 709–10.

77. Ibid., 727–29.

78. Ibid., 741 (notes omitted).

79. Shiffrin offered no evidence that contract undermines promise, other than her own supposition. And that supposition would seem to admit of empirical verification, were it accurate.

80. See Kimel, "Personal Autonomy," 97 (citing J. Raz, "Promises and Obligations," in *Law, Morality, and Society*, ed. P. M. S. Hacker and Joseph Raz [Oxford, UK: Clarendon, 1977], 210, 227–28).

81. Kimel, *From Promise to Contract*, 28.

82. See Bruce Waller, "The Stubborn Illusion of Moral," in *Exploring the Illusion of Free Will and Moral Responsibility*, ed. Gregg Caruso (Lanham, MD: Lexington, 2013), 81–85, responding to Fischer's "plateau model," which stated that

we are all "roughly equal" in our various capacities to avoid bad acts and act as one should; and so when there are differences in behavior, those differences are by and large just and fair bases to reward and punish accordingly.

83. In *Thinking, Fast and Slow* (New York: Farrar, Straus and Giroux, 2011), Daniel Kahneman described the benefits of slow thinking, when one is more deliberative and more logical. He concluded that this mode of thinking can shape our judgments and act as a safeguard against biases (heuristics that mislead). (Note added.)

84. Waller, "Stubborn Illusion of Moral," 82–83 (citation omitted).

85. Barnett, "Contract Is Not Promise," 45.

86. Ibid. (citing David Hume, *An Enquiry concerning the Principles of Morals*, ed. Eugene Freeman [New York: Liberal Arts Press, 1957], 30n1: "If the secret direction of the intention, said every man of sense, could invalidate a contract; where is our security? And yet a metaphysical schoolman might think, that where an intention was supposed to be requisite, if that intention really had no place, no consequence ought to follow, and no obligation be imposed.").

87. Barnett, "Contract Is Not Promise," 45.

88. This practical necessity can be described as dubious because he ignores completely and conveniently the manipulability of ostensible consent identified by situationist commentators. See chapter 6, "Making Sense of Situation."

89. Barnett, "Contract Is Not Promise," 48–49.

90. Barnett, "Contract Is Not Promise," 49 (emphasis added). See also Bar-Gill, *Seduction by Contract*, on thinking behind that conclusion and actual creation of consumer contracts.

91. Barnett, "Contract Is Not Promise," 52.

92. The reporters of the Consumer Contracts Restatement compiled a compendium of cases for the latest draft of the Restatement Reporter's Notes. "In choosing to follow the passive contracting approach in the restatement, the methodology utilized by the reporters relied on quantitative statistical analysis of precedents, asking which approach has garnered greater following among courts. For this purpose, a database of all published decisions was collected and analyzed. For each decision, various identifiers were coded, including outcome, prior cases cited, principles followed, and many others." See American Law Institute, *Restatement of the Law, Consumer Contracts: Preliminary Draft No. 2* (St. Paul, MN: American Law Institute, 2015), 4–5. For example, in section 2, "Adoption of Standard Contract Terms," the reporters collected 103 cases; and in section 3, "Modification of Standard Contract Terms," the total is 88 cases.

93. The reporters detailed the mechanism designed to safeguard consumers in the Consumer Contracts Restatement (ibid., § 2). Reporter's Notes at 22: "The requirement of assent, when it is meaningfully imposed, protects consumers from harsh terms. The consumer would simply withhold assent if offered harsh terms. While it is implausible, in many cases, for consumers to review the terms in ad-

vance, those consumers interested in proper review of the standard terms may do so after the purchase, and may terminate the contract within a reasonable time."

94. Ibid., § 3–8, 5: "Because §§ 2–3 provide for prima facie enforcement of the standard contract terms, the Restatement balances their effect with a set of rules that rely on ex post scrutiny by courts to limit the risk of abuse." Section 4 provides for protection against illusory promises, § 5 considers unconscionable terms, § 6 deals with deception, § 7 reiterates rules on precontractual representations, and § 8 stipulates the effects of derogation from mandatory rules.

95. See AT&T Mobility LLC v. Concepcion, 563 U.S. 333, 352 (U.S. 2011), ruling that the Federal Arbitration Act preempts state law rulings that certain consumer contracts of adhesion should not be enforced.

96. See Omri Ben-Shahar and Carl E. Schneider, "The Failure of Mandated Disclosure," *University of Pennsylvania Law Review* 159 (2011): 671: "some direct as well as indirect evidence suggests that almost no consumers read boilerplate, even when it is fully and conspicuously disclosed"; Florencia Marotta-Wurgler, "Will Increased Disclosure Help? Evaluating the Recommendations of the ALI's 'Principles of the Law of Software Contracts,'" *University of Chicago Law Review* 78 (2011): 183: "Depending on the methodology, I estimate that moving from browsewraps to clickwraps would increase shoppers' readership rates by 0.04 percent to 1.32 percent relative to a baseline readership rate of around 0 percent. An average estimate of the effect across six methodologies is 0.36 percent"; Bar-Gill, *Seduction by Contract*, 1: "That no one reads the fine print is old news"; ibid., 2–3: "The prevalence of contracts and prices that cannot be fully explained within a rational-choice framework proves the robustness of the biases and misperceptions driving the behavioral economics theory"; ibid., 3: "imagine an imperfectly rational consumer trying to choose among several . . . complex, multidimensional contracts. The task is a daunting one. Many consumers will simply avoid it. Markets don't work well when consumers do not shop for the best deal."

97. American Law Institute, *Consumer Contracts* [web page], https://www.ali .org/projects/show/consumer-contracts/.

Chapter Eight

1. Jerry A. Fodor, *Psychosemantics: The Problem of Meaning in the Philosophy of Mind* (Cambridge, MA: MIT Press, 1987), xii.

2. See generally A. C. Grayling, *The Age of Genius: The Seventeenth Century and the Birth of the Modern Mind* (New York: Bloomsbury, 2016).

3. Morse argued that new advances in neuroscience must fit into folk psychology criteria and will not lead to a legal paradigm shift. Stephen J. Morse, "Law, Responsibility, and the Sciences of the Brain/Mind," in *Oxford Handbook of Law, Regulation, and Technology*, ed. Roger Brownsword, Eloise Scotford, and Karen Yeung

(Oxford, UK: Oxford University Press, 2016); Morse, "Criminal Law and Common Sense: An Essay on the Perils and Promise of Neuroscience," *Marquette Law Review* 99 (2015): 39–74. See also Morse, "Neuroscience, Free Will, and Criminal Responsibility," in *Free Will and the Brain: Neuroscientific, Philosophical, and Legal Perspectives*, ed. Walter Glannon (Cambridge: Cambridge University Press, 2015), arguing that neuroscience has little to contribute to advances in criminal law policy and doctrine; Morse, "Lost in Translation? An Essay on Law and Neuroscience," in *Law and Neuroscience*, Current Legal Issues 13, ed. Michael Freeman (Oxford, UK: Oxford University Press, 2011), 529–62, arguing that the folk psychology view is correct and fits with new findings in neuroscience; and Morse, "Brain Overclaim Syndrome and Criminal Responsibility: A Diagnostic Note," *Ohio State Journal of Criminal Law* 3 (2006): 397–412, suggesting that "brain overclaim syndrome" afflicts those consumed by new discoveries in neuroscience.

4. See, e.g., Eyal Zamir, "Law and Behavioral Economics," in *Encyclopedia of the Philosophy of Law and Social Philosophy*, eds. Mortimer Sellers and Stephan Kirste (New York: Springer, forthcoming); Christine Jolls, Cass R. Sunstein, and Richard H. Thaler, "A Behavioral Approach to Law and Economics," *Stanford Law Review* 50 (1998): 1471–1550; Russell Korobkin, "What Comes after Victory for Behavioral Law and Economics?," *University of Illinois Law Review* 2011 (2011): 1653–74.

5. The so-called scientific revolution of the seventeenth century was largely about refuting Aristotelian scientific conclusions, upon which much theology of the time was founded. Notwithstanding the striking confidence of Aristotle's pronouncements (and the Church's enlistment of them as the basis of Roman Catholicism, and perhaps all of Christianity), Aristotle was very wrong about so much that would lend itself to empirical confirmation or refutation now, and would have then too.

> Aristotle held that the universe is constructed out of five elements. The heavens are made out of ether, or quintessence, which is translucent and unchanging, neither hot nor cold, dry nor damp. The heavens stretch outward from the Earth, which is at the center of the universe, as a series of material spheres carrying the moon, the sun and the planets, and then, above them all, is the starry firmament. The universe is thus spherical and finite; moreover, it is oriented: It has a top, a bottom, a left, and a right. Aristotle never thinks in terms of space in the abstract (as geometers already did) but always in terms of place. He denied the very possibility of an empty space, a vacuum. Empty space was a contradiction in terms. . . . Aristotle does occasionally mention quantities. Thus he says if you have two heavy objects, the heavier one will fall faster than the lighter one, and if the heavy one is twice as heavy it will fall twice as fast. . . . Aristotle took the view that harder substances are denser and heavier than softer substances; it followed that ice is

> heavier than water. . . . The mathematicians followed Archimedes;
> the philosophers [and the Church, too—ed.] followed Aristotle.

See David Wootton, *The Invention of Science: A New History of the Scientific Revolution* (New York: Harper, 2015), 69–72. With how wrong he was about so much, one might wonder at Aristotle's prominence. That may be explained, though, by the great temporal power of the Church. Disagreement with Aristotelian precepts was grounds for execution (ibid.).

6. Fodor, *Psychosemantics*, xii.

7. Morse, "Law, Responsibility, and the Sciences of the Brain/Mind," 5; Morse, "Criminal Law and Common Sense," 42.

8. David Wootton, *The Invention of Science: A New History of the Scientific Revolution* (New York: Harper), 6–7 (citations omitted).

9. Ibid., 10–12.

10. See, e.g., Scott v. Sandford, 60 U.S. 393, 407 (1857), reasoning that because African Americans historically "had no rights which the white man was bound to respect," they are not protected under the Bill of Rights; Bradwell v. State, 83 U.S. 130, 141 (1872): "The natural and proper timidity and delicacy which belongs to the female sex evidently unfits it for many of the occupations of civil life"; Goesaert v. Cleary, 335 U.S. 464 (1948), upholding law prohibiting female bar owners because it may cause moral and social problems.

11. Examples of racism can be found in the Bible. Deut. 7:1–2: God tells the Israelites to "smite [other tribes], and utterly destroy them; make no covenant with them, nor show mercy unto them." Sexism, too, can be found. See, e.g., I Cor. 14:34–35: "if [women] will learn anything, let them ask their husbands at home: for it is a shame for women to speak in the church"; I Timothy 2:12: "But I suffer not a woman to teach, nor to usurp authority over the man, but to be in silence"; I Cor. 11:8–9: "For the man is not of the woman: but the woman of the man. Neither was the man created for the woman; but the woman for the man."

12. Bowers v. Hardwick, 106 S. Ct. 2841, 2846-47 (1986), upholding Georgia law banning sodomy; Lawrence v. Texas, 123 S. Ct. 2472, 2490 (2003) (Scalia, J. dissenting), likening sodomy to bigamy, adultery, incest, prostitution, and bestiality.

13. Stephen Wolfram, *A New Kind of Science* (Champaign, IL: Wolfram Media, 2002), 750–53.

14. Companies such as No Lie MRI and Cephos developed fMRI lie-detection technology for commercial and courtroom use. Owen D. Jones, Jeffrey D. Schall, and Francis X. Shen, *Law and Neuroscience* (New York: Wolters Kluwer, 2014), 20. Cephos has since discontinued this service after a federal court ruled their test is inadmissible for lack of sufficient reliability. Julia Calderone, "There Are Some Big Problems with Brain-Scan Lie Detectors," *Tech Insider*, April 19, 2016, http://www.tech insider.io/dr-oz-huizenga-fmri-brain-lie-detector-2016-4; U.S. v. Semrau, 693 F.3d 510 (6th Cir. 2012), wherein No Lie MRI evidence was ruled inadmissible on particular facts (Center for Science and the Law, *The Many Dangers of Brain-Based Lie Detection*, 2012).

15. Greene and Cohen are exemplars of this type of thinking. (Note in original.)

16. Morse, "Law, Responsibility, and the Sciences of the Brain/Mind," 9; Morse, "Criminal Law and Common Sense," 44.

17. Compare Michael S. Moore, "A Natural Law Theory of Interpretation," *Southern California Law Review* 58 (1985): 286: "there is a right answer to moral questions, a moral reality if you like. . . . [R]eal morals . . . have a necessary place in the interpretation of any legal text"; Moore, "Moral Reality," *Wisconsin Law Review* 1982 (1982): 1061–1156, canvassing views of skeptics of moral realism; Moore, "Moral Reality Revisited," *Michigan Law Review* 90 (1992): 2424–2533, describing moral realist thesis and making a positive case for moral realism.

18. See chapter 3, "Evolution of the Doctrine."

19. See, e.g., Michael L. Radelet and Ronald L. Akers, "Deterrence and the Death Penalty: The Views of the Experts," *Journal of Criminal Law and Criminology* 87, no. 1 (1996): 1–16, canvassing expert views on capital punishment and concluding it has little to no deterrent effect; William J. Bowers and Glenn L. Pierce, "The Illusion of Deterrence in Isaac Ehrlich's Research on Capital Punishment," *Yale Law Journal* 85, no. 2 (1975): 187–208, arguing that capital punishment has no deterrent effect and criticizing results of a popular study suggesting otherwise; John K. Cochran, Mitchell B. Chamlin, and Mark Seth, "Deterrence or Brutalization? An Impact Assessment of Oklahoma's Return to Capital Punishment," *Criminology* 32 (1994): 107–34, finding no evidence of deterrent effect in an Oklahoma death penalty case.

20. Though solitary confinement of juveniles has been described as a teaching tool, intended to modify behavior Christopher Bickel, "The Scene of the Crime: Children in Solitary Confinement," in *The Marion Experiment: Long-Term Solitary Confinement and the Supermax Movement*, ed. Stephen C. Richards (Carbondale, IL: Southern Illinois University Press, 2015), 132. Further resistance by juveniles in confinement is used to rationalize even further punishment and confinement (ibid.).

21. At least eighteen states have enacted laws that allow indeterminate sentencing for sexual predators on the basis of findings of dangerousness. Christopher Slobogin, "The Civilization of the Criminal Law," *Vanderbilt Law Review* 58 (2005): 121–70. Such statutes have been upheld by the US Supreme Court: Kansas v. Hendricks, 521 U.S. 346, 371 (1997).

22. Morse, "Lost in Translation?," 560–62. Mr. Oft displayed abrupt behavioral changes, ultimately leading to a conviction for child molestation, after a tumor developed in his orbitofrontal lobe, the part of the brain involved in social behavior. Upon removal of the tumor, his behavior returned to normal. See chapter 1, notes 43–50 and accompanying text.

23. Morse, "Lost in Translation?," 560: "Oft's desires may have been mechanistically caused, but acting on them was intentional action. An abnormal cause for his behaviour does not mean that he could not control his actions. . . . We can

reasonably infer that Oft had difficulty controlling behaviour . . . [b]ut this is true of all paedophiles and we do not excuse them."

24. For an example of the kind of empirical neuroscience that could render Mr. Oft as no longer a threat to society, see Jeffrey M. Burns and Russell H. Swerdlow, "Right Orbitofrontal Tumor with Pedophilia Symptom and Constructional Apraxia Sign," *JAMA Neurology* 60 (2003): 437–40.

25. See, e.g., 42 U.S.C.A. § 264(d)(1), permitting apprehension and examination of individuals with certain communicable diseases; 42 U.S.C.A. § 265, empowering the surgeon general to prohibit "the introduction of persons and property" into the United States from countries with known communicable diseases. See also chapter 2, "Conceptions of Capacity and the Curious Case of Psychopathy."

26. See chapter 3, "Blame, Desert, and Culpability," discussing Moore's conceptions of retribution, guilt, and desert.

27. See chapter 2, " 'Responsibility,' Retribution, and Deterrence."

28. See chapter 3, "The Rationalization of Emotion."

29. Morse, "Law, Responsibility, and the Sciences of the Brain/Mind," 9; Morse, "Criminal Law and Common Sense," 44.

30. Michael S. Moore, *Placing Blame: A Theory of the Criminal Law* (Oxford, UK: Oxford University Press, 2010), 147–49.

31. Morse, "Law, Responsibility, and the Sciences of the Brain/Mind," 9; Morse, "Criminal Law and Common Sense," 44; Morse, "Neuroscience, Free Will, and Criminal Responsibility," 271, acknowledging that criminal law is not the only area of law "in peril" and contract law is similarly based on folk psychology.

32. Morse, "Law, Responsibility, and the Sciences of the Brain/Mind," 12. See also Morse, "Criminal Law and Common Sense," 51–52: "the law presupposes folk psychology, even when we most habitually follow the legal rules. Unless people are capable of understanding and then using legal rules to guide their conduct, the law is powerless to affect human behavior" (citation omitted).

33. Morse, "Law, Responsibility, and the Sciences of the Brain/Mind," 13; Morse, "Criminal Law and Common Sense," 52 (emphasis added).

34. See Bruce Waller, *Against Moral Responsibility* (Cambridge, MA: MIT Press, 2011), 117–18: "the supposition that those with disadvantaged starting points will be compensated by later good fortune—'after all, luck averages out in the long run'—is absurd" (quoting Dennett, *Elbow Room*, 95); Neil Levy, *Hard Luck: How Luck Undermines Free Will and Moral Responsibility* (New York: Oxford University Press, 2011), 199: "Constitutive luck, good and bad tends to ramify, not even out. . . . Chance events that are genuinely lucky and that actually compensate for constitutive luck are rare and extraordinary."

35. Small differences in starting rules can yield great differences in system output. See Wolfram, *New Kind of Science*, 551; Judith Rich Harris, *No Two Alike: Human Nature and Human Individuality* (New York: W. W. Norton, 2007), 84, discussing findings on the effect of differences in home environment on differences

in adult personality; chapter 7, "The Folk Psychology of Assent and Misunderstanding Human Agency."

36. Morse, "Law, Responsibility, and the Sciences of the Brain/Mind," 14.

37. Ibid.

38. Ibid., 14–15.

39. Morse conceded that psychopaths are not morally blameworthy and the law should be reformed to excuse them Stephen J. Morse, "Psychopathy and Criminal Responsibility," *Neuroethics* 1 (2008): 205–12; see also the discussions in chapter 3 and chapter 2.

40. Morse, "Law, Responsibility, and the Sciences of the Brain/Mind," said that "In some cases, the capacity for control is poor characterologically; in other cases it may be undermined by variables that are not the defendant's fault, such as mental disorder. The meaning of this capacity is fraught" (17). But he offered us no measure of such "fault" other than, we may presume, the simplistic assumptions of folk psychology: "the capacity for control will once again be a folk psychological capacity" (ibid., 18).

41. Morse has avowed suspicion that deterministic theories are true, though he believes nobody can know with certainty. Stephen J. Morse, "Waiting for Determinism to Happen," *Legal Essays*, 1999, http://people.virginia.edu/~dll2k/morse.pdf.

42. Of the 59 percent of philosophers considered to be compatibilists, only 35 percent fully accept compatibilism and 24 percent "lean toward" compatibilism. David Bourget and David J. Chalmers, "What Do Philosophers Believe?," *Philosophical Studies* 170 (2013): 476 (cited in Morse, "Criminal Law and Common Sense," 49). See also Eddy Nahmias, "Is Free Will an Illusion?," in *Moral Psychology*, vol. 4, *Free Will and Moral Responsibility*, ed. Walter Sinnott-Armstrong (Cambridge, MA: MIT Press, 2014), 22–23n6, finding 59 percent of philosophers to be compatibilists (citing http://philpapers.org/surveys/results.pl).

43. Kadri Vihvelin, *Causes, Laws, and Free Will: Why Determinism Doesn't Matter* (New York: Oxford University Press, 2013); R. Jay Wallace, *Responsibility and the Moral Sentiments* (Cambridge, MA: Harvard University Press, 1994). (Note in original.)

44. Morse, "Law, Responsibility, and the Sciences of the Brain/Mind," 19.

45. See Gregg D. Caruso, ed., *Exploring the Illusion of Free Will and Moral Responsibility* (Lanham, MD: Lexington, 2013), 1. See also John M. Doris, *Lack of Character: Personality and Moral Behavior* (New York: Cambridge University Press, 2002), using experimental social psychology and other scientific research to challenge traditional assumptions about character and ethics; and Kwame Anthony Appiah, *Experiments in Ethics* (Cambridge, MA: Harvard University Press, 2008), asserting that modern empirical research can contribute to morality and ethics.

46. Morse cited no source for this phrase, but it appears to come from another compatibilist. Compare Moore, "Causation and the Excuses," *California Law Review* 73 (1985): 1091. (Note added.)

47. Morse, "Law, Responsibility, and the Sciences of the Brain/Mind," 21.

48. The French phrase "*tout comprendre, c'est tout pardoner*" (to understand all is to forgive all) gets it wrong. See, e.g., Nicholas Rescher, *Free Will: A Philosophical Reappraisal* (New Brunswick, NJ: Transaction, 2009), 108.

49. Morse, "Law, Responsibility, and the Sciences of the Brain/Mind," 21.

50. Philosopher Herbert Hart distinguishes several classifications of the word *responsibility*, including "causal responsibility." H. L. A. Hart, "Postscript: Responsibility and Retribution," in Hart, ed., *Punishment and Responsibility* (Oxford, UK: Oxford University Press, 1968), 211–12.

51. Bruce Waller, *The Stubborn System of Moral Responsibility* (Cambridge, MA: MIT Press, 2014), 101.

52. Morse, "Law, Responsibility, and the Sciences of the Brain/Mind," 11.

53. See Craig Haney, *Reforming Punishment: Psychological Limits to the Pains of Imprisonment* (Washington, DC: American Psychological Association, 2006), 8, finding that, by way of associated negative psychological effects, "prison paradoxically may serve to increase the amount of crime that occurs"; Sharon Shalev, *Supermax: Controlling Risk through Solitary Confinement* (Portland, OR: Willan, 2009, 207–17), noting solitary confinement's ultimate ineffectiveness in meeting its objectives.

54. As well as many similar "state of mind" nouns, e.g., "consent." See chapter 7, "Promise, Bargain, and Agreement."

55. Eric A. Posner, "Economic Analysis of Contract Law after Three Decades: Success or Failure?," *Yale Law Journal* 112 (2003): 864–68, arguing that economic analysis of contract law fails by assuming that individuals perform complex efficiency calculations.

56. The Nobel Prize in Physiology or Medicine 2000 was jointly awarded to Arvid Carlsson, Paul Greengard, and Eric R. Kandel for "discoveries concerning signal transduction in the nervous system." http://www.nobelprize.org/nobel_prizes/medicine/laureates/2000.

57. Eric R. Kandel, *In Search of Memory: The Emergence of a New Science of Mind* (New York: W. W. Norton, 2006), 145, studied memory in giant marine snails (*Aplysia californica*).

58. Ibid., xii–xiii.

59. See Francis Crick, *The Astonishing Hypothesis: The Scientific Search for the Soul* (New York: Scribner's, 1994).

Bibliography

Abraham, Kenneth S. *The Forms and Functions of Tort Law*. Eagan, MN: Foundation Press, 2007.

Adler, Matthew D., and Eric A. Posner. "Implementing Cost-Benefit Analysis When Preferences Are Distorted." In *Cost-Benefit Analysis: Legal, Economic, and Philosophical Perspectives*, edited by Matthew D. Adler and Eric A. Posner, 1105–48. Chicago: University of Chicago Press, 2000.

Alces, Peter A. "Contract Reconceived." *Northwestern University Law Review* 96 (2001): 39–97.

———. "The Death of Consent?" In *Comparative Contract Law: British and American Perspectives*, edited by Larry A. DiMatteo and Martin Hogg, 30–60. Oxford, UK: Oxford University Press, 2016.

———. "Guerrilla Terms." *Emory Law Journal* 56 (2007): 1511–62.

———. "The Moral Impossibility of Contract." *William and Mary Law Review* 48 (2007): 1647–72.

———. "Naturalistic Contract." In *Commercial Contract Law: Transatlantic Perspectives*, edited by Larry A. DiMatteo, Qi Zhou, Séverine Saintier, and Keith Rowley, 85–115. Cambridge: Cambridge University Press, 2013.

———. "Regret and Contract 'Science.'" *Georgetown Law Journal* 89 (2000): 143–70.

———. *A Theory of Contract Law: Empirical Insights and Moral Psychology*. New York: Oxford University Press, 2011.

Alces, Peter A., and Michael Greenfield. "They Can Do What!? Limitations on the Use of Change-of-Terms Clauses." *Georgia State Law Review* 26 (2010): 1099–45.

Alexander, Larry. "Theory's a What Comes Natcherly." *San Diego Law Review* 37 (2000): 777–82.

Alexander, Larry, and Michael Moore. "Deontological Ethics." *Stanford Encyclopedia of Philosophy*, edited by Edward N. Zalta. https://plato.stanford.edu/entries/ethics-deontological/.

Allen, Tom, and Robin Widdison. "Can Computers Make Contracts?" *Harvard Journal of Law and Technology* 9 (1996): 25–52.

Allman, John, and James Woodward. "What Are Moral Intuitions and Why Should We Care about Them? A Neurobiological Perspective." *Philosophical Issues* 18 (2008): 164–85.

"Alone and Afraid: Children Held in Solitary Confinement and Isolation in Juvenile Detention and Correctional Facilities." *American Civil Liberties Union*, June 2014, 3–4. https://www.aclu.org/files/assets/Alone%20and%20Afraid%20COMPLETE%20FINAL.pdf.

American Law Institute. *Consumer Contracts* [web page]. https://www.ali.org/projects/show/consumer-contracts/.

———. *Model Penal Code: Sentencing.* https://www.ali.org/projects/show/sentencing/.

———. "Model Penal Code: Sentencing Foreword." Tentative Draft No. 1. St. Paul, MN: American Law Institute, 2007.

———. *Principles of the Law of Software Contracts.* St. Paul, MN: American Law Institute, 2010.

———. *Principles of the Law of Software Contracts: Summary Overview to Topic 2.* St. Paul, MN: American Law Institute, 2010.

———. *Restatement (First) of Torts.* St. Paul, MN: American Law Institute, 1934.

———. *Restatement of the Law, Consumer Contracts: Preliminary Draft No. 2.* St. Paul, MN: American Law Institute, 2015.

———. *Restatement (Second) of Contracts.* St. Paul, MN: American Law Institute, 1981.

———. *Restatement (Second) of Torts.* St. Paul, MN: American Law Institute, 1965.

———. *Restatement (Third) of Torts: Physical and Emotional Harm.* St. Paul, MN: American Law Institute, 2012.

American Psychiatric Association. *Diagnostic and Statistical Manual of Mental Disorders.* 4th ed. Washington, DC: American Psychiatric Association, 2000. Cited as *DSM-IV.*

———. *Diagnostic and Statistical Manual of Mental Disorders.* 5th ed. Washington, DC: American Psychiatric Publishing, 2013. Cited as *DSM-V.*

Anderson, Steven W., Antoine Bechara, Hanna Damasio, Daniel Tranel, and Antonio R. Damasio. "Impairment of Social and Moral Behavior Related to Early Damage in Human Prefrontal Cortex." *Nature Neuroscience* 2 (1999): 1032–37.

Apkarian, A. Vania. "A Brain Signature for Acute Pain." *Trends in Cognitive Science* 17 (2013): 309–10.

Apollonio, Dorie E., and Ruth E. Malone. "Marketing to the Marginalised: Tobacco Industry Targeting of the Homeless and Mentally Ill." *Tobacco Control* 14 (2005): 409–15.

Appiah, Kwame Anthony. *Experiments in Ethics.* Cambridge, MA: Harvard University Press, 2008.

Aristotle. *Nicomachean Ethics.* Edited by Roger Crisp. Cambridge: Cambridge University Press, 2000.

Arpaly, Nomy. *Unprincipled Virtue*. New York: Oxford University Press, 2002.

Ascoli, Giorgio A. *Trees of the Brain Roots of the Mind*. Cambridge, MA: MIT Press, 2015.

Bach-y-Rita, Paul. "Brain Plasticity as a Basis for Recovery of Function in Humans." *Neuropsychologia* 28 (1990): 547–54.

Badhwar, Neera Kapur. "Friendship, Justice and Supererogation." *American Philosophical Quarterly* 22 (1985): 123–31.

Bagaric, Mirko, and Julie Clarke. *Torture: When the Unthinkable Is Permissible*. Albany: State University of New York Press, 2007.

Baird, Douglas G. "The Boilerplate Puzzle." *Michigan Law Review* 104 (2006): 933–52.

Bakos, Yannis, Florencia Marotta-Wurgler, and David R. Trossen. "Does Anyone Read the Fine Print? Consumer Attention to Standard-Form Contracts." *Journal of Legal Studies* 43 (2014): 1–45.

Barber, Brad M., Terrance Odean, and Lu Zheng. "Out of Sight, Out of Mind: The Effects of Expenses on Mutual Fund Flows." *Journal of Business* 78 (2005): 2095–2120.

Bar-Gill, Oren. *Seduction by Contract: Law, Economics, and Psychology in Consumer Markets*. Oxford, UK: Oxford University Press, 2012.

Bar-Gill, Oren, and Omri Ben-Shahar. Reporter's Notes to *Restatement of the Law Third, Consumer Contracts*. http://www.ballardspahr.com/~/media/files/alerts/2012 -12-06-outline.pdf.

Barnett, Randy E. "Contract Is Not Promise; Contract Is Consent." In *Philosophical Foundations of Contract Law*, edited by Gregory Klass, George Letsas, and Prince Saprai, 42–57. Oxford, UK: Oxford University Press, 2014.

Batthyany, Alexander. "Mental Causation and Free Will after Libet and Soon: Reclaiming Conscious Agency." In *Irreducibly Conscious*, edited by Alexander Batthyany and Avshalom Elitzur, 135–60. Heidelberg: Universitätsverlag Winter, 2009.

Baumgartner, Thomas, M. Heinrichs, A. Vonlanthen, U. Fischbacher, and E. Fehr. "Oxytocin Shapes the Neural Circuitry of Trust and Trust Adaptation in Humans." *Neuron* 58, no. 4 (2008): 639–50.

Bayne, Tim. "Libet and the Case for Free Will Scepticism," in *Free Will and Modern Science*, edited by Richard Swinburne, 551–564. Oxford, UK: Oxford University Press, 2011.

Bechara, Antoine. "The Role of Emotion in Decision-Making: Evidence from Neurological Patients with Orbitofrontal Damage." *Brain and Cognition* 55 (2004): 30–40.

Bechara, Antoine, Hanna Damasio, and Antonio R. Damasio. "Emotion, Decision Making and the Orbitofrontal Cortex." *Cerebral Cortex* 10, no. 3 (2000): 295–307.

Bechara, Antoine, Hanna Damasio, Daniel Tranel, and Antonio R. Damasio. "Deciding Advantageously before Knowing the Advantageous Strategy." *Science* 275 (1997): 1293–95.

Bellia, Anthony J. Jr. "Contracting with Electronic Agents." *Emory Law Journal* 50 (2001): 1047–94.

Benforado, Adam, and Jon Hanson. "The Great Attributional Divide: How Divergent Views of Human Behavior Are Shaping Legal Policy." *Emory Law Journal* 57 (2008): 311–408.

Benforado, Adam, Jon Hanson, and David Yosifon. "Broken Scales: Obesity and Justice in America." *Emory Law Journal* 53 (2004): 1645–1806.

Bennett, Jonathan. "Accountability." In *Philosophical Subjects*, edited by Zak van Stratten, 89–103. Oxford, UK: Oxford University Press, 1980.

Bennett, Maxwell R., Daniel Dennett, Peter Hacker, and John Searle. *Neuroscience and Philosophy: Brain, Mind, and Language.* New York: Columbia University Press, 2007.

Bennett, Maxwell R., and P. M. S. Hacker. *History of Cognitive Neuroscience.* Malden, MA: Wiley-Blackwell, 2008.

———. *Philosophical Foundations of Neuroscience.* Malden, MA: Blackwell, 2003.

Ben-Shahar, Omri. Foreword to " 'Boilerplate': Foundations of Market Contracts Symposium." *Michigan Law Review* 104 (2006): 821–26.

Ben-Shahar, Omri, and Carl E. Schneider. "The Failure of Mandated Disclosure." *University of Pennsylvania Law Review* 159 (2011): 647–750.

———. *More Than You Wanted to Know: The Failure of Mandated Disclosure.* Princeton, NJ: Princeton University Press, 2014.

Berker, Selim. "The Normative Insignificance of Neuroscience." *Philosophy and Public Affairs* 37 (2009): 293–329.

Bern, Roger C. "Terms Later Contracting: Bad Economics, Bad Morals, and a Bad Idea for a Uniform Law, Judge Easterbrook Notwithstanding." *Journal of Law and Policy* 12 (2004): 641–796.

Bernheim, B. Douglas, and Antonio Rangel. "Addiction and Cue-Triggered Decision Processes." *American Economic Review* 94 (2004): 1558–90.

Bickel, Christopher. "The Scene of the Crime: Children in Solitary Confinement." In *The Marion Experiment: Long-Term Solitary Confinement and the Supermax Movement*, edited by Stephen C. Richards, 129–59. Carbondale: Southern Illinois University Press, 2015.

Bisiach, Edoardo. "The (Haunted) Brain and Consciousness." In *Consciousness in Contemporary Science*, edited by Anthony J. Marcel and Edoardo Bisiach, 464–84. New York: Oxford University Press, 1988.

Blair, R. J. "Facial Expressions, Their Communicatory Functions and Neurocognitive Substrates." *Philosophical Transactions of the Royal Society of London B: Biological Sciences* 358 (2003): 561–72.

Blakemore, Sarah-Jayne. "The Mysterious Workings of the Adolescent Brain." *TED*, June 2012. http://www.ted.com/talks/sarah_jayne_blakemore_the_mysterious _workings_of_the_adolescent_brain?language=en.

Blakemore, Sarah-Jayne, and Suparna Choudhury. "Development of the Adolescent Brain: Implications for Executive Function and Social Cognition." *Journal of Child Psychology and Psychiatry* 47 (2006): 296–312.

Blakeslee, Sandra. "Cells That Read Minds." *New York Times*, January 10, 2006. http://www.nytimes.com/2006/01/10/science/10mirr.html?pagewanted=all.

Blumenthal, Jeremy A. "Law and the Emotions: The Problems of Affective Forecasting." *Indiana Law Journal* 80 (2005): 154–237.

Boardman, J. D., S. Menard, M. E. Roettger, K. E. Knight, B. B. Boutwell, and A. Smolen. "Genes in the Dopaminergic System and Delinquent Behaviors across the Life Course: The Role of Social Controls and Risks." *Criminal Justice and Behavior* 41 (2014): 713–31.

Boston University CTE Center. *What Is CTE?* http://www.bu.edu/cte/about/what-is-cte/.

Bouchard, Thomas J., and Matt McGue. "Genetic and Environmental Influences on Human Psychological Differences." *Developmental Neurobiology* 54 (2003): 4–45.

Bourget, David, and David J. Chalmers. "What Do Philosophers Believe?" *Philosophical Studies* 170 (2013): 465–500.

Bowers, William J., and Glenn L. Pierce. "The Illusion of Deterrence in Isaac Ehrlich's Research on Capital Punishment." *Yale Law Journal* 85, no. 2 (1975): 187–208.

Braithwaite, John. "Restorative Justice: Assessing Optimistic and Pessimistic Accounts." *Crime and Justice* 25 (1999): 1–128.

Breese, G. R., K. Chu, C. V. Dayas, D. Funk, D. J. Knapp, and D. F. Koob. "Stress Enhancement of Craving during Sobriety: A Risk for Relapse." *Alcoholism: Clinical and Experimental Research* 29 (2005): 185–95.

Burns, Jeffrey M., and Russell H. Swerdlow. "Right Orbitofrontal Tumor with Pedophilia Symptom and Constructional Apraxia Sign." *JAMA Neurology* 60 (2003): 437–40. http://archneur.jamanetwork.com/article.aspx?articleid=783830.

Calabresi, Guido. *The Costs of Accidents: A Legal and Economic Analysis*. New Haven, CT: Yale University Press, 1970.

Calderone, Julia. "There Are Some Big Problems with Brain-Scan Lie Detectors." *Tech Insider*, April 19, 2016. http://www.techinsider.io/dr-oz-huizenga-fmri-brain-lie-detector-2016-4.

Caruso, Gregg D. *Exploring the Illusion of Free Will and Moral Responsibility*. Lanham, MD: Lexington, 2013.

Caspi A., J. McClay, T. Moffitt, J. Mill, and J. Martin. "Role of Genotype in the Cycle of Violence in Maltreated Children." *Science* 297 (2002): 851–54.

Cavanagh, Sarah Rose. "Female Sexual Desire: An Evolutionary Biology Perspective." *Psychology Today*, June 19, 2013. https://www.psychologytoday.com/blog/once-more-feeling/201306/female-sexual-desire-evolutionary-biology-perspective.

Center for Science and the Law. *The Many Dangers of Brain-Based Lie Detection*. Houston, TX: Center for Science and the Law, 2012. http://www.neulaw.org/blog/1034-class-blog/4131-the-many-dangers-of-brain-based-lie-detection.

Centers for Disease Control. "Cigarette Smoking among Adults and Trends in Smoking Cessation—United States." *Morbidity and Mortality Weekly Report* 58 (2009): 1227–32.

Chein, Jason, Dustin Albert, Lia O'Brien, Kaitlyn Uckert, and Laurence Steinberg. "Peers Increase Adolescent Risk Taking by Enhancing Activity in the Brain's Reward Circuitry." *Developmental Science* 14 (2011): F1–F10.

Chen, Ronald, and Jon D. Hanson. "Categorically Biased: The Influence of Knowledge Structures on Law and Legal Theory." *Southern California Law Review* 77 (2004): 1103–1254.

———. "The Illusion of Law: The Legitimating Schemas of Modern Policy and Corporate Law." *Michigan Law Review* 103 (2004): 1–149.

Christen, Markus, and Sabine Müller. "Effects of Brain Lesions on Moral Agency: Ethical Dilemmas in Investigating Moral Behavior." *Current Topics in Behavioral Neuroscience* 19 (2014): 159–88.

Churchland, Patricia S. *Braintrust: What Neuroscience Tells Us about Morality.* Princeton, NJ: Princeton University Press, 2011.

Churchland, Paul M. "Eliminative Materialism and the Propositional Attitudes." *Journal of Philosophy* 78 (1981): 67–90.

Coase, Ronald H. "Biographical." In *Nobel Lectures: Prize Lectures in Economic Sciences.* Vol. 3, *1991–1995*, edited by Torsten Persson, 7–10. Singapore: World Scientific, 1997.

Cochran, John K., Mitchell B. Chamlin, and Mark Seth. "Deterrence or Brutalization? An Impact Assessment of Oklahoma's Return to Capital Punishment." *Criminology* 32 (1994): 107–34.

Coghill, R. C. "Pain: Neuroimaging." In *Encyclopedia of Neuroscience*, edited by Larry R. Squire, 409–14. New York: Academic Press, 2009.

Cohen, Mark A. "The Monetary Value of Saving a High-Risk Youth." *Quantitative Criminology* 14 (1998): 5–34.

Cohen, Mark A., and Alex R. Piquero. "New Evidence on the Monetary Value of Saving a High-Risk Youth." *Journal of Quantitative Criminology* (2009): 25–49.

Colapinto, John. "Brain Games: The Marco Polo of Neuroscience." *New Yorker*, May 11, 2009, 76.

Colby, Ann, and Lawrence Kohlberg. *The Measurement of Moral Judgment.* Vol. 2, *Standard Issue Scoring Manual.* New York: Cambridge University Press, 1987.

Coleman, Jules L. *Markets, Morals, and the Law.* New York: Oxford University Press, 2002.

———. "The Mixed Conception of Corrective Justice." *Iowa Law Review* 77 (1991): 427–44.

———. *The Practice of Principle: In Defence of a Pragmatist Approach to Legal Theory.* Oxford, UK: Oxford University Press, 2001.

———. *Risks and Wrongs.* Oxford, UK: Oxford University Press, 2002.

Coleman, Jules L., Scott Hershovitz, and Gabriel Mendlow. "Theories of the

Common Law of Torts." In *Stanford Encyclopedia of Philosophy*, edited by Edward N. Zalta. http://plato.stanford.edu /entries/tort-theories/#CorJus.

Connell, Simon. "What Is the Place of Corrective Justice in Criminal Justice?" *Waikato Law Review* 19 (2011): 134–44.

Connolly, Gregory N., Hillel R. Alpert, Geoffrey Ferris Wayne, and Howard Koh. "Trends in Nicotine Yield in Smoke and Its Relationship with Design Characteristics among Popular US Cigarette Brands, 1997–2005." *Tobacco Control* 16 (2007): e5.

Cooper, Geoffrey M. *The Cell: A Molecular Approach: The Molecular Composition of Cells.* 2nd ed. Sunderland, UK: Sinauer Associates, 2000.

Costigan, George P. "Implied-in-Fact Contracts and Mutual Assent." *Harvard Law Review* 33 (1920): 376–400.

Craswell, Richard. "Contract Law, Default Rules, and the Philosophy of Promising." *Michigan Law Review* 88 (1989): 489–529.

Crick, Francis. *The Astonishing Hypothesis: The Scientific Search for the Soul.* New York: Scribner's, 1994.

Cullen, Francis T., and Karen E. Gilbert. *Reaffirming Rehabilitation.* Cincinnati, OH: Anderson, 1982.

Cummings, K. Michael, C. P. Morley, J. K. Horan, C. Steger, and N.-R. Leavell. "Marketing to America's Youth: Evidence from Corporate Documents." *Tobacco Control* 11 (2002): i5–i17.

Damasio, Antonio R. *Descartes' Error: Emotion, Reason, and the Human Brain.* New York: Harper-Perennial, 1995.

Davies, Paul S. "Skepticism concerning Human Agency." In *Neuroscience and Legal Responsibility*, edited by Nicole A. Vincent, 113–30. New York: Oxford University Press, 2013.

Daw, Nathaniel D. "Advanced Reinforcement Learning." In *Neuroeconomics: Decision Making and the Brain*, edited by Paul W. Glimcher and Ernst Fehr, 299–320. Cambridge, MA: Academic, 2013.

Daw, Nathaniel D., and Philippe N. Tobler. "Value Learning through Reinforcement: The Basics of Dopamine and Reinforcement Learning." In *Neuroeconomics: Decision Making and the Brain*, edited by Paul W. Glimcher and Ernst Fehr, 283–98. Cambridge, MA: Academic, 2013.

Dawkins, Richard. "Straf is wetenschappelijk achterhaald." *NRC-Handelsblad*, January 14, 2006.

De Martino, Benedetto, Dharshan Kumaran, Ben Seymour, and Raymond J. Dolan. "Frames, Biases, and Rational Decision-Making in the Human Brain." *Science* 313 (2006): 684–87.

Dennett, Daniel C. *Elbow Room: The Varieties of Free Will Worth Wanting.* Cambridge, MA: MIT Press, 1984.

———. "Intentional Systems." *Journal of Philosophy* 68 (1971): 87–106.

———. Review of *Against Moral Responsibility*, by Bruce Waller. *Naturalism* (2012).

http://www.naturalism.org/resources/book-reviews/dennett-review-of-against
-moral-responsibility.

———. "The Self as a Responding—and Responsible—Artifact." *Annals of the New York Academy of Science* 1001 (2003): 39–50.

Dick, Philip K. *Collected Stories of Phillip K. Dick*. Vol. 4, *The Minority Report*. Burton, MI: Subterranean, 2013.

Dobbs, Dan B. *The Law of Torts*. St. Paul, MN: West Group, 2000.

Dodge, William S. "The Case for Punitive Damages in Contracts." *Duke Law Journal* 48 (1999): 629–99.

Dolinko, David. "Some Thoughts about Retributivism." *Ethics* 101 (1991): 537–39.

———. "Three Mistakes of Retributivism." *UCLA Law Review* 39 (1992): 1623–58.

Doris, John M. *Lack of Character: Personality and Moral Behavior*. New York: Cambridge University Press, 2002.

Dosenbach, Nico U. F., Binyam Nardos, Alexander L. Cohen, Damien A. Fair, Jonathan D. Power, Jessica A. Church, Steven M. Nelson, Gagan S. Wig, Alecia C. Vogel, Christina N. Lessov-Schlaggar, Kelly Anne Barnes, Joseph W. Dubis, Eric Feczko, Rebecca S. Coalson, John R. Pruett Jr., Deanna M. Barch, Steven E. Petersen, and Bradley L. Schlaggar "Prediction of Individual Brain Maturity Using fMRI." *Science* 329 (2010): 1358–61.

Dumontheil, Iroise, Ian A. Apperly, and Sarah-Jayne Blakemore. "Online Usage of Theory of Mind Continues to Develop in Late Adolescence." *Developmental Science* 13, no. 2 (2010): 331–38.

Dunn, Barnaby D., Tim Dalgleish, and Andrew D. Lawrence. "The Somatic Marker Hypothesis: A Critical Evaluation." *Neuroscience and Biobehavioral Reviews* 30 (2006): 239–71.

Dunnell, Mark B. *Dunnell Minnesota Digest*. New York: LexisNexis, 2014.

Edwards, Matthew A. "The Virtue of Mandatory Disclosure." *Notre Dame Journal of Law, Ethics and Public Policy* 28 (2014): 47–78.

Eich, Eric, Ian A. Brodkin, John L. Reeves, and Anuradha F. Chawla. "Questions concerning Pain." In *Well-Being: The Foundations of Hedonic Psychology*, edited by Daniel Kahneman, Edward Deiner, and Norbert Schwarz, 155–68. New York: Russell Sage Foundation, 1999.

Epstein, Richard A. "Unconscionability: A Critical Reappraisal." *Journal of Law and Economics* 18 (1975): 239–315.

Esfeld, Michael, and Christian Sachse. "Theory Reduction by Means of Functional Sub-types." *International Studies in the Philosophy of Science* 21 (2007): 1–17.

Evans, Jonathan St. B. T. "Dual-Processing Accounts of Reasoning, Judgment, and Social Cognition." *Annual Review of Psychology* 59 (2008): 255–78.

Eysenck, H. J. *Crime and Personality*. 3rd ed. London: Routledge and Kegan Paul, 1977.

Fainaru, Steve, and Mark Fainaru-Wada. "Youth Football Participation Drops." *ESPN.com*, November 14, 2013. http://espn.go.com/espn/otl/story/_/page/pop warner/pop-warner-youth-football-participation-drops-nfl-concussion-crisis -seen-causal-factor.

Fallon, James. *The Psychopath Inside: A Neuroscientist's Personal Journey into the Dark Side of the Brain*. New York: Current, 2013.

Farahany, Nita A. "Cruel and Unequal Punishments." *Washington University Law Review* 86 (2009): 859–915.

Farnsworth, E. Allan. *Changing Your Mind: The Law of Regretted Decisions*. New Haven, CT: Yale University Press, 1998.

———. "Ingredients in the Redaction of the 'Restatement (Second) of Contracts.'" *Columbia Law Review* 81 (1981): 1–12.

Farnsworth, Ward, and Mark F. Grady. *Torts: Cases and Questions*. 2nd ed. Austin, TX: Wolters Kluwer Law and Business, 2009.

Farrell, Ian P., and Justin F. Marceau. "Taking Voluntariness Seriously." *Boston College Law Review* 54 (2013): 1545–1612.

Federal Trade Commission. N.d. *Payday Lending* [web page]. https://www.ftc.gov /news-events/media-resources/consumer-finance/payday-lending.

Feigel, Herbert. *The "Mental" and the "Physical."* Minneapolis: University of Minnesota Press, 1967.

Fingarette, Herbert. "Punishment and Suffering." *Proceedings and Addresses of the American Philosophical Association* 50 (1977): 499–525.

Fleischman, John. *Phineas Gage: A Gruesome but True Story about Brain Science*. Boston: Houghton Mifflin, 2002.

Fletcher, George P. *Rethinking Criminal Law*. Oxford, UK: Oxford University Press, 2000.

Fodor, Jerry A. *Psychological Explanation*. New York: Random House, 1968.

———. *Psychosemantics: The Problem of Meaning in the Philosophy of Mind*. Cambridge, MA: MIT Press, 1987.

———. "Special Sciences (Or, The Disunity of Science as a Working Hypothesis)." *Synthese* 28 (1974): 97–115.

———. "Special Sciences: Still Autonomous after All These Years." *Noûs Supplement: Philosophical Perspectives* 31 (1997): 149–63.

Foot, Philippa. "The Problem of Abortion and the Doctrine of Double Effect." *Oxford Review* 5 (1967): 1–6.

Fowler, Lucy. "Gender and Jury Deliberations: The Contributions of Social Science." *William and Mary Journal of Women and the Law* 12 (2005): 1–48.

Fox, Stuart. "Laws Might Change as the Science of Violence Is Explained." *Live Science*, June 7, 2010. http://www.livescience.com/6535-laws-change-science-violence -explained.html.

Frankena, William K. *Ethics*. Englewood Cliffs, NJ: Prentice Hall, 1973.

Fried, Charles. *Contract as Promise: A Theory of Contractual Obligation*. Cambridge, MA: Harvard University Press, 1981.

Gabaix, Xavier, and David Laibson. "Shrouded Attributes, Consumer Myopia, and Information Suppression in Competitive Markets." *Quarterly Journal of Economics* 121 (2006): 505–40.

Gao, Y., A. Raine, P. H. Venables, M. E. Dawson, and S. A. Mednick. "Association of Poor Childhood Fear Conditioning and Adult Crime." *American Journal of Psychiatry* 167 (2010): 56–60.

Garvin, Larry T. "Small Business and the False Dichotomies of Contract Law." *Wake Forest Law Review* 40 (2005): 295–388.

Gautam, Kirti A., S. Pooja, S. N. Sankhwar, P. L. Sankhwar, A. Goel, and S. Rajender. "c.29C>T Polymorphism in the Transforming Growth Factor-β1 (TGFB1) Gene Correlates with Increased Risk of Urinary Bladder Cancer." *Cytokine* 75 (2015): 344–48.

Gendreau, Paul, N. L. Freedman, G. J. Wilde, and G. D. Scott. "Changes in EEG Alpha Frequency and Evoked Response Latency during Solitary Confinement." *Journal of Abnormal Psychology* 79, no. 1 (1972): 54–59.

Gibbard, Allen. "Normative and Recognitional Concepts." *Philosophy and Phenomenological Research* 64 (2002): 151–67.

Giedd, Jay N., Jonathan Blumenthal, Neal O. Jeffries, F. X. Castellanos, Hong Liu, Alex Zijdenbos, Tomás caron Paus, Alan C. Evans, and Judith L. Rapoport "Brain Development during Childhood and Adolescence: A Longitudinal MRI Study." *Nature Neuroscience* 2 (1999): 861–63.

Giere, Ronald N. *Scientific Perspectivism*. Chicago: University of Chicago Press, 2006.

Gigerenzer, Gerd. "Moral Intuition = Fast and Frugal Heuristics?" In *Moral Psychology*, vol. 2, edited by Walter Sinnott-Armstrong, 1–26. Cambridge, MA: MIT Press, 2008.

Ginther, Matthew R, Francis X. Shen, Richard J. Bonnie, Morris B. Hoffman, Owen D. Jones, René Marois, and Kenneth W. Simons. "The Language of *Mens Rea*." *Vanderbilt Law Review* 67 (2014): 1327–72.

Glannon, Walter. "Brain, Behavior, and Knowledge." *Neuroethics* 4 (2011): 191–94.

Glimcher, Paul W., and Ernst Fehr, eds. *Neuroeconomics: Decision Making and the Brain*. Cambridge, MA: Academic, 2013.

Gold, Joshua I., and Michael N. Shadlen. "The Neural Basis of Decision Making." *Annual Review of Neuroscience* 30 (2007): 535–74.

Goldberg, John C. P., and Benjamin C. Zipursky. "Civil Recourse Defended: A Reply to Posner, Callabresi, Rustad, Chamallas, and Robinette." *Indiana Law Journal* 88 (2013): 569–609.

———. "The Moral of *MacPherson*." *University of Pennsylvania Law Review* 146 (1998): 1733–1846. http://scholarship.law.upenn.edu/penn_law_review/vol146/iss6/1.

———. "The Myths of *MacPherson*." *Journal of Tort Law* 9, no. 1 (2016): 91–117.

———. *The Oxford Introduction to U.S. Law: Torts*. New York: Oxford University Press, 2010.

———. *Tort Law: Responsibilities and Redress*. New York: Aspen, 2004.

———. "Torts as Wrongs." *Texas Law Review* 88 (2010): 917–86.

Goldman, Alvin I. "Theory of Mind." In *Oxford Handbook of Philosophy of*

Cognitive Science, edited by Eric Margolis, Richard Samuels, and Stephen P. Stich, 402–24. Oxford, UK: Oxford University Press, 2012.

Gordon, James D. III. "Dialogue about the Doctrine of Consideration." *Cornell Law Review* 75 (1990): 987–1006.

Gould, Stephen Jay. Introduction to *Dance of the Tiger: A Novel of the Ice Age*, edited by Björn Kurtén, xi–xix. New York: Random House, 1980.

Grassian, Stuart. "Psychiatric Effects of Solitary Confinement." *Washington University Journal of Law and Policy* 22 (2006): 325–84.

Grayling, A. C. *The Age of Genius: The Seventeenth Century and the Birth of the Modern Mind*. New York: Bloomsbury, 2016.

Greene, Joshua D. "The Cognitive Neuroscience of Moral Judgment." In *The Cognitive Neurosciences IV*, edited by Michael S. Gazzaniga, 987–1000. Cambridge, MA: MIT Press, 2009.

———. "From Neural 'Is' to Moral 'Ought': What Are the Moral Implications of Neuroscientific Moral Psychology?" *Nature Reviews Neuroscience* 4 (2003): 846–50.

———. *Moral Tribes: Emotion, Reason, and the Gap between Us and Them*. New York: Penguin Press, 2013.

———. "The Secret Joke of Kant's Soul." *Moral Psychology* 3 (2008): 35–80.

Greene, Joshua D., and Jonathan Cohen. "For the Law, Neuroscience Changes Nothing and Everything." *Philosophical Transactions of the Royal Society of London B: Biological Sciences* 359 (2004): 1775–85.

Greene, Joshua D., Sylvia A. Morelli, Kelly Lowenberg, Leigh E. Nystrom, and Jonathan D. Cohen. "Cognitive Load Selectively Interferes with Utilitarian Moral Judgment." *Cognition* 107 (2008): 1144–54.

Greene, Joshua D., Leigh E. Nystrom, Andrew D. Engell, John M. Darley, and Jonathan D. Cohen. "The Neural Bases of Cognitive Conflict and Control in Moral Judgment." *Neuron* 44 (2004): 389–400.

Greene, Joshua D., Brian R. Sommerville, Leigh E. Nystrom, John M. Darley, and Jonathon D. Cohen. "An fMRI Investigation of Emotional Engagement in Moral Judgment." *Science* 293 (2001): 2105–8.

Greenstone, Gerry. "The History of Bloodletting." *British Columbia Medical Journal* 52 (2010): 12–14.

Grey, David. "Punishment as Suffering." *Vanderbilt Law Review* 63 (2010): 1619–93.

Guenther, Lisa. *Solitary Confinement: Social Death and Its Afterlives*. Minneapolis: University of Minnesota Press, 2013.

Guo, G., M. E. Roettger, and J. C. Shih. "Contributions of the DAT1 and DRD2 Genes to Serious and Violent Delinquency among Adolescents and Young Adults." *Human Genetics* 121 (2007): 125–36.

Guskiewicz, Kevin M., Michael McCrea, Stephen W. Marshall, Robert C. Cantu, Christopher Randolph, William Barr, James A. Onate, and James P. Kelly.

"Cumulative Effects Associated with Recurrent Concussion in Collegiate Foot-ball Players." *Journal of the American Medical Association* 290 (2003): 2549–55.

Güth, Werner, Rolf Schmittberger, and Bernd Schwarze. "An Experimental Anal-ysis of Ultimatum Bargaining." *Journal of Economic Behavior and Organiza-tion* 3 (1982): 367–88.

Haney, Craig. *Reforming Punishment: Psychological Limits to the Pains of Impris-onment.* Washington, DC: American Psychological Association, 2006.

Hanson, Jon D., and Douglas A. Kysar. "Taking Behavioralism Seriously: The Problem of Market Manipulation." *New York University Law Review* 74 (1999): 630–749.

———. "Taking Behavioralism Seriously: Some Evidence of Market Manipula-tion." *Harvard Law Review* 112 (1999): 1420–1572.

Hanson, Jon D., and David Yosifon. "The Situational Character: A Critical Realist Perspective on the Human Animal." *Georgetown Law Journal* 93 (2004): 1–182.

———. "The Situation: An Introduction to the Situational Character, Critical Re-alism, Power Economics, and Deep Capture." *University of Pennsylvania Law Review* 152 (2003): 129–218.

Hare, Robert. "Focus on Psychopathy." *FBI Law Enforcement Bulletin*, July 2012. https://leb.fbi.gov/2012/july/focus-on-psychopathy.

———. *Hare Psychopathy: Checklist—Revised (PCL-R).* North Tonawanda, NY: Multi-Health Systems, 2003. Cited as *Hare PCL.*

———. "This Charming Psychopath: How to Spot Social Predators before They Attack." *Psychology Today*, January 1, 1994. https://www.psychologytoday.com /articles/199401/charming-psychopath.

Harris, Judith Rich. *No Two Alike: Human Nature and Human Individuality.* New York: W. W. Norton, 2007.

Harris, Sam. *The End of Faith.* New York: W. W. Norton, 2004.

———. *Free Will.* New York: Free Press, 2012.

———. *Letter to a Christian Nation.* New York: Knopf Doubleday Publishing Group, 2006.

———. *The Moral Landscape: How Science Can Determine Human Values.* New York: Free Press, 2010.

Hart, H. L. A. "Legal Responsibility and Excuses." In Hart, ed., *Punishment and Responsibility.*

———. "Postscript: Responsibility and Retribution." In Hart, ed., *Punishment and Responsibility*, 211–12.

———, ed. *Punishment and Responsibility.* Oxford, UK: Oxford University Press, 1968.

Heinzen, Hanna, D. Kohler, N. Godt, F. Geiger, and C. Huchzermeier. "Psychop-athy, Intelligence and Conviction History." *International Journal of Law and Psychiatry* 34, no. 5 (2011): 336–40.

Herbert, Bob. "In America; Tobacco Dollars." *New York Times*, November 28, 1993. http://www.nytimes.com/1993/11/28/opinion/in-america-tobacco-dollars .html.

Hermalin, Benjamin E., Avery W. Katz, and Richard Craswell. "Contract Law." In *Handbook of Law and Economics*, edited by A. Mitchell Polinsky and Steven Shavell, 15, 3–138. London: Elsevier, 2007.

Hibbeln, J. R., J. M. Davis, C. Steer, P. Emmett, I. Rogers, C. Williams, and J. Golding. "Maternal Seafood Consumption in Pregnancy and Neurodevelopmental Outcomes in Childhood (ALSPAC Study): An Observational Cohort Study." *Lancet* 369 (2007): 578–85.

Hillman, Robert A. *Principles of Contract Law*. St. Paul, MN: West Academic, 2004.

———. "Rolling Contracts." *Fordham Law Review* 71, no. 3 (2002): 743–60.

Hillman, Robert A., and Jeffrey Rachlinski. "Standard-Form Contracting in the Electronic Age." *New York University Law Review* 77, no. 3 (2002): 429–95.

Holmes, Eric Mills. "The Four Phases of Promissory Estoppel." *Seattle University Law Review* 20 (1996): 45–79.

Hrdy, Sarah Blaffer. 2009. *Mothers and Others: The Evolutionary Origins of Mutual Understanding*. Cambridge, MA: Belknap.

Hughes, Virginia. "Science in Court: Head Case." *Nature* 464 (2010): 340–42.

Hull, Richard. *Deprivation and Freedom: A Philosophical Enquiry*. New York: Routledge, 2007.

Hume, David. *An Enquiry concerning the Principles of Morals*. Edited by Eugene Freeman. New York: Liberal Arts Press, 1957.

———. *A Treatise of Human Nature*, edited by L. A. Selby-Bigge. London: Oxford University Press, 1896.

Hunter, Howard O. *Modern Law of Contracts*. St. Paul, MN: Thompson Reuters, 2016.

Hurd, Heidi M. "The Innocence of Negligence." *Contemporary Readings in Law and Social Justice* 8, no. 2 (2016): 48–95.

Hurt, Richard D., and Channing R. Robertson. "Prying Open the Door to the Tobacco Industry's Secrets about Nicotine: The Minnesota Tobacco Trial." *Journal of the American Medical Association* 280 (1998): 1173–81.

Husak, Douglas. "Rethinking the Act Requirement." *Cardozo Law Review* 28 (2007): 2437–60.

Hyman, Mark. "Why Kids under 14 Should Not Play Tackle Football." *Time*, November 6, 2012. http://ideas.time.com/2012/11/06/why-kids-under-14-should-not-play-tackle-football/.

Hyser, Sarah. "Two Steps Forward, One Step Back: How Federal Courts Took the 'Fair' Out of the Fair Sentencing Act of 2010." *Penn State Law Review* 117 (2012): 503–35.

"iCloud Terms and Conditions." *Apple* [web page]. http://www.apple.com/legal/internet-services/icloud/en/terms.html.

Iribarren, C., J. H. Markovitz, D. R. Jacobs, P. J. Schreiner, and M. Daviglus. "Dietary Intake of N-3, N-6 Fatty Acids and Fish: Relationship with Hostility in Young Adults—the CARDIA Study." *European Journal of Clinical Nutrition* 58 (2004): 24–31.

Ishikawa, Sharon H., A. Raine, T. Lencz, S. Bihrle, and L. Lacasse. "Autonomic Stress Reactivity and Executive Functions in Successful and Unsuccessful Criminal Psychopaths from the Community." *Journal of Abnormal Psychology* 110 (2001): 423–32.

Jain, Sarah S. Lochlann. "'Come Up to the Kool Taste': African American Upward Mobility and the Semiotics of Smoking Menthols." *Public Culture* 15 (2003): 295–322.

Jiang, Zihong, Gregory R. Rompala, Shuquin Zhang, Rita M. Cowell, and Kazu Nakazawa. "Social Isolation Exacerbates Schizophrenia-like Phenotypes via Oxidative Stress in Cortical Interneurons." *Biological Psychiatry* 73 (2013): 1024–34.

Jolls, Christine, Cass R. Sunstein, and Richard H. Thaler. "A Behavioral Approach to Law and Economics." *Stanford Law Review* 50 (1998): 1471–1550.

Jones, Owen D., Jeffrey D. Schall, and Francis X. Shen. *Law and Neuroscience.* New York: Wolters Kluwer, 2014.

Joyce, Richard. *The Evolution of Morality.* Cambridge, MA: MIT Press, 2006.

Kable, Joseph W. "Valuation, Intertemporal Choice, and Self-Control." In *Neuroeconomics: Decision Making and the Brain*, edited by Paul W. Glimcher and Ernst Fehr, 173–89. Cambridge, MA: Academic, 2013.

Kahneman, Daniel. *Thinking, Fast and Slow.* New York: Farrar, Straus and Giroux, 2011.

Kahneman, Daniel, Paul Slovic, and Amos Tversky. *Judgment under Uncertainty: Heuristics and Biases.* Cambridge: Cambridge University Press, 1982.

Kalant, Harold. "What Neurobiology Cannot Tell Us about Addiction." *Addiction* 105 (2010): 780–89.

Kandel, Eric R. *In Search of Memory: The Emergence of a New Science of Mind.* New York: W. W. Norton, 2006.

Kandel, Eric R., and J. H. Schwartz. "Molecular Biology of Learning." *Science* 218 (1982): 433–43.

Kant, Immanuel. *Critique of Pure Reason.* Translated by F. Max Muller. New York: Macmillan, 1922. First published in c. 1781.

———. *Groundwork of the Metaphysics of Morals.* Edited by Mary Gregor and Jens Timmermann. Cambridge: Cambridge University Press, 2011.

Kaplan, Jonas T., and Marco Iacoboni. "Getting a Grip on Other Minds: Mirror Neurons, Intention Understanding, and Cognitive Empathy." *Social Neuroscience* 1 (2006): 175–83.

Kar, Robin Bradley. "Contract as Empowerment." *University of Chicago Law Review* 83 (2016): 101–74.

Kathol, Mary H., Timothy E. Moore, Georges Y. El-Khoury, William T. C. Yuh, and William J. Montgomery. "Magnetic Resonance Imaging of Athletic Soft Tissue Injuries." *Iowa Orthopaedic Journal* 9 (1989): 44–50.

Keeton, W, William Lloyd Prosser, Dan B. Dobbs, Robert E. Keeton, and David G. Owen. *Prosser and Keeton on the Law of Torts § 65.* 5th ed. Eagan, MN: West Group, 1984.

Kessler, David A. "The Control and Manipulation of Nicotine in Cigarettes." *Tobacco Control* 3 (1994): 362–69.

Kiani, Roozbeh, and Michael N. Shadlen. "Representation of Confidence Associated with a Decision by Neurons in the Parietal Cortex." *Science* 324, no. 5928 (2009): 759–64.

Kiehl, K. A. "A Cognitive Neuroscience Perspective on Psychopathy: Evidence for Paralimbic System Dysfunction." *Psychiatry Research* 142 (2006): 107–28.

Kiehl, K. A., A. M. Smith, R. D. Hare, A. Mendrek, B. B. Forster, J. Brink, and P. F. Liddle. "Limbic Abnormalities in Affective Processing by Criminal Psychopaths as Revealed by Functional Magnetic Resonance Imaging." *Biological Psychiatry* 50 (2001): 677–84.

Kimel, Dori. *From Promise to Contract: Towards a Liberal Theory of Contract.* Portland, OR: Hart, 2003.

———. "Personal Autonomy and Change of Mind in Promise and in Contract." In *Philosophical Foundations of Contract Law*, edited by Gregory Klass, George Letsas, and Prince Saprai, 96–115. Oxford, UK: Oxford University Press, 2014.

Klass, Gregory. Introduction to *Philosophical Foundations of Contract Law*, edited by Gregory Klass, George Letsas, and Prince Saprai, 1–16. Oxford, UK: Oxford University Press, 2014.

Klass, Gregory, George Letsas, and Prince Saprai. *Philosophical Foundations of Contract Law*. Oxford, UK: Oxford University Press, 2014.

Klebs, Chandler, and George Ortega. "Exploring the Illusion of Free Will and Moral Responsibility." *Causal Consciousness*, 2012. www.causalconciousness.com.

Knutson, Brian, Scott Rick, G. Elliott Wimmer, Drazen Prelec, and George Loewenstein. "Neural Predictors of Purchases." *Neuron* 53 (2007): 147–56.

Koenigs, Michael, Liane Young, Ralph Adolphs, Daniel Tranel, Fiery Cushman, Marc Hauser, and Antonio Damasio. "Damage to the Prefrontal Cortex Increases Utilitarian Moral Judgments." *Nature* 446 (2007): 908–11.

Kolber, Adam J. "Pain Detection and the Privacy of Subjective Experience." *American Journal of Law and Medicine* 33 (2007): 433–56.

———. "Punishment and Portfolio of Beliefs." Unpublished manuscript, 2016.

———. "The Subjective Experience of Punishment." *Columbia Law Review* 109 (2009): 182–236.

———. "Will There Be a Neurolaw Revolution?" *Indiana Law Journal* 89 (2014): 808–45.

Korobkin, Russell B. "What Comes after Victory for Behavioral Law and Economics?" *University of Illinois Law Review* 2011 (2011): 1653–74.

Korobkin, Russell B., and Thomas S. Ulen. "Law and Behavioral Science: Removing the Rationality Assumption from Law and Economics." *California Law Review* 88 (2000): 1051–1144.

Kosfeld, Michael, Markus Heinrichs, Paul J. Zak, Urs Fischbacher, and Ernst Fehr. "Oxytocin Increases Trust in Humans." *Nature* 435 (2005): 673–76. http://econstor.eu/bitstream/10419/95 865/1/782390358.pdf.

Kosfeld, Michael, and Ulrich Schüwer. "Add-on Pricing in Retail Financial Markets and the Fallacies of Consumer Education." *Review of Finance* 21(2017): 1189–1216.

Label, Catherine, and Christian Beaulieu. "Longitudinal Development of Human Brain Wiring Continues from Childhood into Adulthood." *Journal of Neuroscience* 31 (2011): 10937–47.

Lanphear, Bruce P., and Klaus J. Roghmann. "Pathways of Lead Exposure in Urban Children." *Environmental Law Research* 74 (1997): 67–73.

Larsen, Richard W. *Bundy: The Deliberate Stranger*. Englewood Cliffs, NJ: Prentice Hall, 1980.

Lawrence, William H. "Rolling Contracts Rolling over Contract Law." *San Diego Law Review* 41 (2004): 1099–1122.

Leff, Arthur Allen. "Contract as Thing." *American University Law Review* 19 (1970): 131–57.

Leknes, Siri, and Tracey, Irene. "A Common Neurobiology for Pain and Pleasure." *Nature Reviews Neuroscience* 9 (2008): 314–20.

Lelling, Andrew E. "Eliminative Materialism, Neuroscience and the Criminal Law." *University of Pennsylvania Law Review* 141 (1993): 1471–1564.

Lempert, Karolina M., and Elizabeth A. Phelps. "Neuroeconomics of Emotion and Decision Making." In *Neuroeconomics: Decision Making and the Brain*, edited by Paul W. Glimcher and Ernst Fehr, 219–36. Cambridge, MA: Academic, 2013.

Leshner, Alan I. "Addiction Is a Brain Disease, and It Matters." *Science* 278 (1997): 45–47.

Levin, E. D., A. Wilkerson, J. P. Jones, N. C. Christopher, and S. J. Briggs. "Prenatal Nicotine Effects on Memory in Rats: Pharmacological and Behavioral Challenges." *Developmental Brain Research* 97 (1996): 207–15.

Levine, Joseph. "Conceivability and the Metaphysics of Mind." *Noûs* 32 (1998): 449–80.

———. "On Leaving Out What It's Like." In *Consciousness: Psychological and Philosophical Essays*, edited by Martin Davies and Glyn W. Humphreys, 121–36. Oxford, UK: Blackwell, 1993.

Levy, Ken. "Why Retributivism Needs Consequentialism: The Rightful Place of Revenge in the Criminal Justice System." *Rutgers Law Review* 66 (2014): 629–84.

Levy, Neil. *Hard Luck: How Luck Undermines Free Will and Moral Responsibility*. New York: Oxford University Press, 2011.

———. *Neuroethics: Challenges for the 21st Century*. Cambridge: Cambridge University Press, 2007.

Libet, Benjamin. "Unconscious Cerebral Initiative and the Role of Conscious Will in Voluntary Action." *Behavioral and Brain Sciences* 8 (1985): 529–66.

Libet, Benjamin, Curtis A. Gleason, Elmwood W. Wright, and Dennis K. Pearl. "Time of Conscious Intention to Act in Relation to Onset of Cerebral Activity (Readiness Potential): The Unconscious Initiation of a Freely Voluntary Act." *Brain* 106 (1983): 623–42.

Lilienfeld, Scott O., and Ashley Watts. "Not All Psychopaths Are Criminals—Some Psychopathic Traits Are Actually Linked to Success." *The Conversation*, January 26, 2016. https://theconversation.com/not-all-psychopaths-are-criminals-some -psychopathic-traits-are-actually-linked-to-success-51282.

Lilienfeld, Scott O., Ashley L. Watts, and Sarah Francis Smith. "Successful Psychopathy." *Current Directions in Psychological Science* 24 (2015): 298–303.

Ling, Pamela M., and Stanton A. Glantz. "Why and How the Tobacco Industry Sells Cigarettes to Young Adults: Evidence from Industry Documents." *American Journal of Public Health* 92 (2002): 908–16.

Litton, Paul. "Responsibility Status of the Psychopath." *Rutgers Law Journal* 39 (2008): 349–92.

Loewenstein, George. "Affect Regulation and Affective Forecasting." In *Handbook of Emotion Regulation*, edited by James J. Gross, 180–203. New York: Guilford Press, 2007.

Lorenzen, Ernest G. "Causa and Consideration in the Law of Contracts." *Yale Law Journal* 28 (1919): 621–46.

Lowe, Brian M. *Emerging Moral Vocabularies: The Creation and Establishment of New Forms of Moral and Ethical Meanings*. Lanham, MD: Lexington, 2006.

Lu, Yi, Geraldine T. Klein, and Michael Y. Wang. "Can Pain Be Measured Objectively?" *Neurosurgery* 73 (2013): N24–N25.

Luna, Beatriz, Keith R. Thulborn, Douglas P. Munoz, Elisha P. Merriam, Krista E. Garver, Nancy J. Minshew, Matcheri S. Keshavan, Christopher R. Genovese, William F. Eddy, and John A. Sweeney. "Maturation of Widely Distributed Brain Function Subserves Cognitive Development." *NeuroImage* 13 (2001): 786–93.

Mackie, J. L. "Morality and the Retributive Emotions." *Criminal Justice Ethics* 1 (1982): 3–10.

———. "Retribution: A Test Case for Ethical Objectivity." In *Philosophy of Law*, edited by Joel Feinberg and Hyman Gross, 676–84. Belmont, CA: Wadsworth, 1991.

Mackor, Anne Ruth. "What Can Neuroscience Say about Responsibility?" In *Neuroscience and Legal Responsibility*, edited by Nicole A. Vincent, 53–84. New York: Oxford University Press, 2013.

Macmillan, Malcolm. *An Odd Kind of Fame: Stories of Phineas Gage*. Cambridge, MA: MIT Press, 2002.

Main, Chris J. "The Nature of Chronic Pain." In *Malingering and Illness Deception*, edited by Peter W. Halligan, Christopher Bass, and David A. Oakley, 171–83. New York: Oxford University Press, 2003.

Makinodan, Manabu, Kenneth M. Rosen, Susomo Ito, and Gabriel Corfas. "A Critical Period for Social Experience-Dependent Oligodendrocyte Maturation and Myelination." *Science* 337 (2012): 1357–60.

Mann, Ronald J. *Charging Ahead: The Growth and Regulation of Payment Card Markets*. Cambridge: Cambridge University Press, 2006.

"The Many Dangers of Brain-Based Lie Detection." *Center for Science and Law*, November 19, 2012. http://www.neulaw.org/blog/1034-class-blog/4131-the-many -dangers-of-brain-based-lie-detection.

Margolis, Joseph. *Persons and Minds*. Boston: D. Reidel, 1978.

Markou, Athina. "Neurobiology of Nicotine Dependence." *Philosophical Transactions of the Royal Society of London B: Biological Sciences* 363 (2008): 3159–68.

Marotta-Wurgler, Florencia. "Are 'Pay Now, Terms Later' Contracts Worse for Buyers?" *Journal of Legal Studies* 38 (2009): 309–43.

———. "Will Increased Disclosure Help? Evaluating the Recommendations of the ALI's 'Principles of the Law of Software Contracts.'" *University of Chicago Law Review* 78 (2011): 165–86.

Marotta-Wurgler, Florencia, and Robert Taylor. "Set in Stone? Change and Innovation in Consumer Standard-Form Contracts." *New York University Law Review* 88 (2013): 240–85.

Maughan, B., A. Taylor, A. Caspi, and T. E. Moffit. "Prenatal Smoking and Early Childhood Conduct Problems." *Archives of General Psychiatry* 61 (2004): 836–43.

McClure, Samuel M., D. I. Laibson, G. Loewenstein, and J. D. Cohen. "Separate Neural Systems Value Immediate and Delayed Monetary Rewards." *Science* 306 (2004): 503–7.

McGeer, Victoria. "Co-reactive Attitudes and the Making of Moral Community." In *Emotions, Imagination, and Moral Reasoning*, edited by Robyn Langdon and Catriona Mackenzie, 299–325. New York: Psychology Press, 2012.

McKee, Ann C., Robert C. Cantu, Christopher J. Nowinski, E. Tessa Hedley-Whyte, Brandon E. Gavett, Andrew E. Budson, Veronica E. Santini, et al. "Chronic Traumatic Encephalopathy in Athletes: Progressive Tauopathy after Repetitive Head Injury." *Journal of Neuropathology and Experimental Neurology* 68 (2009): 709–35.

McNamara, R. K., and S. E. Carlson. "Role of Omega-3 Fatty Acids in Brain Development and Function: Potential Implications for the Pathogenesis and Prevention of Psychopathology." *Prostaglandins, Leukotrienes and Essential Fatty Acids* 75 (2006): 329–49.

Meier, Barry. "U.S. Brings First Charges in Inquiry on Tobacco Companies." *New York Times*, January 8, 1998, A16.

Mele, Alfred R. *Effective Intentions: The Power of Conscious Will*. Oxford, UK: Oxford University Press, 2009.

Merzenich, Michael M., and William M. Jenkins. "Reorganization of Cortical Representations of the Hand Following Alterations of Skin Inputs Induced by Nerve Injury, Skin Island Transfers, and Experience." *Journal of Hand Therapy* 6 (1993): 89–104.

Mesulam, M. Marsel. *Principles of Behavioral and Cognitive Neurology*. 2nd ed. New York: Oxford University Press, 2000.

Metzger, Michael B., and Michael J. Phillips. "Emergence of Promissory Estoppel as an Independent Theory of Recovery." *Rutgers Law Review* 35 (1983): 472–557.

Miao, Chun-Hui. "Consumer Myopia, Standardization and Aftermarket Monopolization." *European Economic Review* 54 (2010): 931–46.

Mielke, H. W., and S. Zahran. "The Urban Rise and Fall of Air Lead (Pb) and the Latent Surge and Retreat of Societal Violence." *Environment International* 43 (2012): 48–55.

Milgate, Deborah E. "The Flame Flickers, but Burns On: Modern Judicial Application of the Ancient Heat of Passion Defense." *Rutgers Law Review* 51 (1998): 193–228.

Mill, John Stuart. *On Liberty*. New York: D. Appleton, 1863.

Mills, Kathryn L., François Lalonde, Liv S. Clasen, Jay N. Giedd, and Sarah-Jayne Blakemore. "Developmental Changes in the Structure of the Social Brain in Late Childhood and Adolescence." *Social Cognitive and Affective Neuroscience* 123 (2014): 123–31.

Minority Report. Paramount Pictures. Directed by Steven Spielberg. Distributed by Twentieth Century Fox, 2002.

Moffitt, Terrie E. "Adolescence-Limited and Life-Course Persistent Antisocial Behavior: A Developmental Taxonomy." *Psychology Review* 100 (1993): 674–701.

Moore, G. E. *Principia Ethica*. Cambridge: Cambridge University Press, 1903.

Moore, Michael S. *Act and Crime: The Philosophy of Action and Its Implications for Criminal Law*. Oxford, UK: Oxford University Press, 1993.

———. "Causation and the Excuses." *California Law Review* 73 (1985): 1091–1150.

———. "Moral Reality." *Wisconsin Law Review* 6 (1982): 1061–1156.

———. "Moral Reality Revisited." *Michigan Law Review* 90 (1992): 2424–2533.

———. "A Natural Law Theory of Interpretation." *Southern California Law Review* 58 (1985): 286–398.

———. *Placing Blame: A Theory of the Criminal Law*. Oxford, UK: Oxford University Press, 2010.

———. "Responsible Choices, Desert-Based Legal Institutions, and the Challenges of Contemporary Neuroscience." In *New Essays in Political and Social Philosophy*, edited by Ellen Frankel Paul, Fred D. Miller Jr., and Jeffrey Paul, 233–79. Cambridge: Cambridge University Press, 2012.

Morris, Herbert. "Persons and Punishment." *Monist* 52 (1968): 475–501.

Morse, Stephen J. "Avoiding Irrational NeuroLaw Exuberance: A Plea for Neuromodesty." *Mercer Law Review* 62 (2011): 837–60.

———. "Brain Overclaim Syndrome and Criminal Responsibility: A Diagnostic Note." *Ohio State Journal of Criminal Law* 3 (2006): 397–412.

———. "Common Criminal Law Compatibalism." In *Neuroscience and Legal Responsibility*, edited by Nicole A. Vincent, 27–52. New York: Oxford University Press, 2013.

————. "Criminal Law and Common Sense: An Essay on the Perils and Promise of Neuroscience." *Marquette Law Review* 99 (2015): 39–74.

————. "Deprivation and Desert." In *From Social Justice to Criminal Justice: Poverty and the Administration of Criminal Law*, edited by William C. Heffernan and John Kleinig, 114–60. New York: Oxford University Press, 2000.

————. "The Ethics of Forensic Practice: Reclaiming the Wasteland." *Journal of the American Academy of Psychiatry and the Law* 36 (2008): 206–17.

————. "Failed Explanations and Criminal Responsibility." *Virginia Law Review* 68 (1982): 971–1083.

————. Introduction to *A Primer on Criminal Law and Neuroscience*, edited by Stephen J. Morse and Adina L. Roskies, xv–xxii. New York: Oxford University Press, 2013.

————. "Law, Responsibility, and the Sciences of the Brain/Mind." In *Oxford Handbook of Law, Regulation, and Technology*, edited by Roger Brownsword, Eloise Scotford, and Karen Yeung. Oxford, UK: Oxford University Press, 2016.

————. "Laws Might Change as the Science of Violence Is Explained" [interview by Stuart Fox]. LiveScience, June 7, 2010.

————. "Lost in Translation? An Essay on Law and Neuroscience." In *Law and Neuroscience*, Current Legal Issues 13, edited by Michael Freeman, 529–62. Oxford, UK: Oxford University Press, 2011.

————. "Neuroscience, Free Will, and Criminal Responsibility." In *Free Will and the Brain: Neuroscientific, Philosophical, and Legal Perspectives*, edited by Walter Glannon, 251–86. Cambridge: Cambridge University Press, 2015.

————. "New Neuroscience, Old Problems." In *Neuroscience and the Law: Brain, Mind, and the Scales of Justice*, edited by Brent Garland, 177–81. New York: Dana, 2004.

————. "Psychopathy and Criminal Responsibility." *Neuroethics* 1 (2008): 205–12.

————. "Reason, Results, and Criminal Responsibility." *University of Illinois Law Review* (2004): 363–444.

————. "Waiting for Determinism to Happen." *Legal Essays*, unpublished, 1999. http://people.virginia.edu/~dll2k/morse.pdf.

Moser, Rosemarie S., Philip Schatz, and Barry D. Jordan. "Prolonged Effects of Concussion in High School Athletes." *Neurosurgery* 57 (2005): 300–306.

Mukherjee, Raja A. S., S. Hollins, and Jeremy Turk. "Low Level Alcohol Consumption and the Fetus." *British Medical Journal* 330 (2005): 375–76.

Mullenix, Linda S. "Gaming the System: Protecting Consumers from Unconscionable Contractual Forum-Selection and Arbitration Clauses." *Hastings Law Journal* 66 (2015): 719–60.

Murray, John E. Jr. "The Dubious Status of the Rolling Contract Formation Theory." *Duquesne Law Review* 50 (2012): 35–82.

Myers, G. J., P. W. Davidson, C. Cox, C. F. Shamlaye, and D. Palumbo. "Prenatal

Methylmercury Exposure from Ocean Fish Consumption in the Seychelles Child Development Study." *Lancet* 361 (2003): 1686–92.

Nabavi, Sadegh, Rocky Fox, Christophe D. Poux, John Y. Lin, Roger Y. Tisen, and Roberto Malinow. "Engineering a Memory with LTD and LTP." *Nature* 511 (2014): 348–52. http://www.nature.com/nature/journal/vaop/ncurrent/full/nature13294.html.

Nadelhoffer, Thomas. "Neural Lie Detection, Criterial Change, and Ordinary Language." *Neuroethics* 4 (2011): 205–15.

Nadelhoffer, Thomas, and Walter P. Sinnott-Armstrong. "Is Psychopathy a Mental Disease?" In *Neuroscience and Legal Responsibility*, edited by Nicole A. Vincent, 229–55. New York: Oxford University Press, 2013.

Nahmias, Eddy. "Is Free Will an Illusion?" In *Moral Psychology*. Vol. 4, *Free Will and Moral Responsibility*, edited by Walter Sinnott-Armstrong, 1–58. Cambridge, MA: MIT Press, 2014.

———."Why We Have Free Will." *Scientific American*, January 2015, 77–79.

Naqvi, Nasir H., David Rudrauf, Hanna Damasio, and Antoine Bechara. "Damage to the Insula Disrupts Addiction to Cigarette Smoking." *Science* 315 (2007): 531–34.

National Institute of Neurological Disorders and Stroke. *NINDS Arachnoid Cyst Information Page.* http://www.ninds.nih.gov/disorders/all-disorders/arachnoid-cysts-information-page.

Nestler, Eric J., and Robert C. Malenka. "The Addicted Brain." *Scientific American* 290 (2004): 78–85.

Neumann, Craig S., David S. Schmitt, Rachel Carter, Iva Embley, and Robert D. Hare. "Psychopathic Traits in Females and Males across the Globe." *Behavioral Science and the Law* 30 (2012): 557–74.

Nevins-Saunders, Elizabeth. "Not Guilty as Charged: The Myth of *Mens Rea* for Defendants with Mental Retardation." *UC Davis Law Review* 45 (2012): 1419–86.

Nygaard, Richard Lowell. "Crime, Pain, and Punishment." *Dickinson Law Review* 102 (1998): 355–82.

Olender, Jack H. "Showing Pain and Suffering." In *American Jurisprudence Trials*, vol. 5, edited by Thompson West, 921. Rochester, MN: Lawyers Co-operative, 1966.

Omalu, Bennet I., Steven T. DeKosky, Ryan L. Minster, M. Ilyas Kamboh, Ronald L. Hamilton, and Cyril H. Wecht. "Chronic Traumatic Encephalopathy in a National Football League Player." *Neurosurgery* 57 (2005): 128–34.

Oman, Nathan B. "The Failure of Economic Interpretations of the Law of Contract Damages." *Washington and Lee Law Review* 64 (2007): 829–76.

Panksepp, J. Review of "Looking for Spinoza: Joy, Sorrow, and the Feeling Brain," by Antonio Damasio. *Consciousness and Emotion* 4 (2003): 111, 126–27.

Pardo, Michael S., and Dennis Patterson. *Minds, Brains, and Law: The Conceptual Foundations of Law and Neuroscience.* New York: Oxford University Press, 2013.

————. "Philosophical Foundations of Law and Neuroscience." *University of Illinois Law Review*, no. 4 (2010): 1211–50.

Pellman, Elliot J. "Background on the National Football League's Research on Concussion in Professional Football." *Neurosurgery* 53 (2003): 797–98.

Pereboom, Derk. *Living without Free Will*. Cambridge: Cambridge University Press, 2001.

Perry, Stephen. "Comment on Coleman: Corrective Justice." *Indiana Law Journal* 67 (1992): 381–409.

————. "The Distributive Turn: Mischief, Misfortune and Tort Law." In *Analyzing Law: New Essays in Legal Theory*, edited by Brian Bix, 141–61. New York: Oxford University Press, 1998.

————. "The Moral Foundations of Tort Law." *Iowa Law Review* 77 (1992): 449–511.

————. "Political Authority and Political Obligation." In *Oxford Studies in the Philosophy of Law 2*, edited by Leslie Green and Brian Leiter, 1–74. New York: Oxford University Press, 2012.

————. "Responsibility for Outcomes, Risk, and the Law of Torts." In *Philosophy and the Law of Torts*, edited by Gerald Postema, 72–130. Cambridge: Cambridge University Press, 2001.

————. "Torts, Rights, and Risks." In *Philosophical Foundations of the Law of Torts*, edited by John Oberdiek, 38–64. New York: Oxford University Press, 2014.

Pettit, Philip. "Responsibility Incorporated." *Ethics* 117 (2007): 171–201.

Pinker, Stephen. *The Better Angels of Our Nature: Why Violence Has Declined*. New York: Viking, 2012.

Plassmann, Hilke, John O'Doherty, Baba Shiv, and Antonio Rangel. "Marketing Actions Can Modulate Neural Representations of Experienced Pleasantness." *Proceedings of the National Academy of Sciences* 105 (2008): 1050–54.

Plomin, Robert, Michael J. Owen, and Peter McGuffin. "The Genetic Basis of Complex Human Behaviors." *Science* 264 (1994): 1733–39.

Polinsky, A. Mitchell. *An Introduction to Law and Economics*. New York: Wolters Kluwer Law and Business, 1989.

Pollock, Frederick. *The Law of Torts*. 11th ed. London: Stevens and Sons, 1920.

Popper, Karl R. *Objective Knowledge*. Oxford, UK: Oxford University Press, Clarendon, 1972.

Pop Warner League. *Ages and Weights*. http://www.popwarner.com/football/foot ballstructure.htm.

Posner, Eric A. "Economic Analysis of Contract Law after Three Decades: Success or Failure?" *Yale Law Journal* 112 (2003): 829–80.

————. "ProCD v. Zeidenberg and Cognitive Overload in Contractual Bargaining." *University of Chicago Law Review* 77 (2010): 1181–94.

Posner, Richard A. *Economic Analysis of Law*. 8th ed. New York: Aspen, 2010.

———. "Justices Should Use More than Their Gut and 'Brain Science' to Decide Cases." *Slate*, June 26, 2012. http://www.slate.com/articles/news_and_politics /the_breakfast_table/features/2012/_supreme_court_year_in_review/supreme _court_year_in_review_the_justices_should_use_more_than_their_emotions_to _decide_how_to_rule_.html.

———. "Let Us Never Blame a Contract Breaker." *Michigan Law Review* 107 (2009): 1349–64.

Potenza, Marc N. "What Integrated Interdisciplinary and Translational Research May Tell Us about Addiction." *Addiction* 105 (2010): 792–96.

Price, Donald D. "Psychological and Neural Mechanisms of the Affective Dimension of Pain." *Science* 288 (2000): 1769–72.

Priest, George L. "A Theory of the Consumer Product Warranty." *Yale Law Journal* 90 (1981): 1292–1352.

Prosser, William L. *Handbook of the Law of Torts*. St. Paul, MN: West, 1941.

Putnam, Hillary. *Meaning and the Moral Sciences*. London: Routledge and Kegan Paul, 1978.

———. "Robots: Machines or Artificially Created Life?" *Journal of Philosophy* 61 (1964): 668–91.

Quinton, A. M. "On Punishment." *Analysis* 14 (1954): 133–42.

Rabadi, Meheroz H., and Barry D. Jordan. "The Cumulative Effect of Repetitive Concussion in Sports." *Clinical Journal of Sports Medicine* 11 (2001): 194–98.

Radelet, Michael L., and Ronald L. Akers. "Deterrence and the Death Penalty: The Views of the Experts." *Journal of Criminal Law and Criminology* 87, no. 1 (1996): 1–16.

Radin, Margaret Jane. "Boilerplate: Foundations of Market Contracts." *Michigan Law Review* 104 (2006): 820–26.

———. *Boilerplate: The Fine Print, Vanishing Rights, and the Rule of Law*. Princeton, NJ: Princeton University Press, 2013.

———. "Boilerplate Today: The Rise of Modularity and the Waning of Consent." *Michigan Law Review* 104 (2006): 1223–34.

Raine, Adrian, P. A. Brennan, and D. P. Farrington. *Biosocial Bases of Violence: Conceptual and Theoretical Issues*. New York: Plenum, 1997.

Raine, Adrian, W. S. Laufer, Y. Yang, K. L. Narr, and A. W. Toga. "Increased Executive Functioning, Attention, and Cortical Thickness in White-Collar Criminals." *Human Brain Mapping* 33 (2012): 2932–40.

Raine, Adrian, and P. H. Venables. "Classical Conditioning and Socialization—A Biosocial Interaction." *Personality and Individual Differences* 2 (1981): 273–83.

Raine, Adrian. *The Anatomy of Violence: The Biological Roots of Crime*. New York: Pantheon, 2013.

Raine, Adrian, and Yaling Yang. "Neural Foundations to Moral Reasoning and Antisocial Behavior." *Social Cognitive and Affective Neuroscience* 1 (2006): 203–13.

Ramachandran, Vilanyan S. *A Brief Tour of Human Consciousness*. New York: Pi Press, 2005.

Ramachandran, Vilanyan S., D. Rogers-Ramachandran, and S. Cobb. "Touching the Phantom Limb." *Nature* 377 (1995): 489–90.

Ramachandran, Vilanyan S., D. Rogers-Ramachandran, and M. Stewart. "Perceptual Correlates of Massive Cortical Reorganization." *Science* 258 (1992): 1159–60.

Random House Unabridged Dictionary. New York: Random House, 2001.

Ravenscroft, Ian. N.d. "Folk Psychology as a Theory." In *Stanford Encyclopedia of Philosophy*, edited by Edward N. Zalta. http://plato.stanford.edu/archives/fall 2010/entries/folkpsych-theory/.

Rawls, John. "Two Concepts of Rules." *Philosophical Review* 64 (1955): 3–32.

Raz, J. "Promises and Obligations." In *Law, Morality, and Society*, edited by P. M. S. Hacker and Joseph Raz, 210–28. Oxford, UK: Clarendon, 1977.

Redish, A. David. *The Mind within the Brain: How We Make Decisions and How Those Decisions Go Wrong*. Oxford, UK: Oxford University Press, 2013.

Reisel, Daniel. "Towards a Neuroscience of Morality." In *The Psychology of Restorative Justice*, edited by Theo Gavrielides, 49–64. New York: Routledge, 2015.

Renthal, William, Tiffany L. Carle, Ian Maze, Herbert E. Covington III, Hoang-Trang Truoang, Imran Alibhai, Arvind Kumar, et al. "ΔFosB Mediates Epigenetic Desensitization of the c-fos Gene after Chronic Amphetamine Exposure." *Journal of Neuroscience* 28 (2008): 7344–49.

Rescher, Nicholas. *Free Will: A Philosophical Reappraisal*. New Brunswick, NJ: Transaction, 2009.

Richards, Stephen C., ed. *The Marion Experiment: Long-Term Solitary Confinement and the Supermax Movement*. Carbondale: Southern Illinois University Press, 2015.

Ridge, Michael. N.d. "Moral Non-Naturalism." In *Stanford Encyclopedia of Philosophy*, edited by Edward N. Zalta. http://plato.stanford.edu/archives/spr2010 /entries/moral-non-naturalism/.

Robinson, Michael E., Roland Straud, and Donald D. Price. "Pain Measurement and Brain Activity: Will Neuroimages Replace Pain Ratings?" *Journal of Pain* 14 (2013): 323–27.

Robinson, Paul H. "Punishing Dangerousness: Cloaking Preventative Detention as Criminal Justice." *Harvard Law Review* 114 (2001): 1429–56.

Robinson, Paul H., and Markus D. Dubber. "The American Model Penal Code: A Brief Overview." *New Criminal Law Review* 10 (2007): 319–41.

Rolls, E. T., B. J. Everitt, and A. Roberts. "The Orbitofrontal Cortex." *Philosophical Transactions of the Royal Society of London B: Biological Sciences* 351 (1996): 1433–43.

Ronson, Jon. *The Psychopath Test: A Journey through the Madness Industry*. New York: Riverhead, 2012.

Rosen, Jeffery. "The Brain on the Stand." *New York Times Magazine*, March 11, 2007.

Roskies, Adina, and Walter Sinnott-Armstrong. "Between a Rock and a Hard Place: Thinking about Morality." *Scientific American*, July 29, 2008. http://www.scientificamerican.com/article/thinking-about-morality/.

Rossi, Peter H., and Richard A. Berk. *Just Punishments: Federal Guidelines and Public Views Compared*. New York: Aldine de Gruyter, 1997.

Rush, Benjamin. "A Defence of Blood-Letting, as a Remedy for Certain Diseases." In *Medical Inquiries and Observations*, edited by Benjamin Rush, 183–254. Philadelphia: Kimber and Richardson, 1812.

Sacks, Oliver. *Awakenings*. New York: Harper Perennial, 1973.

———. *Hallucinations*. New York: Vintage, 2013.

———. *A Leg to Stand On*. New York: Summit Books, 1984.

———. *The Man Who Mistook His Wife for a Hat*. New York: Summit Books, 1985.

———. *The Mind's Eye*. New York: Knopf, 2010.

———. *Seeing Voices*. Berkeley: University of California Press, 1989.

Salmond, John W. *Salmond on the Law of Torts*. London: Sweet and Maxwell, 1973.

Santarelli, Luca, Michael Saxe, Cornelius Gross, Alexandre Surget, Fortunato Battaglia, Stephanie Dulawa, Noelia Weisstaub, et al. "Requirement of Hippocampal Neurogenesis for the Behavioral Effects of Antidepressants." *Science* 301 (2003): 805–9.

Santos, Laurie R., and Michael L. Platt. "Evolutionary Anthropological Insights into Neuroeconomics: What Non-Human Primates Can Tell Us about Human Decision-Making Strategies." In *Neuroeconomics: Decision Making and the Brain*, edited by Paul W. Glimcher and Ernst Fehr, 109–46. Cambridge, MA: Academic, 2013.

Sapolsky, Robert M. "The Frontal Cortex and the Criminal Justice System." *Philosophical Transactions of the Royal Society of London B: Biological Sciences* 359 (2004): 1787–96.

Saver, Jeffrey L., and Antonio Damasio. "Preserved Access and Processing of Social Knowledge in a Patient with Acquired Sociopathy Due to Ventromedial Frontal Damage." *Neuropsychology* 29 (1991): 1241–49.

Scanlon, T. M. *What We Owe to Each Other*. Cambridge, MA: Harvard University Press, 1998.

Scerri, Eric R., and Lee McIntyre. "The Case for the Philosophy of Chemistry." *Synthese* 111 (1997): 213–32.

Schmithorst, V. J., and W. Yuan. "White Matter Development during Adolescence as Shown by Diffusion MRI." *Brain and Cognition* 72 (2010): 16–25.

Schopenhauer, Arthur. *On the Basis of Morality*. London: Swan Sonnenschein, 1903.

Schroeder, Christopher H. "Corrective Justice and Liability for Increasing Risks." *UCLA Law Review* 37 (1990): 439–78.

———. "Thousands of Prison Terms in Crack Cases Could Be Eased." *New York Times*, June 30, 2011.

Scott, Hal S. "The Risk Fixers." *Harvard Law Review* 91 (1978): 737–92.

Scott, Robert E. "The Rise and Fall of Article 2." *Louisiana Law Review* 62 (2002): 1009–64.

Scull, Andrew. "Mind, Brain, Law and Culture." *Brain* 130 (2007): 585–91.

Sedgwick, Sally S. "Hegel's Critique of the Subjective Idealism of Kant's Ethics." *Journal of the History of Philosophy* 26 (1998): 89–105.

Seidelson, David E. "Reasonable Expectations and Subjective Standards in Negligence Law: The Minor, the Mentally Impaired, and the Mentally Incompetent." *George Washington Law Review* 50 (1981): 67–92.

Shalev, Sharon. *Supermax: Controlling Risk through Solitary Confinement.* Portland, OR: Willan, 2009.

Shavell, Steven. "An Economic Analysis of Altruism and Deferred Gifts." *Journal of Legal Studies* 20 (1991): 401–22.

Sheline, Yvette I., P. W. Wang, M. H. Gado, J. G. Csernansky, and M. W. Vannier. "Hippocampal Atrophy in Recurrent Major Depression." *Proceedings of the National Academy of Sciences* 93 (1996): 3908–13.

Shiffrin, Seana Valentine. "The Divergence of Contract and Promise." *Harvard Law Review* 120 (2007): 708–53.

Sifferd, Katrina L. "Translating Scientific Evidence into the Language of the 'Folk': Executive Function as Capacity-Responsibility." In *Neuroscience and Legal Responsibility*, edited by Nicole A. Vincent, 183–204. New York: Oxford University Press, 2013.

Simpson, A. Rae. N.d. "Brain Changes." *MIT Young Adult Development Project.* http://hrweb.mit.edu/worklife/youngadult/brain.html.

Singer, Peter. "Ethics and Intuitions." *Journal of Ethics* 9 (2005): 331–52.

Singer, Tania, Ben Seymour, John P. O'Doherty, Klaas E. Stephan, Raymond J. Dolan, and Chris D. Frith. "Empathic Neural Responses Are Modulated by the Perceived Fairness of Others." *Nature* 439 (2006): 466–69.

Slobogin, Christopher. "The Civilization of the Criminal Law." *Vanderbilt Law Review* 58 (2005): 121–70.

Smith, Angela. "Control, Responsibility, and Moral Assessment." *Philosophical Studies* 138 (2008): 367–92.

———. "Responsibility for Attitudes: Activity and Passivity in Mental Life." *Ethics* 115 (2005): 236–71.

Smith, Trenton G., and Attila Tasnádi. "The Economics of Information, Deep Capture and the Obesity Debate." *American Journal of Agricultural Economics* 96 (2014): 533–41.

Solomon, Jason. "Civil Recourse as Social Equity." *Florida State University Law Review* 39 (2011): 243–72.

Soon, Chun Siong, Marcel Brass, Hans-Jochen Heinze, and John-Dylan Haynes. "Unconscious Determinants of Free Decisions in the Human Brain." *Nature Neuroscience* 11 (2008): 543–45.

Sowell, Elizabeth R., Paul M. Thompson, Colin J. Holmes, Terry L. Jernigan, and

Arthur W. Toga. "*In Vivo* Evidence for Post-Adolescent Brain Maturation in Frontal and Striatal Regions." *Nature Neuroscience* 2 (1999): 859–61.

Stark, Debra Pogrund. "Ineffective in Any Form: How Confirmation Bias and Distractions Undermine Improved Home-Loan Disclosures." *Yale Law Journal Online* 122 (2013): 377–400.

Steinberg, Laurence, Elizabeth Cauffman, Jennifer Woolard, Sandra Graham, and Marie Banich. "Are Adolescents Less Mature than Adults? Minors' Access to Abortion, the Juvenile Death Penalty, and the Alleged APA 'Flip-Flop.'" *American Psychologist* 64 (2009): 583–94.

Strawson, Galen. *Freedom and Belief.* New York: Oxford University Press, 2010.

Strawson, Peter F. "Freedom and Resentment." In *Freedom and Resentment and Other Essays.* London: Harper and Row, 1974.

Sullivan, Timothy J. "Punitive Damages in the Law of Contract: The Reality and the Illusion of Legal Change." *Minnesota Law Review* 61 (1977): 207–52.

Sunstein, Cass R., ed. *Behavioral Law and Economics.* New York: Cambridge University Press, 2000.

———. "Moral Heuristics." *Behavioral and Brain Sciences* 28 (2005): 531–73.

Swayze, V. W., V. P. Johnson, J. W. Hanson, J. Piven, Y. Sato, J. N. Giedd, D. Mosnik, and N. C. Andreasen. "Magnetic Resonance Imaging of Brain Anomalies in Fetal Alcohol Syndrome," *Pediatrics* 99 (2006): 232–40.

Tehrani, J. A., P. A. Brennan, S. Hodgins, and S. A. Mednick. "Mental Illness and Criminal Violence." *Social Psychiatry and Psychiatric Epidemiology* 33 (1998): S81–S85.

Teicher, M. H., Susan L. Andersen, Ann Polcari, Carl M. Anderson, Carryl P. Navalta, and Dennis M. Kim. "The Neurobiological Consequences of Early Stress and Childhood Maltreatment." *Neuroscience and Biobehavioral Review* 27 (2003): 33–44.

Terasaki, Michael. "Do End User License Agreements Bind Normal People?" *Western State University Law Review* 41 (2014): 467–89.

Thomas, M. E. "Confessions of a Sociopath." *Psychology Today*, May 7, 2013.

Toledo, Esteban, Alyssa Lebela, Lino Beccarr, Anna Minister, Clas Linman, Nasim Maleki, David W. Dodick, and David Borsook. "The Young Brain and Concussion: Imaging as a Biomarker for Diagnosis and Prognosis." *Neuroscience and Biobehavioral Reviews* 36 (2009): 1510–31.

Tomberlin, James. *Mind Causation and World.* Oxford, UK: Blackwell, 1997.

Toro, R., G. Leonard, J. V. Lerner, R. M. Lerner, M. Perron, G. B. Pike, L. Richer, et al. "Prenatal Exposure to Maternal Cigarette Smoking and the Adolescent Cerebral Cortex." *Neuropsychopharmacology* 33 (2008): 1019–27.

Tranel, D., and B. T. Hyman. "Neuropsychological Correlates of Bilateral Amygdala Damage." *Archives of Neurology* 47 (1990): 349–55.

Treede, Rolf-Detlef, Daniel R. Kenshalo, Richard H. Gracely, and Anthony K. P. Jones. "The Cortical Representation of Pain," *Pain* 79, nos. 2–3 (1999): 105–11.

Tropp, Burton E. *Molecular Biology: Genes to Proteins.* 4th ed. Burlington, VT: Jones and Bartlett Learning, 2012.

Tversky, Amos, and Daniel Kahneman. "Judgment under Uncertainty: Heuristics and Biases." *Science* 185 (1974): 1124–31.

Twining, William. *Karl Llewellyn and the Realist Movement.* 2nd ed. Cambridge: Cambridge University Press, 2012.

Van Gulick, Robert. "Nonreductive Materialism and the Nature of Intertheoretical Constraint." In *Emergence or Reduction? Essays on the Prospects of Nonreductive Physicalism,* edited by A. Beckermann, H. Flohr, and J. Kim, 157–79. Berlin: Walter de Gruyter, 1992.

Van Riel, Raphael, and Robert Van Gulick. "Scientific Reduction." In *The Stanford Encyclopedia of Philosophy,* edited by Edward N. Zalta, http://plato.stanford.edu/archives/spr2016/entries/scientific-reduction.

Varden, Helga. "Kant and Lying to the Murderer at the Door . . . One More Time: Kant's Legal Philosophy and Lies to Murderers and Nazis." *Journal of Social Philosophy* 41 (2010): 403–21.

Veit, R., M. Lotze, S. Sewing, H. Missenhardt, T. Gaber, and N. Birbaumer. "Aberrant Social and Cerebral Responding in a Competitive Reaction Time Paradigm in Criminal Psychopaths." *NeuroImage* 49 (2010): 3365–72.

Velmans, Max. *Understanding Consciousness.* New York: Routledge, 2009.

Vihvelin, Kadri. *Causes, Laws, and Free Will: Why Determinism Doesn't Matter.* New York: Oxford University Press, 2013.

Vogel, Howard J. "The Restorative Justice Wager: The Promise and Hope of a Value-Based, Dialogue-Driven Approach to Conflict Resolution for Social Healing." *Cardozo Journal of Conflict Resolution* 8 (2007): 565–610.

Vrca, A., V. Bozikov, Z. Brzovic, R. Fuchs, and M. Malinar. "Visual Evoked Potentials in Relation to Factors of Imprisonment in Detention Camps." *International Journal of Legal Medicine* 109 (1996): 114–17.

Wager, Tor D., Lauren Y. Atlas, Martin A. Lindquist, Mathieu Roy, Choong-Wan Woo, and Ethan Kross. "An fMRI-Based Neurologic Signature of Physical Pain." *New England Journal of Medicine* 368 (2013): 1388–97.

Wallace, David L. "Addiction Postulates and Legal Causations, or Who's in Charge, Person or Brain?" *Journal of American Academy of Psychiatry and the Law* 41 (2013): 92–97.

Wallace, Deanna L., Vincent Vialou, Loretta Rios, Tiffany L. Carle-Florence, Sumana Chakravarty, Arvind Kumar, Danielle L. Graham, et al. "The Influence of ΔFosB in the Nucleus Accumbens on Natural Reward-Related Behavior." *Journal of Neuroscience* 28 (2008): 10272–77.

Wallace, R. Jay. *Responsibility and the Moral Sentiments.* Cambridge, MA: Harvard University Press, 1994.

Waller, Bruce N. *Against Moral Responsibility.* Cambridge, MA: MIT Press, 2011.

———. "The Stubborn Illusion of Moral." In *Exploring the Illusion of Free Will and*

Moral Responsibility, edited by Gregg Caruso, 82–85. Lanham, MD: Lexington, 2013.

———. *The Stubborn System of Moral Responsibility*. Cambridge, MA: MIT Press, 2014.

Wang, Y., Ning Sun, Suping Li, Qiaorong Du Bachelor, Yong Xu Bachelor, Zhifeng Liu, and Kerang Zhang. "A Genetic Susceptibility Mechanism for Major Depression: Combinations of Polymorphisms Defined the Risk of Major Depression and Subpopulations." *Medicine* 94 (2015): e778.

Webber, Ann L., and Joanne Wood. "Amblyopia: Prevalence, Natural History, Functional Effects and Treatment." *Clinical and Experimental Optometry* 88 (2005): 365–75.

Wegner, Daniel M. *The Illusion of Conscious Will*. Cambridge, MA: MIT Press, 2002.

———. "Précis of the Illusion of Conscious Will." *Behavioral and Brain Sciences* 27 (2004): 649–92.

Weinrib, Ernest J. "Corrective Justice." *Iowa Law Review* 77 (1992): 403–9.

———. *Corrective Justice*. New York: Oxford University Press, 2012.

———. "Corrective Justice in a Nutshell." *University of Toronto Law Journal* 52 (2002): 349–56.

———. "Correlativity, Personality, and the Emerging Consensus on Corrective Justice." *Theoretical Inquiries in Law* 2 (2001): 107–59.

———. *The Idea of Private Law*. Oxford, UK: Oxford University Press, 2012.

———. "Toward a Moral Theory of Negligence Law." *Law and Philosophy* 2 (1983): 37–62.

Wertheimer, Ellen. "The Smoke Gets in Their Eyes: Product Category Liability and Alternative Feasible Designs in the Third Restatement." *Tennessee Law Review* 61 (1994): 1429–54.

Wessman, Mark B. "Should We Fire the Gatekeeper? An Examination of the Doctrine of Consideration." *University of Miami Law Review* 43 (1993): 45–117.

Westen, Peter. "Individualizing the Reasonable Person in Criminal Law." *Criminal Law and Philosophy* 2 (2008): 137–62.

Whalen, Paul J. "Fear, Vigilance, and Ambiguity: Initial Neuroimaging Studies of the Human Amygdala." *Current Directions in Psychological Science* 7, no. 6 (1998): 177–88.

Wheatley, C. S. Jr. "Future Pain and Suffering as Elements of Damages for Physical Injury." *American Law Reports* 81 (1932): 423–630.

White, G. Edward. *Tort Law in America: An Intellectual History*, exp. ed. New York: Oxford University Press, 2003.

Wilkinson-Ryan, Tess. "A Psychological Account of Consent to Fine Print." *Iowa Law Review* 99 (2014): 1745–84.

Williams, Bernard. *Ethics and the Limits of Philosophy*. Cambridge, MA: Harvard University Press, 1985.

Wilson, Margo, and Martin Daly. "Do Pretty Women Inspire Men to Discount the Future?" *Proceedings of the Royal Society of London B: Biological Sciences* 271 (2004): 177–79.

Winkielman, Piotr, Kent C. Berridge, and Julia L. Wilbarger. "Unconscious Affective Reactions to Masked Happy versus Angry Faces Influence Consumption Behavior and Judgments of Value." *Personality and Social Psychology Bulletin* 31 (2005): 121–35.

Wolf, Susan. *Freedom within Reason.* Oxford, UK: Oxford University Press, 1990.

Wolfram, Stephen. *A New Kind of Science.* Champaign, IL: Wolfram Media, 2002.

Woodward, James, and John Allman. "Moral Intuition: Its Neural Substrates and Normative Significance." *Journal of Physiology-Paris* 101 (2007): 179–202.

Wootton, David. *The Invention of Science: A New History of the Scientific Revolution.* New York: Harper, 2015.

Wright, Lord. "Ought the Doctrine of Consideration to Be Abolished from the Common Law?" *Harvard Law Review* 49 (1936): 1225–53.

Yaffe, Gideon. "Criminal Attempts." *Yale Law Journal* 124 (2014): 1–247.

Yang, Tianming, and Michael N. Shadlen. "Probabilistic Reasoning by Neurons." *Nature* 447 (2007): 1075–80.

Yang, Y., A. Raine, K. L. Narr, P. Colletti, and A. W. Toga. "Localization of Deformations within the Amygdala in Individuals with Psychopathy." *Archives of General Psychiatry* 66, no. 9 (2009): 986–94.

Zamir, Eyal. Forthcoming. "Law and Behavioral Economics." In *Encyclopedia of the Philosophy of Law and Social Philosophy*, edited by Stephan Sellers and Mortimer Kirste. New York: Springer.

Zipursky, Benjamin. "Civil Recourse, Not Corrective Justice." *Georgetown Law Journal* 91 (2003): 694–756.

———. "*Palsgraf*, Punitive Damages, and Preemption." *Harvard Law Review* 125 (2012): 1757–97.

Zirin, Dave. "Jovan Belcher's Murder-Suicide: Did the Kansas City Chiefs Pull the Trigger?" *The Nation*, January 6, 2014. http://www.thenation.com/blog/177787/jovan-belchers-murder-suicide-did-kansas-city-chiefs-pull-trigger#.

Index